Energy, Environment, and Sustainability

Series Editor

Avinash Kumar Agarwal, Department of Mechanical Engineering, Indian Institute of Technology Kanpur, Kanpur, Uttar Pradesh, India

AIMS AND SCOPE

This books series publishes cutting edge monographs and professional books focused on all aspects of energy and environmental sustainability, especially as it relates to energy concerns. The Series is published in partnership with the International Society for Energy, Environment, and Sustainability. The books in these series are edited or authored by top researchers and professional across the globe. The series aims at publishing state-of-the-art research and development in areas including, but not limited to:

- Renewable Energy
- Alternative Fuels
- Engines and Locomotives
- Combustion and Propulsion
- Fossil Fuels
- Carbon Capture
- Control and Automation for Energy
- Environmental Pollution
- Waste Management
- Transportation Sustainability

Review Process

The proposal for each volume is reviewed by the main editor and/or the advisory board. The chapters in each volume are individually reviewed single blind by expert reviewers (at least four reviews per chapter) and the main editor.

Ethics Statement for this series can be found in the Springer standard guidelines here https://www.springer.com/us/authors-editors/journal-author/journal-author-helpdesk/before-you-start/before-you-start/1330#c14214

More information about this series at http://www.springer.com/series/15901

Akhilendra Pratap Singh · Avinash Kumar Agarwal
Editors

Novel Internal Combustion Engine Technologies for Performance Improvement and Emission Reduction

Springer

Editors
Akhilendra Pratap Singh
Department of Mechanical Engineering
Indian Institute of Technology Kanpur
Kanpur, Uttar Pradesh, India

Avinash Kumar Agarwal
Department of Mechanical Engineering
Indian Institute of Technology Kanpur
Kanpur, Uttar Pradesh, India

ISSN 2522-8366 ISSN 2522-8374 (electronic)
Energy, Environment, and Sustainability
ISBN 978-981-16-1581-8 ISBN 978-981-16-1582-5 (eBook)
https://doi.org/10.1007/978-981-16-1582-5

This Springer imprint is published by the registered company Springer Nature Singapore Pte Ltd.
The registered company address is: 152 Beach Road, #21-01/04 Gateway East, Singapore 189721, Singapore

Preface

Energy demand has been rising remarkably due to the increasing population and urbanization. Global economy and society are significantly dependent on the energy availability because it touches every facet of human life and activities. Transportation and power generation are two major examples. Without the transportation by millions of personalized and mass transport vehicles and availability of 24×7 power, human civilization would not have reached contemporary living standards.

The International Society for Energy, Environment and Sustainability (ISEES) was founded at Indian Institute of Technology Kanpur (IIT Kanpur), India, in January 2014 with an aim to spread knowledge/awareness and catalyse research activities in the fields of energy, environment, sustainability and combustion. The society's goal is to contribute to the development of clean, affordable and secure energy resources and a sustainable environment for the society and to spread knowledge in the above-mentioned areas and create awareness about the environmental challenges, which the world is facing today. The unique way adopted by the society was to break the conventional silos of specializations (engineering, science, environment, agriculture, biotechnology, materials, fuels, etc.) to tackle the problems related to energy, environment and sustainability in a holistic manner. This is quite evident by the participation of experts from all fields to resolve these issues. The ISEES is involved in various activities such as conducting workshops, seminars, conferences, etc. in the domains of its interests. The society also recognizes the outstanding works done by the young scientists and engineers for their contributions in these fields by conferring them awards under various categories.

Fourth International Conference on 'Sustainable Energy and Environmental Challenges' (IV-SEEC) was organized under the auspices of ISEES from 27 to 29 November 2019, at NEERI, Nagpur. This conference provided a platform for discussions between eminent scientists and engineers from various countries including India, USA, China, Italy, Mexico, South Korea, Japan, Sweden, Greece, Czech Republic, Germany, Netherland and Canada. In this conference, eminent speakers from all over the world presented their views related to different aspects of energy, combustion, emissions and alternative energy resource for sustainable development and cleaner environment. The conference presented one high-voltage plenary talk by Mrs. Rashmi Urdhwareshe, Director, Automotive Research Association of India (ARAI), Pune.

The conference included 28 technical sessions on topics related to energy and environmental sustainability including 1 plenary talk, 25 keynote talks and 54 invited talks from prominent scientists, in addition to 70+ contributed talks, and 80+ poster presentation by students and researchers. The technical sessions in the conference included Fuels, Engine Technology and Emissions, Coal and Biomass Combustion/Gasification, Atomization and Sprays, Combustion and Modelling, Alternative Energy Resources, Water and Water and Wastewater Treatment, Automobile and other Environmental Applications, Environmental Challenges and Sustainability, Nuclear Energy and Other Environmental Challenges, Clean Fuels and Other Environmental Challenges, Water Pollution and Control, Biomass and Biotechnology, Waste to Wealth, Microbiology, Biotechnological and Other Environmental Applications, Waste and Wastewater Management, Cleaner Technology and Environment, Sustainable Materials and Processes, Energy, Environment and Sustainability, Technologies and Approaches for Clean, Sensors and Materials for Environmental, Biological Processes and Environmental Sustainability. One of the highlights of the conference was the rapid fire poster sessions in (i) engine/fuels/emissions, (ii) environment and (III) biotechnology, where 50+ students participated with great enthusiasm and won many prizes in a fiercely competitive environment. 300+ participants and speakers attended this 3-day conference, where 12 ISEES books published by Springer, Singapore under a special dedicated series 'Energy, environment and sustainability' were released. This was third time in a row that such significant and high-quality outcome has been achieved by any society in India. The conference concluded with a panel discussion on 'Balancing Energy Security, Environmental Impacts & Economic Considerations: Indian Perspective', where the panelists were Dr. Anjan Ray, CSIR-IIP Dehradun; Dr. RR Sonde, Thermax Ltd.; Prof. Avinash Kumar Agarwal, IIT Kanpur; Dr. R Srikanth, National Institute of Advanced Studies, Bengaluru; and Dr. Rakesh Kumar, NEERI Nagpur. The panel discussion was moderated by Prof. Ashok Pandey, Chairman, ISEES. This conference laid out the roadmap for technology development, opportunities and challenges in energy, environment and sustainability domain. All these topics are very relevant for the country and the world in present context. We acknowledge the support received from various funding agencies and organizations for the successful conduct of the Fourth ISEES conference IV-SEEC, where these books germinated. We would therefore like to acknowledge SERB, Government of India (special thanks to Dr. Sandeep Verma, Secretary); NEERI Nagpur (special thanks to Dr. Rakesh Kumar, Director); CSIR and our publishing partner Springer (special thanks to Swati Mehershi).

The editors would like to express their sincere gratitude to a large number of authors from all over the world for submitting their high-quality work in a timely manner and revising it appropriately at a short notice. We would like to express our special thanks to Dr. Pravesh Chandra Shukla, Dr. Nirendra Nath Mustafi, Prof. V. Ganeshan, Dr. Joonsik Hwang, Dr. Rajan Kumar, Dr. Vikram Kumar, Dr. Nikhil Sharma, Dr. Dhinesh Balasubramanian, Krishna Kumar, Ashutosh Jena, Ankur Kalwar, Utkarsha Sonawane, Harsimran Singh, Sulav Kafle, Manojit Pal,

Sam Joe, Arun Lal, Akash Rai, Ayush Tripathi and Tushar Kakkar, who reviewed various chapters of this book and provided their valuable suggestions to improve the manuscripts.

This book covers different aspects of internal combustion (IC) engines including engine performance and emissions and presents various solutions to resolve the issues of IC engines. In this book, solution strategies have been divided into three categories, namely advanced combustion techniques, after-treatment techniques and alternative fuels. In each section, a detailed discussion about each strategy has been explained using a number of case studies. Few chapters based on review of technologies present the status of different combustion and emission control technologies and their suitability for different types of IC engines. A few novel technologies for Spark Ignition (SI) engines have been also included in this book, which makes this book a complete solution for both diesel as well as gasoline engines. This book summarizes different issues involved with automotive sectors and presents available solutions strategically to resolve these issues. Overall, this book will be useful for both engine researchers and energy experts. We hope that the book would be of great interest to the professionals and post-graduate students involved in fuels, IC engines, engine instrumentation and environmental research.

Kanpur, India

Akhilendra Pratap Singh
Avinash Kumar Agarwal

Contents

About the Editors

Dr. Akhilendra Pratap Singh is currently working at the Indian Institute of Technology (IIT) Kanpur, India. He received his Master's and Ph.D. in Mechanical Engineering from IIT Kanpur, India in 2010 and 2016, respectively. He worked at the Engine Research Center, University of Wisconsin at Madison, USA as a postdoctoral fellow (2017–2018). His areas of research include advanced low temperature combustion, optical diagnostics with special reference to engine endoscopy and PIV, combustion diagnostics, engine emissions measurement, particulate characterization and their control, and alternative fuels. Dr. Singh has edited 10 books and authored 27 book chapters, 45+ research articles in international journals and conferences. He has been awarded with ISEES Best Ph.D. Thesis Award (2017), SERB Indo-US Postdoctoral Fellowship (2017) and IEI Young Engineer Award (2017). He is a member of numerous professional societies, including SAE, ASME, and ISEES.

Prof. Avinash Kumar Agarwal joined the Indian Institute of Technology (IIT) Kanpur, India in 2001 after working as a post-doctoral fellow at the Engine Research Center, University of Wisconsin at Madison, USA. His interests are IC engines, combustion, alternate and conventional fuels, lubricating oil tribology, optical diagnostics, laser ignition, HCCI, emissions and particulate control, and large bore engines. Prof. Agarwal has published 290+ peer reviewed international journal and conference papers, 42 edited books, 78 books chapters and has 10000+ Scopus and 15300+ Google scholar citations. He is a Fellow of SAE (2012), Fellow of ASME (2013), Fellow of ISEES (2015), Fellow of INAE (2015), Fellow of NASI (2018), Fellow of Royal Society of Chemistry (2018), and a Fellow of American Association of Advancement in Science (2020). He is recipient of several prestigious awards such as Clarivate Analytics India Citation Award-2017 in Engineering and Technology, NASI-Reliance Industries Platinum Jubilee Award-2012; INAE Silver Jubilee Young Engineer Award-2012; Dr. C. V. Raman Young Teachers Award: 2011; SAE Ralph R. Teetor Educational Award -2008; INSA Young Scientist Award-2007; UICT Young Scientist Award-2007; INAE Young Engineer Award-2005. Prof. Agarwal received Prestigious Shanti Swarup Bhatnagar Award-2016 in Engineering Sciences. For his outstanding contributions, Prof. Agarwal is conferred upon Sir J C Bose National Fellowship (2019) by SERB.

Part I
General

Chapter 1
Introduction to Novel Internal Combustion Engine Technologies for Performance Improvement and Emission Reduction

Akhilendra Pratap Singh and Avinash Kumar Agarwal

Abstract In the last two decades, efficient and clean internal combustion (IC) engines have become the main requirement of society. To fulfil this demand, a lot of research is going on throughout the world. Few researchers have already focussed on advanced engine technologies for improving engine efficiency; however, current emission standards are pushing researchers to carry out some fundamental investigations to reduce the pollutant formation inside the combustion chamber and the development of advanced after-treatment systems to cut down the emissions from tail pipe. This book is based on these two aspects. The first section of this book covers several fundamental studies related to laser induced fluorescence (LIF) for pollutant formation, particle image velocimetry (PIV) for in-cylinder air flow characteristics and fuel spray investigations. The second section of this book is based on performance improvement and emission reduction by using after-treatment technologies and other techniques such as blending of nano-additives. Overall this book is based on the current measures of engine technologies used for performance improvement and emission reduction.

Keywords Alternative fuels · Advanced combustion techniques · Engine performance · Emission reduction

Transportation is the major requirement in the modern society, which fulfils the daily requirement as well as affects the growth of any country. Among all modes of transportation, road transportation is an important one, which is used for both passenger transport as well as freight transport. In road transport vehicles, mineral diesel-fuelled compression ignition (CI) engines and gasoline-fuelled spark ignition (SI) engines are the most used engines. However, conventional engines are now facing performance and emission related issues, which need to be upgraded using fundamental researches. To fulfil the current emission standards, it is necessary to

A. P. Singh · A. K. Agarwal (✉)
Department of Mechanical Engineering, Indian Institute of Technology Kanpur, Kanpur, India
e-mail: akag@iitk.ac.in

A. P. Singh and A. K. Agarwal (eds.), *Novel Internal Combustion Engine Technologies for Performance Improvement and Emission Reduction*, Energy, Environment, and Sustainability, https://doi.org/10.1007/978-981-16-1582-5_1

utilize some after-treatment technology so that pollutants from on-road vehicles can be reduced. This book covers all these aspects in different sections.

The first section of this book is based on several fundamental studies to enhance engine performance and to reduce the harmful pollutant formation during combustion. The first chapter of this section is based on laser diagnostic techniques, which have been extensively used in internal combustion (IC) engine research. These techniques can be used to explore the fundamental insights about the in-cylinder processes such as spray characteristics, fuel–air mixing, combustion, pollutant formation, etc. without affecting/altering the underlying physics. This chapter discusses about the 'laser-induced fluorescence (LIF)', which is capable of imaging the temperature-field and in-situ species concentration during combustion. A detailed discussion about the fundamentals of LIF, design, and development, current status and future trends for the application of this technology in IC engine research is the innovative aspect of this chapter, which provides necessary information for a beginner. Inclusion of several case studies based of LIF application in IC engines is another important aspect of this chapter, which makes it suitable reference for practitioners of this technology. The second chapter of this section is based on Particle Image Velocimetry (PIV), which has evolved as a pioneering tool for the investigation of intake air flow structure, flow interaction and in-cylinder air flow. PIV is a powerful tool for getting important information, which affects both engine performance and exhaust emissions. However, there are several challenges for the adaptation of PIV in IC engines. Getting an appropriate optical access and selection of seeds according to the dynamic nature of ambient conditions are some of the practical difficulties. This chapter deals with such issues and provides several feasible solutions to resolve these issues. To present the importance of this technique, this chapter also includes several case studies in which the impact of flow structures on combustion and pollutant formation has been also discussed. The next chapter of this section is based on spray characteristics of Di-methyl ether (DME), which has a great potential to be utilized as an alternative fuel in IC engines. DME can resolve the energy as well as emissions issues of IC engines; however, a lot of research related to fuel characteristics of DME is still needed. Among different fuel characteristics, spray characteristics are the important one because it has a dominating effect on fuel–air mixture preparation. This chapter deals with various factors related to spray characteristics of DME such as fuel injection pressure (FIP), injector geometry, fuel properties, and ambient conditions. This chapter provides the results of spray characteristics of DME under different ambient temperature and pressure conditions and their associated effects on in-cylinder mixture formation and engine combustion. The next chapter of this section is based on the effect of spray characteristics on fuel–air mixing and in-cylinder combustion. For carrying out spray investigations in engine like conditions, this study was focused on the development of constant volume combustion chamber (CVCC). Developments of CVCC and getting optical access for fuel spray and combustion visualization involve a number of complications. This chapter deals with these complications and provides feasible solutions to these issues. This chapter starts with design of combustion chambers, followed by explanation of prominent optical techniques. Several case studies and particle examples given in this chapter are unique, which help in better understanding of spray atomization and combustion in different operating conditions.

The second section of this book is based on the techniques, which can be used for performance improvement and emission reduction. For controlling the exhaust emissions, the use of after-treatment systems has been discussed in this section. The first chapter of this section is based on the waste heat recovery methods such as turbo charging, turbo compounding, organic Rankine cycle, and thermoelectric generators, to enhance the thermal efficiency of the IC engine. This chapter provides the summary of these techniques along with important features. This chapter also discusses about several advanced technologies such as engine downsizing, advanced engine controls, variable valve timing, variable geometry engine design, advanced fuel injection, advanced compression ignition engines, advanced SI engines, alternative fuels, which have been demonstrated as an important technique for engine performance improvement. The second chapter of this section is based on catalytic converter, which is installed in the engine exhaust manifold of modern vehicles for emission reduction. This chapter gives insights into the flow behaviour along with the transient temperature field across the catalytic converter, which helps understand the overall working of the catalytic converter. Also, catalyst temperature distributions are discussed based on engine operating conditions in order to understand the effect of exhaust gas temperature on the catalytic converter efficiency. The next chapter of this section is based on the adaptation of after-treatment systems to meet the stringent emissions norms effectively. In this chapter, a number of after-treatment systems such as Lean NOx Traps (LNT), Selective Catalytic Reduction (SCR) catalysts, and Diesel Particulate Filter (DPF) are discussed in detail. This chapter also includes various challenges and their solutions to meet the emission norms using exhaust gas after-treatment devices. The next chapter of this section is based on the impact of the after-treatment system on the structure (i.e. morphology and nanostructure) and chemical characteristics of the exhaust PM. This review chapter will discuss the different theories and experimental work provided in the literature regarding the impact of after-treatment systems on the PM characteristics. Special attention will be given to the impact of alcohols and other oxygenated fuels on this mechanism compared to conventional diesel fuel. The last chapter of this section is based on the combustion, performance and emission characteristics of a diesel engine fuelled with hybrid nanoparticles. The investigation was carried out using cerium oxide (CeO_2), aluminium oxide (Al_2O_3), titanium dioxide (TiO_2) nanoparticles as an additive to diesel fuel. This chapter reported a significant reduction in exhaust species.

Specific topics covered in the monograph include:

- Introduction to novel internal combustion engine technologies for performance improvement and emission reduction
- Application of laser-induced fluorescence technique in internal combustion engine investigations
- Challenges and opportunities of particle imaging velocimetry as a tool for internal combustion engine diagnostics

- Dimethyl ether spray characteristics for compression ignition engines
- Spray chamber designs, and optical techniques for fundamental spray investigations
- Efficiency improvement of internal combustion engines over time
- Evolution of catalytic converters for spark-ignition engines to control emissions
- Engine emission control devices for particulate matter and oxides of nitrogen: challenges and emerging trends
- Variation of soot structure along with the exhaust after-treatment system—Impact of fuel type on the soot/catalyst interactions
- Effect of hybrid nanoparticle on DI diesel engine performance, combustion and emission studies

A comprehensive study on the utilization of producer gas as IC engine fuel.

The topics are organized in three different sections: (i) General, (ii) Advanced Technologies for Internal Combustion Engines, (iii) Technologies for Performance Improvement and Emissions Reduction.

Part II
Advanced Technologies for Internal Combustion Engines

Chapter 2
Application of Laser-Induced Fluorescence Technique in Internal Combustion Engine Investigations

Tushar Kakkar, Ashutosh Jena, and Avinash Kumar Agarwal

Abstract Laser diagnostic techniques have evolved as a pioneering tool for understanding the combustion and fluid dynamics in internal combustion (IC) engines. These are non-intrusive techniques, which provide fundamental insights about the in-cylinder processes such as spray characteristics, fuel-air mixing, combustion, pollutant formation, etc. without affecting/altering the underlying physics. This chapter is based on one such laser diagnostic technique namely 'laser-induced fluorescence (LIF)', which is capable of imaging the temperature-field and in situ species concentration during combustion. This chapter discusses a brief overview of optical diagnostics techniques, fundamentals of LIF, design, and development, current status and future trends for the application of this technology in IC engine research. This chapter also includes a comprehensive literature review of the applications of LIF in combustion investigations with a special emphasis on IC engines. Several examples and case studies have also been included in this chapter for a better understanding.

Keywords Laser-induced fluorescence · Optical diagnostics · IC engines · Equivalence ratio · Emission formation

2.1 Introduction

In the last few decades, advanced combustion technologies have been developed to reduce the formation of harmful pollutants in the engine along with an optimum trade-off between engine-out emissions and engine performance. These advanced combustion techniques are based on optimum engine control parameters as well as a selection of advanced control strategies such as fuel injection parameters, intake air pressure and temperature, etc. In many studies, it has already been proven that the in-cylinder temperature/pressure and equivalence ratio of the fuel–air mixture affect the quality of combustion. These physical quantities together control the onset

T. Kakkar · A. Jena · A. K. Agarwal (✉)
Engine Research Laboratory, Department of Mechanical Engineering, Indian Institute of Technology Kanpur, Kanpur 208016, India
e-mail: akag@iitk.ac.in

A. P. Singh and A. K. Agarwal (eds.), *Novel Internal Combustion Engine Technologies for Performance Improvement and Emission Reduction*, Energy, Environment, and Sustainability, https://doi.org/10.1007/978-981-16-1582-5_2

of ignition, flame propagation, and combustion stability, which are important for optimum engine design. Computational fluid dynamics (CFD) analysis has evolved as an important tool for engine development; however, the use of CFD models for successful prediction of complicated combustion phenomenon relies heavily on the generation of accurate experimental data for model validation. Therefore, reliable measurement of in-cylinder processes and temperature distribution in the engine hold the key for futuristic emission compliant high-performance internal combustion (IC) engines.

For this purpose, several optical diagnostic tools have been developed to understand the in-cylinder physics/processes, in which several laser diagnostic techniques such as Laser Rayleigh Scattering (LRS), Raman scattering, Coherent Anti-stokes Raman Scattering (CARS), etc. are used for in-cylinder temperature and species concentration measurements. Laser-induced fluorescence (LIF) is a well-established optical diagnostics technique for in situ measurement of species concentrations and temperature measurement during combustion. It produces relatively stronger signal intensity with high spatial resolution compared to other techniques. In this technique, the emission spectra are red-shifted from absorption, hence, provide relatively easier discrimination of the results. This technique is important for both quantitative and qualitative analyses of different species up to sub-ppm level. This chapter discusses different aspects of LIF technique such as the basic principle of LIF technique, method of measurements, application of LIF in different engine technologies, etc. Several case studies have also been discussed for a better understanding of this technique and its capabilities.

2.2 LIF-Based Diagnostics

2.2.1 Basic Principle

LIF works on the basis of excitation of target species molecules from lower energy state to higher energy state, which emit fluorescence during spontaneous de-excitation. The emitted photon wavelength by the excited species is different from the wavelength of the exciting source. The excitation wavelength coincides with the absorption wavelength. The amount of energy absorbed and the amount of energy emitted depend on the particular species getting excited.

Excited species are often involved in multiple energy interactions, in which a further excitation to a higher energy state is possible due to excitation by absorption of an additional photon. The molecules also undergo inelastic collisions with other molecules. This phenomenon alters the electronic energy of the molecules and is known as quenching. Quenching affects the fluorescence quantum yield (ϕ), which is defined as the ratio of photons emitted during emission to photons absorbed during excitation. Therefore, by analysing the radiation spectra, desired information such as temperature and concentration of species can be obtained. Figure 2.1 shows the

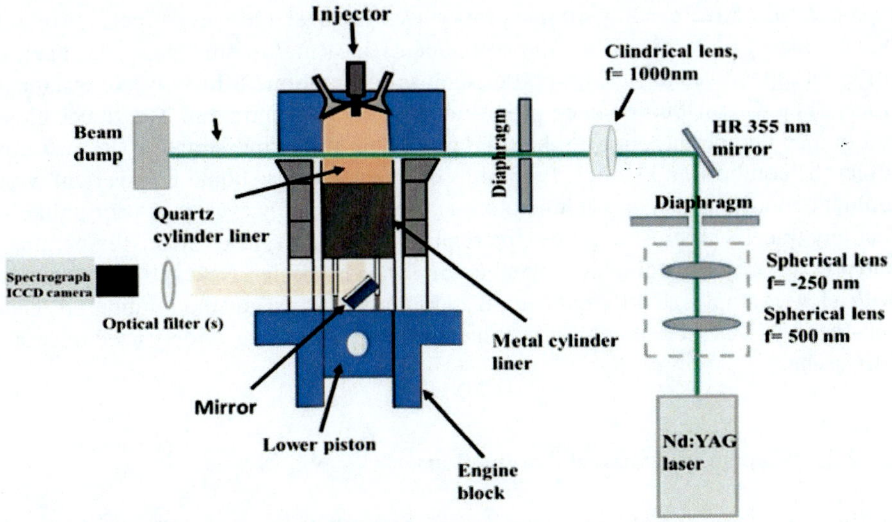

Fig. 2.1 Schematic of the LIF experimental setup

schematic of a LIF experimental setup in an optical engine, in which the laser beam is converted into a thin light sheet with the help of light-sheet optics. The engine cylinder liner is modified to provide laser access to the cylinder. The LIF signals are captured by the high-speed camera using appropriate filters. The piston is also provided with an optical window for enabling the camera to capture the fluorescence signals.

2.2.2 *Tracers*

In LIF technique, single component species having definite spectral behaviour are used as tracer to measure the local distribution of physical quantities like temperature, concentration, etc. Selection of suitable tracers is important for precise measurement in the experimental domain. The signal yield for a specific tracer depends on the environment under study. This sub-section describes the photo-physical aspects and associated challenges for tracers in the LIF studies.

2.2.2.1 Ideal Tracer

Stimulation of suitable wavelength excites the species from one electronic/vibrational state to another. Such species are known as the tracer species, which may be naturally present or sometimes need to be introduced. While natural tracers are inherent to fuels, foreign tracers are mixed with the fuel in required proportions. Ideally, the tracer should have exactly similar properties to that of the parent fuel. It should behave like a replacement for the components in a multi-component fuel having different aspects of physical properties such as evaporation, diffusivity, etc. An ideal tracer should emit fluorescence proportional to its concentration. The tracer must not induce any perturbation to the original system under investigation. It must not affect the combustion kinetics. Properties of tracer species should be invariant with ambient conditions such as temperature and pressure. However, it is very unlikely that any tracer will closely follow the requirements of an ideal tracer. Temperature and pressure dependencies are inevitable for all tracer species. Next sub-section deals with a few examples of different natural and foreign tracers, and their limitation. This sub-section gives more emphasis to toluene, which is the most widely used tracer in LIF studies.

2.2.2.2 Tracers in Practical Applications

During the combustion of fuels, a number of intermediate species and by-products are formed. However, the major part of intermediate species is saturated hydrocarbons, which do not generate any LIF signal in the required range of the spectrum. There are few species such as aromatic hydrocarbons present in fuels, which emit fluorescence signals and act as natural tracers. Similarly, formaldehyde, an intermediate species formed during combustion in the low-temperature regime, is also capable of producing fluorescence signals. It gets further oxidized and vanishes, when the local temperature rises. The Presence of formaldehyde marks a low-temperature region in the combustion chamber and its measurement has been demonstrated elsewhere as well (Hildingsson et al. 2005; Schießl and Maas 2003). By fuel tracer LIF approach, it is confirmed that the dark pockets are lower temperature regimes, where intermediate reactions are yet to start. Since formation of intermediates are highly temperature-sensitive, the temperature variation can be extracted from the fluctuations of the concentration field of these species. For this, first, it is necessary to investigate a correlation between temperature and concentration of intermediate species. Schießl and Maas (2003) observed a strong relationship between temperature and concentration of formaldehyde, which indicates the possibilities of detecting temperature fluctuations by means of measuring spatial changes in $[H_2CO]$. This study demonstrated that the fluctuations in the amplitude of concentrations can be mapped into a temperature interval via the correlation $[H_2CO] = f(T)$. Commercial fuels consist of several species, which act as a natural tracer, however, they are not helpful in the quantitative analysis. Physical properties of such species present in commercial fuels vary with varying ambient conditions, leading to an ambiguity in signal detection. However, some fuels do not emit detectable fluorescence. Peterson et al. (2015) found that iso-octane vapours in the intake flow did not yield any fluorescence. In such conditions, foreign tracers such as rhodamine, ketones etc. are used in a non-fluorescing base liquid (Schulz and Sick 2005). For quantitative analysis, spectroscopic grade fuel should be used for the best results. Thallium chloride and indium chloride can also be used for qualitative temperature analysis of burned gases. An extensive study of a variety of tracers was carried out by Schulz and Sick (2005), in which ketones were suggested as one of the best tracers for LIF. LIF signal obtained from organic molecules like ketones or toluene depends strongly on temperature, which is helpful for simultaneous measurements of in situ species concentration and temperature. Fujikawa et al. (2006) used 3-pentanone as a fluorescence tracer for the LIF technique. They found that the quantum efficiencies of ketones are 1/10th or less than that of aromatic hydrocarbon molecules. Therefore, aromatic hydrocarbons such as toluene are often preferred as a foreign tracer. They demonstrated that toluene (5% w/w) has a comparable intensity for 3-pentanone; however, the quantum efficiency of toluene was relatively higher than ketones. A temperature-dependent shift in toluene emission spectra offers the simplicity of using single line extraction (excitation with a single laser of a particular wavelength), which reduces the experimental cost and complexity (Peterson et al. 2013). 3-pentanone and toluene are also extensively used for experiments under high temperature and pressure conditions. Therefore, 3-pentanone and toluene can also be used as a primary LIF tracer in IC engines. In IC engine experiments, strong temperature sensitivity of toluene also provides important information about the flow structure, however, 3-pentanone provides only noteworthy measurements compared to toluene (Peterson et al. 2015; Gamba et al. 2015). Toluene is compatible with most commercial fuels (iso-octane and n-heptane), which have components similar to toluene (Gessenhardt et al. 2015; Kaiser et al. 2013). Although, high-speed toluene LIF thermometry measurements are accurate and provide important information about the transport of gas temperature, improved precision is still essential to make this technique more effective (Peterson et al. 2013). In most studies, one colour detection technique was used in toluene LIF due to superior temperature sensitivity and precision, however, sometimes, one colour detection technique results in inaccuracy due to local LIF signal variations caused by local toluene mixture heterogeneity (Gamba et al. 2015). Under such conditions, two-colour detection technique is preferred in toluene LIF thermometry measurements. In two-colour toluene thermometry, modest sensitivity and strong quenching of toluene LIF by oxygen reduce the signals in the air (in fired engine operation conditions) with minimum variation in the colour ratio. In some studies, nitrogen addition resolved quenching issues in non-reacting environment (Gessenhardt et al. 2015; Peterson et al. 2014). There are few other aspects such as pressure effect (air/fuel) of toluene, which make its application challenging in LIF measurements (Kaiser et al. 2013).

2.2.3 Material for Optical Access

Detailed information related to optical access is an important aspect, which is by and large missing in the previous studies. In most studies, quartz is the most widely used material for providing optical access into an engine as well as in the constant volume combustion chamber (CVCC) (Schießl and Maas 2003; Peterson et al. 2013; Zhang et al. 2019; Haessler et al. 2012; Li et al. 2013; Scott et al. 2017; Chen et al. 2019).

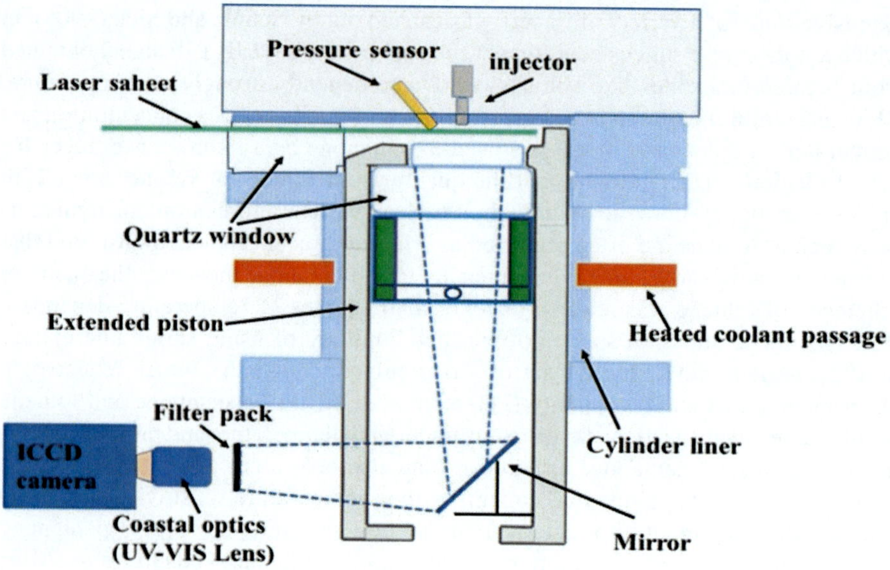

Fig. 2.2 Schematic of an engine with optical access

In last few years, the use of quartz as a suitable material for optical access into the IC engine combustion chamber has increased rapidly due to the adaptation of different laser diagnostic techniques (Fig. 2.2).

There are few other materials such as sapphire, and fused silica which are used for providing the optical access in to metallic engines. Tensile strength of sapphire is comparable to cast iron. It has a Mohs hardness of 9 with favourable mechanical properties, which provides abrasion resistance (Miles 2014). Sapphire has the ability to retain its strength at a very high temperature. Thermal expansion coefficient of sapphire is appropriate to form an interface boundary with stainless steel. This is a unique feature, which reduces thermal stress/thermal shock at the interface of the two materials. Figure 2.3 depicts that a large thermal strain can change the shape of the material, which alters the laser path. This results in a serious issue related to laser focusing and data acquisition (Miles 2014). Relatively, inferior optical property of sapphire compared to quartz is, however, another issue. It has a much higher refractive index, which leads to a high image distortion with a curved sapphire window. Sapphire has a high mass density, which is almost twice of quartz. This further adds to the balancing issue of the elongated piston when used as a piston window material. Itani et al. (2015) demonstrated the use of sapphire in high-pressure, high-temperature, pre-combustion vessel. They used two cylindrical sapphire windows (with a field of view 85×85 mm^2) on opposite sides of the vessel. This vessel has two more windows of 65×10 mm^2 for providing access to laser beams.

Fused silica has also been used in many optical studies, where the thickness of fused silica window ranges between 3 to 70 mm, depending on ambient requirements

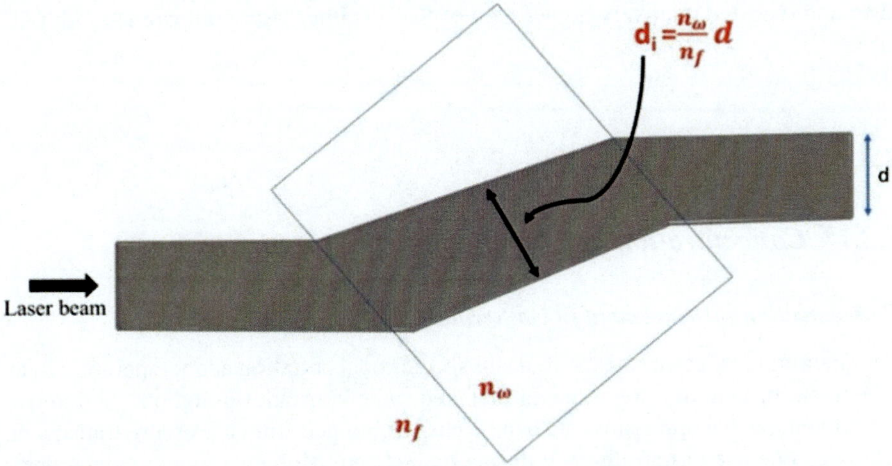

Fig. 2.3 Effect of thermal stresses on optical material shape

(Fujikawa et al. 2006; Li et al. 2006; Burkert et al. 2013; Lind et al. 2017). In order to protect whole optical assembly against the cylinder pressure, Gessenhardt et al. (2015) used fused silica as window material with an arrangement of thin titanium bushing with a press-fit sapphire window of 2.7 mm thickness for pressure sealing. Burkert et al. (2013) also used a high-pressure stainless steel cell, equipped with fused silica windows. These windows were capable of sustaining up to 100 bar pressure and 1000 K temperature.

2.2.4 Laser

Nd:YAG laser with a wavelength in the range 266–355 nm is the most used laser for LIF investigations (Peterson et al. 2013, 2014, 2015; Gessenhardt et al. 2015; Scott et al. 2017; Chen et al. 2019; Lind et al. 2017; Kranz and Kaiser 2019; Strozzi et al. 2019). In few studies, many other types of lasers such as XeCl laser (Schießl and Maas 2003), XeF laser (Burkert et al. 2002), etc. have been used. Fujikawa et al. (2006) carried out experiments using three different lasers namely: KrF excimer laser (248 nm), the fourth harmonic of Nd:YAG laser (266 nm), and the second harmonic of a dye laser. Kaiser et al. (2013) compared the performance of KrF excimer laser with Nd:YAG laser. They reported that KrF excimer laser excited at 248 nm and resulted in more unbiased detection due to different excitation wavelengths compared to fluorescence. KrF excimer laser resulted in 2–3 times longer pulse length compared to Nd:YAG laser. Bessler et al. (2001) and Schulz et al. (1995) also used KrF excimer laser and reported relatively higher gain of KrF excimer laser compared to Nd:YAG laser.

2.3 Applications of LIF

2.3.1 Concentration and Temperature Determination

Concentration measurement in isothermal condition

For quantitative measurements of in situ species concentration and temperature in an IC engine, in-cylinder pressure can be taken to be constant throughout the interrogation volume. The pressure data can be easily obtained from a pressure transducer, which can be used as a reference during calibration. Moreover, under some conditions, the in-cylinder temperature can be approximated as being spatially invariant. Average temperature can be calculated with the knowledge of pressure, assuming adiabatic compression. This would make the measurement of species concentration quite straight forward. Under the isobaric and isothermal conditions, LIF intensity is directly proportional to the species concentration. During a experiment, unknown concentration (n_{exp}) can be determined by using the following equation.

$$n_{exp}=n_{cal}\left(S_{exp}/S_{cal}\right) \tag{2.1}$$

where, S_{exp} is the signal measured during the experiment. The fluorescence signal (S_{cal}) is obtained for a known concentration (n_{cal}). However, this case is not often realized. During transient mixing and evaporation of fuel, spatial gradients in temperature as well as concentration fields exist. Therefore, temperature measurements are essential and primary to correct the fuel concentration measurements.

Temperature measurement

The fluorescence signal intensity is given by Thurber (1999):

$$S_{fl} = \eta \,.\, E \,.\, r_{Tracer} . \sigma\,(\lambda_{ece}, T) . \phi\,(\lambda_{ece},\ p,\ T) \tag{2.2}$$

where

η = optical efficiency factor,
E = laser pulse energy,
r_{Tracer} = tracer partial density,
λ_{ece} = excitation wavelength,
σ = absorption cross-section, and
ϕ =fluorescence quantum yield.

For a certain optical setup and tracer, S_{fl} is a function of pressure, temperature and concentration of tracer only. Thus, for known number density and pressure (isothermal, isobaric conditions), calibration can be done to obtain the temperature.

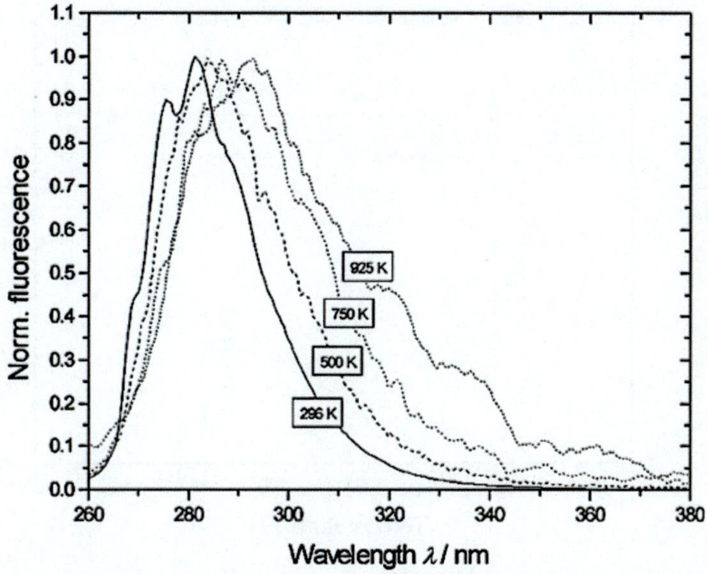

Fig. 2.4 Toluene fluorescence spectra variations with temperature (Koban et al. 2004)

Figure 2.4 depicts the variations in fluorescence yield with temperature for toluene. Calibration was done with 248 nm excitation and nitrogen bath gas pressure of 1 bar for a temperature range of 300–950 K (Koban et al. 2004). However, in many applications, the number density is unknown due to heterogeneous distribution. In such a case, the temperature-dependent red-shift of tracer spectrum can be exploited. Figure 2.4 shows the emission spectra of toluene for 248 nm excitation and 1 bar pressure. The peak of the spectra moves rightwards as the temperature increases. For a given temperature, fluorescence signals at two different wavelengths in the spectrum are detected. The ratio of intensities is determined to eliminate the requirement of number density (r_{Tracer}). Calibration is done at multiple temperature points (at constant pressure) and final calibration curve is obtained by curve fitting. Similarly, multiple calibration curves are required for the measurements at different crank angle positions in the engine because in-cylinder pressure varies with crank angle.

$$\frac{(S_f)_{\lambda 1}}{(S_f)_{\lambda 2}} = f(T) \tag{3}$$

Figure 2.5 shows the signal ratio for the calibration of toluene LIF with 266 nm excitation. The signals were obtained at 280 nm and 335 nm respectively (Koban et al. 2004). This method of temperature measurement with single-colour excitation and two-colour detection is known as 'two-colour detection' technique.

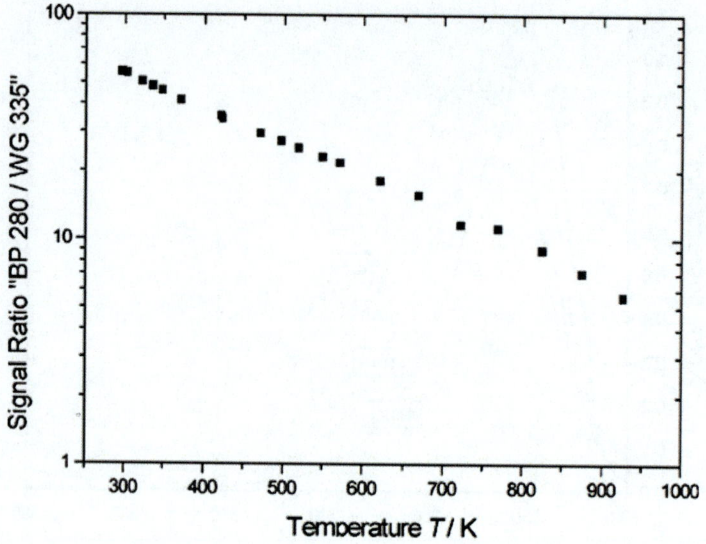

Fig. 2.5 Signal ratio of toluene for different temperatures (Koban et al. 2004)

For in-cylinder measurements, the LIF signal distribution is obtained for a particular crank angle. The data are then normalized and converted to temperature distribution by comparing the signal ratio at each pixel with the calibration curve. Once the temperature is determined from the calibration curve, the unknown concentration can be obtained by putting the value of temperature in the Eq. 2.2. The calibration is preferably done in the same setup to obtain a high degree of accuracy. Table 2.1 documents a chronicle of different studies related to temperature and concentration measurements using LIF technique in engines and CVCCs.

2.4 LIF Case Studies

This section discusses various applications of LIF studies in different engines and CVCCs. A brief summary on the experimental methods and experimental setup details have been provided along with important results of a large number of experimental studies.

Table 2.1 Summary of LIF studies for different applications

Engine	Combustion technology	Fuel with tracer percentage	Physical quantity measured	References
CI	Conventional CI	Heptane with iso-octane and toluene (65%, 25%, 10% v/v)	Temperature	Hildingsson et al. (2005)
		Heptane with ketone (2% v/v)	Temperature	Schulz et al. (1995)
		Diesel and n-heptane with CH_2O	Temperature	Donkerbroek et al. (2010)
			Species concentration	Donkerbroek et al. (2010)
		Diesel with natural occurring tracer (formaldehyde, OH, NO)	Temperature	Verbiezen et al. (2007)
		PRF77 and acetone (10% v/v)	Species concentration	Shi et al. (2020)
			Temperature	Shi et al. (2020)
	HCCI (CAI)	Nitrogen with iso-octane and toluene (50-50% v/v)	Temperature	Kaiser et al. (2013)
		Ethanol with acetone (10% v)	Temperature	Nygren et al. (2002)
			Species concentration	Nygren et al. (2002)
		Iso-octane with n-heptane (50% v/v), (30%, 70%) (natural tracer OH)	Species concentration	Li et al. (2013), Zhao (2004)
		n-heptane with naturally occur tracer (CH_2O, OH)	Temperature	Brackmann et al. (2003)
			Species concentration	Brackmann et al. (2003), Schrewe and Ghandhi (2007)
	RCCI	Iso-octane with n-heptane and toluene (70%, 30% and 1% v/v)	Temperature	Tang et al. (2017)
		Polyoxymethylene dimethyl ethers, iso-octane, n-heptane with cetane and (natural tracer OH)	Temperature	Tang et al. (2020)
			Species concentration	Tang et al. (2020)
	PCC/PCCI/GCI/PPC	Iso-octane with n-heptane and toluene (70%, 30% and 1% v/v)	Temperature	Tang et al. (2017)

(continued)

Table 2.1 (continued)

Engine	Combustion technology	Fuel with tracer percentage	Physical quantity measured	References
		PRF70 fuel and toluene (1% v)	Temperature	Raman et al. (2020)
			Species concentration	Raman et al. (2020)
SI	GDI/DISI	Toluene (2.75% v)	Temperature	Peterson et al. (2015), Peterson et al. (2013), Peterson et al. (2014)
		Iso-octane with toluene (7-3% v/v)	Species concentration	Scott et al. (2017)
		Formaldehyde (H_2CO)	Temperature	Schießl and Maas (2003) Lachaux and Musculus (2007)
			Species concentration	Schießl and Maas (2003), Lachaux and Musculus (2007)
		NO	Temperature	Schulz et al. (1995)
		n-pentane, iso-octane, n -undecane (36%, 46%, 18% v/v) with tracer (DFB (0.02%),1MN (0.1%), ANI (0.4))	Temperature	Kranz and Kaiser (2019)
		Iso-octane with natural occurring tracer (CH_2O,)	Temperature	Brackmann et al. (2003)
			Species concentration	Bessler et al. (2001), Brackmann et al. (2003)
		Gasoline with biacetyl	Species concentration	Deschamps et al. (1996)
		Gasoline surrogate + ethanol (20% and 85%) with tracer	Species concentration	Bardi et al. (2019)
		Iso-octane with triethylamine (TEA) and acetone (0.25% and 2.5% v/v)	Temperature	Lind et al. (2015)
			Species concentration	Lind et al. (2015)

(continued)

Table 2.1 (continued)

Engine	Combustion technology	Fuel with tracer percentage	Physical quantity measured	References
		Butanes with Isopentanes and acetone (5% v/v)	Temperature	Williams et al. (2010)
			Species concentration	Williams et al. (2010)
		Iso-octane with n-octane and toluene (2% v/v)	Temperature	Williams et al. (2010)
			Species concentration	Williams et al. (2010)
		Iso-dodecane with n-Decanes and TMB (2% v/v)	Temperature	Williams et al. (2010)
			Species concentration	Williams et al. (2010)
		Iso-pentane, iso-octane with n-nonane (20%, 70%, 10% v/v) and tracer foxylene/acetone (1:5, v:v), toluene/acetone (1:2, v:v), p-xylene/3-pentanone(1:5 v:v)	Temperature	Krämer et al. (1998)
			Species concentration	Krämer et al. (1998)
		DMF with MF and OH	Species concentration	Ma et al. (2014)
		Oil and pyrromethene 567(0.1 mmol/L)	Temperature	Cheong et al. (2020)
	PFI	Iso-octane with toluene (4:1 v/v)	Temperature	Gessenhardt et al. (2015)
		Iso-octane with toluene (5% v)	Temperature	Fujikawa et al. (2006)
		Hexane with 3-pentanone (25%v) and iso-octane with DMA (10%)	Species concentration	Li et al. (2006)
		Iso-octane with natural occurring tracer	Temperature	Peterson et al. (2019)
			Species concentration	Peterson et al. (2019)
CVCC	Calibration cell/combustion chamber shell	GTL-diesel with 1-MN (0.5%v)	Temperature	Lind et al. (2017)
		Formaldehyde (H_2CO)	Temperature	Haessler et al. (2012), Burkert et al. (2002)
		Nitrogen with acetone (23% v)	Species concentration	Zhang et al.(2019)

(continued)

Table 2.1 (continued)

Engine	Combustion technology	Fuel with tracer percentage	Physical quantity measured	References
		Gasoline with p-Difluorobenzene and 1-methylnaphthalene (0.02% and 0.1%)	Temperature	Itani et al. (2015)
		GTL-diesel with natural tracer	Species concentration	Burkert et al. (2013)
		n-heptane with natural tracer	Temperature	Ottenwälder et al. (2018)
			Species concentration	Ottenwälder et al. (2018)
		Gasoline with ethanol (20% and 85 v/v) and p-difluorobenzene and 1-methylnaphtalene	Temperature	Cordier et al. (2020)
			Species concentration	Cordier et al. (2020)
		N-dodecane with Naphthalene and TMPD (90%, 9% and 1% v/v)	Temperature	Chen et al. (2019)
Misc.	Jets in supersonic cross -flow	Nitrogen with toluene (0.5% v)	Temperature	Gamba et al. (2015)
	Reacting flows	OH	Temperature	Malmqvist et al. (2016)
			Species concentration	Malmqvist et al. (2016)
	Corona discharge	O_2, N_2 with NO	Species concentration	Shen et al. (2012)
	LIF in methane/air Flames	Propane	Species concentration	Bessler et al. (2002)
	Turbulent premixed conical flames	OH radical	Species concentration	Deschamps et al. (1996)
	Study on mixing phenomenon inside reactor down-comer	Salted tracer concentration	Species concentration	Zhang et al. (2019)
	Imaging of methyl radical (CH_3) in premixed methane/air flames	Methane and CH	Species concentration	Li et al. (2017)
	Temperature measurements in RCM	Iso-octane and anisole	Temperature	Tran et al. (2015)
	Magnetic resonance concentration measurements	De-aerated water with copper sulphate	Temperature	Banko et al. (2020)

(continued)

Table 2.1 (continued)

Engine	Combustion technology	Fuel with tracer percentage	Physical quantity measured	References
	Laminar premixed low-pressure flame experiments	OH	Species concentration	Schmitt et al. (2020)
	Micro-explosion of emulsion drops	Oil with water and fluorescein (250 and 300 mg/L) and rhodamine B(25,50,100 mg/L)	Temperature	Omar et al. (2020)
			Species concentration	Omar et al. (2020)
	Effects of lubricant oil film adhering on cylinder wall on the post fuel impingement and adhesion of spray penetration, wall impingement behaviour	Oil and rhodamine-B (0.1% v/v)	Temperature	Shibata et al. (2020)
		JIS K2204 (Japanese diesel) with ethanol (5%v/v) and rhodamine-B (0.05%)	Temperature	Shibata et al. (2020)
	Turbulent buoyant jet in the self-similar region	Rhodamine-6G	Temperature	Qin et al. (2020)
			Species concentration	Qin et al. (2020)
	Single-shot 3D flame diagnostic based on volumetric LIF	CH	Temperature	Ma et al. (2017)
			Species concentration	Ma et al. (2017)
	Single-shot measurements in turbulent flows	Nitrogen and I_2 (iodine crystals)	Temperature	Wu et al. (2015)
			Species concentration	Wu et al. (2015)
	Pyrolysis of toluene behind reflected shock waves	Argon and toluene (0.1–2% v/v)	Temperature	Zabeti et al. (2015)

2.4.1 A Case Study of Thermal Stratification in HCCI Engine

Despite having an immense potential of HCCI combustion engines in cutting-down the emissions to meet future emission norms, there are several hurdles in the commercialization of HCCI engine technology. Cycle-to-cycle variations is one such major hurdle, which has been reported by many researchers. Variations in the local temperature field from cycle-to-cycle is one of the main reasons for cyclic variations in HCCI combustion engines. LIF imaging of homogenous toluene seeded intake charge was used to study the fluctuation of the temperature field over cycle averaged mean temperature field (Kaiser et al. 2013). The experimental setup for this study is shown in Fig. 2.6. Other important experimental details are given in Tables 2.2 and 2.3.

A single-cylinder engine was used in this experiment, which consisted of a pent-roof four-valve cylinder head and a flat top piston with an optical window. Toluene was seeded through the intake to achieve homogenous mixing with the inlet flow. This was essential for precise calibration and measurement of the temperature field. Toluene has properties similar to iso-octane and n-heptane. Toluene was mixed with iso-octane fuel (50% v/v) and was then premixed with nitrogen gas using an injector (hollow type) working at a fuel injection pressure of 120 bar, and then introduced

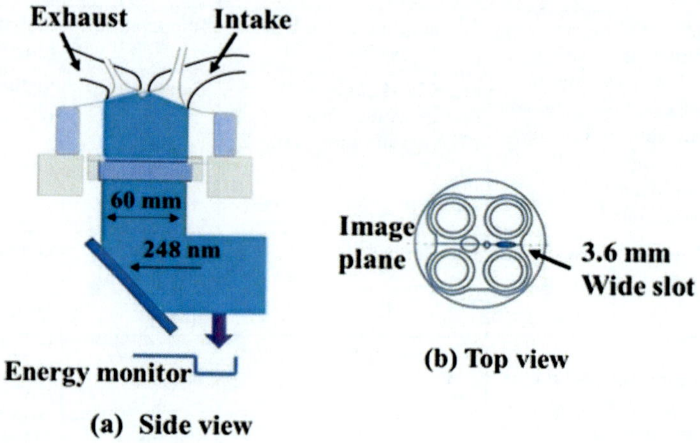

Fig. 2.6 Experimental setup

Table 2.2 Engine geometry and experimental conditions (Kaiser et al. 2013)

Bore	84 mm
Stroke	90 mm
Displacement	499 cm^3
Geometric compression ratio	10
Speed	2000 rpm
Intake pressure and gas	1 bar, nitrogen

Table 2.3 Experimental error sources in LIF measurements (Kaiser et al. 2013)

Error source	Relative signal RMS [%]	
	At BDC	At TDC
Camera noise	0.25	1.1
Energy fluctuations	1.4	1.4
Mode fluctuations	3.5	3.5
Total RMS error [% signal]	3.8	3.9
Total RMS temperature error [K]	1.6	4.0

into the combustion chamber. The KrF excimer laser (248 nm) was used for the excitation.

1 mm thick laser sheet was guided vertically in an upwards direction opposite to the pent-roof cylinder head through the mirror in the extended piston (as shown in Fig. 2.6) and the signal pulse measurement was done by diverting a fraction of laser sheet onto a photo-diode. A CCD camera was employed for signal detection. A fused silica window ring spanning 25 mm from the cylinder head was used to gain optical access.

Figure 2.7 shows the variations in temperature distribution during the compression stroke. The fluctuations shown are relative to the mean bulk temperature measured at each piston position during the compression stroke. Such distribution could have resulted from the inhomogeneous mixing of the charge. It was observed that variations increased as the piston moved closer to the cylinder head. However, no significant gradient in temperature was seen near the central region. Observed fluctuations were located near the wall boundaries. Observed variations at -180 °CA could have resulted from inhomogeneity in the mixing of fuel and air. The location of injector in the port could be crucial in this aspect. During compression, temperature fields become more uniform because of turbulence. Hence variations at TDC have no relation with those present during the intake.

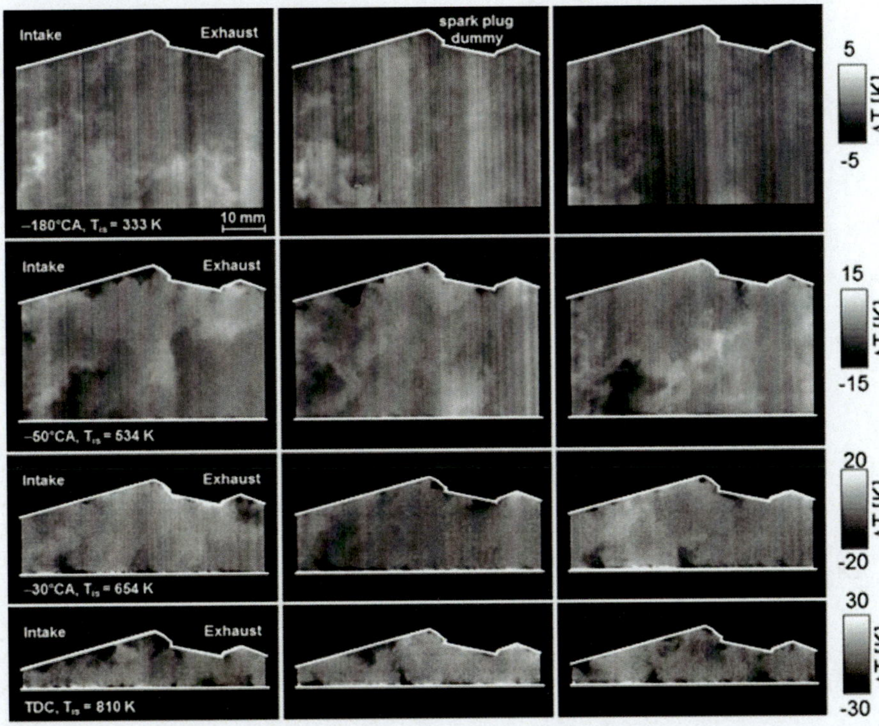

Fig. 2.7 Transition in the spatial structures (Kaiser et al. 2013)

2.4.2 A Case Study of Diesel Spray Diagnostics

The temperature and concentration field of fuel spray are vital to understand fuel–air mixing and ignition delay related phenomenon in CI engines. CVSC and CVCC have emerged as excellent alternatives to optical engines to extract important information about fuel spray and combustion respectively. Lind et al. (Lind et al. 2017) used 1- methyl naphthalene (1-MN) as tracer to understand fuel–air mixing of diesel spray in a high-temperature ambient environment. A study was carried out in a CVCC and it was reported as a pioneering research to employ PLIF concept for such analysis. Schematic of the experimental setup is shown in Fig. 2.8 and important details of spray chamber, operating parameters, etc. are provided in Table 2.4. For spectral fluorescence characterization, fluorescence spectrum was detected by the spectrograph with a blaze wavelength of 300 nm and 300 line/mm. Signal ratio calibration was done through two intensified charge-coupled device (ICCD) cameras. Light scattered from the laser was blocked by providing a 266 nm long-pass filter.

The chamber used in this study had five optical windows made of fused silica. Nd:YAG laser (266 nm, 10 Hz, 16 mJ/cm^3) was used for excitation. Two-colour detection method was adopted for measurement. The effect of laser power fluctuations was taken care during image post-processing. Signals were acquired normal to the plane of illumination. A 50/50 beam splitter enabled imaging using two ICCD

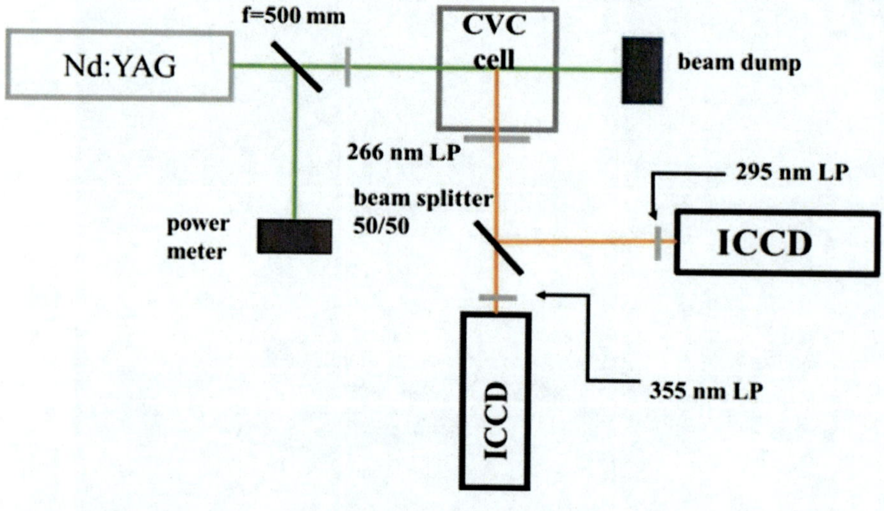

Fig. 2.8 Experiment setup for diesel spray diagnostics using LIF

Table 2.4 Spray chamber operating conditions for LIF investigations (Lind et al. 2017)

	Maximum capacity	Operating condition
Temperature	1000 K	800 K
Pressure	10 MPa	5 MPa

cameras, as shown in the schematic. Signals were acquired for 295 nm and 355 nm using long-pass filters. CVCC was continuously scavenged with nitrogen gas. Non-fluorescent GTL diesel with added tracer (0.5% v/v) was injected using a 3-hole injector.

Figure 2.9 depicts pulse-averaged temperature variations along the spray axis. It can be observed that the temperature decreased with increasing distance from the nozzle exit up to 50 mm and then increased. Reduction in temperature was attributed to evaporative cooling of the spray, which was also supported by increasing density with distance from the nozzle exit. Increased level of turbulence at injector tip promoted turbulent mixing, leading to increased heat transfer. Figure 2.10 shows pulse-to-pulse variations in the temperature and density distributions. Results showed that the spray boundary temperature approached ambient temperature due to hot gas entrainment; however, core region did not show significant variations. This can also be seen in density distributions, where core density was significantly higher than the density at the spray boundary.

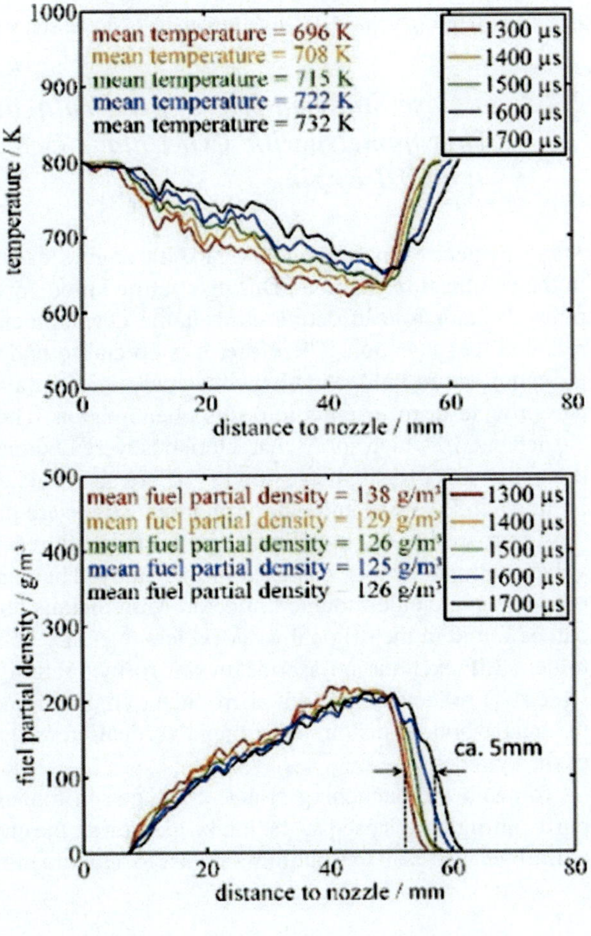

Fig. 2.9 Variations of temperature (top) and density with axial distance from nozzle (Lind et al. 2017)

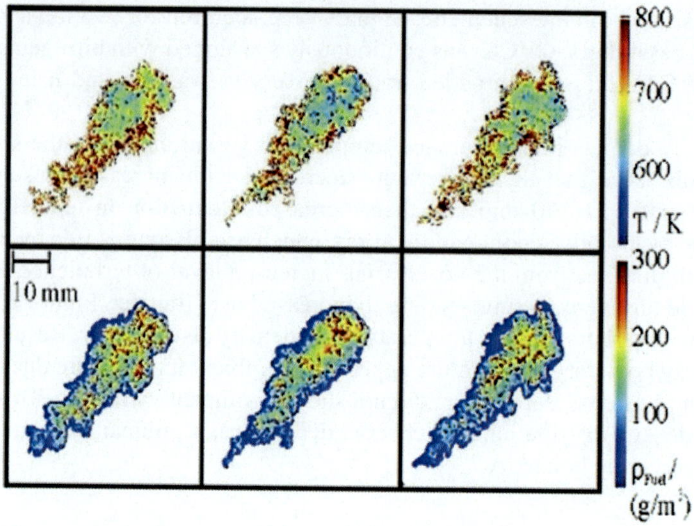

Fig. 2.10 Temperature and density distribution of diesel spray (Lind et al. 2017)

2.4.3 *A Case Study on High-Speed Toluene-LIF Thermometry with Two-Colour Detection Scheme in a GDI Engine*

Spray-induced evaporative cooling in GDI engine leads to a temperature stratification in the combustion chamber. During stratified mode operation, thermal stratification plays a major role in determining flame development and pollutant formation at retarded fuel injection. Therefore, it is crucial to understand the effect of spray on the temperature field. Simultaneous analysis of spray morphology and air motion can provide deep insights into this phenomenon. Therefore, a detailed study was carried out, in which spray characteristics were obtained by Mie scattering and flow dynamics was investigated using PIV (Peterson et al. 2015). The experimental setup is shown in Fig. 2.11 and other important details are given in Table 2.5.

A four-stroke, single-cylinder, spark ignition direct injection (SIDI) engine (four valves with pent-roof) was used in this study. The engine had a centrally mounted spark plug and side-mounted injector. More details about the engine modifications can be found in the original work (Peterson et al. 2015). Nd:YAG laser was use for toluene LIF excitation. Laser beam was formed into light sheet (0.5 mm thickness), which was reflected off by 45° mirror in the engine crankcase and then passed through the quartz-bottom piston, providing a vertical viewing plane within the central axis of the cylinder.

To avoid the quenching effect of oxygen in toluene fluorescence signals, high purity nitrogen was used as the intake gas during the engine operation. Some portion of nitrogen was sent to a bubbling seeder to add toluene vapour (2.75% w/w) in order

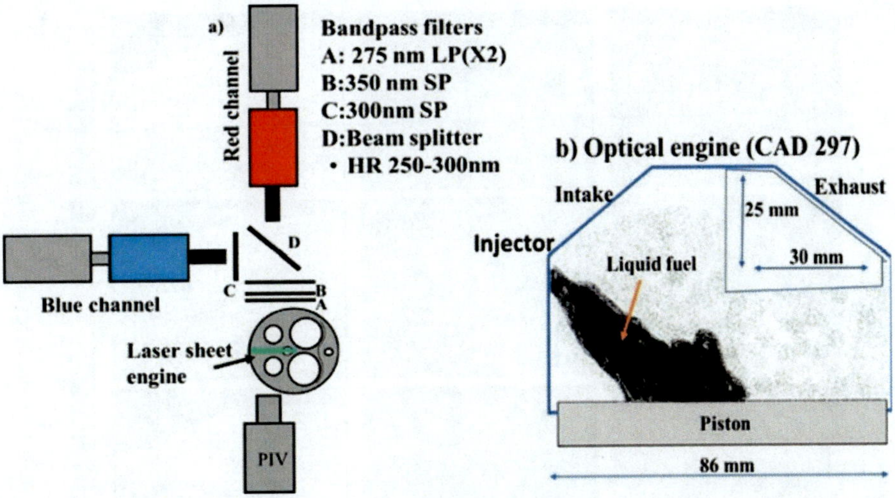

Fig. 2.11 Experimental setup of LIF investigations using high-speed toluene-LIF thermometry with two-colour detection scheme in a GDI engine

Table 2.5 Engine operating condition (Peterson et al. 2015)

Parameters	Value
Engine speed	1000 RPM
Injection timing	294–301° CA bTDC
Intake pressure and temperature	95 kPa, 295 K
Coolant temperature	333 K

to compare the time progression of in-cylinder temperature between motored- and combustion cycles. The area of interest for LIF temperature measurement is shown in Fig. 2.12. Small area marked by a rectangle was used for calibration of LIF signals with temperature. Signal ratio and calibrated temperature variations are shown in Fig. 2.13.

Temperature images were taken within 25 × 30 mm^2 field-of-view to analyse temporal evolution of liquid droplet distribution, which revealed that the temperature stratification originally created by evaporative cooling and bulk-flow motion-driven mixing over time reduces the temperature gradients. A strong counter-clockwise vortex can be observed at 315° CA. The temperature decreased by ~50 K due to spray induced cooling. Low-temperature pocket can be spotted in the bottom right corner at 324° CA. This exhibited that the temperature inhomogeneity diminished well after the end of the injection. Thus, the injection timing can be instrumental in temperature stratification, however, a larger field of view and combustion imaging is required for deeper insights into the after-effects.

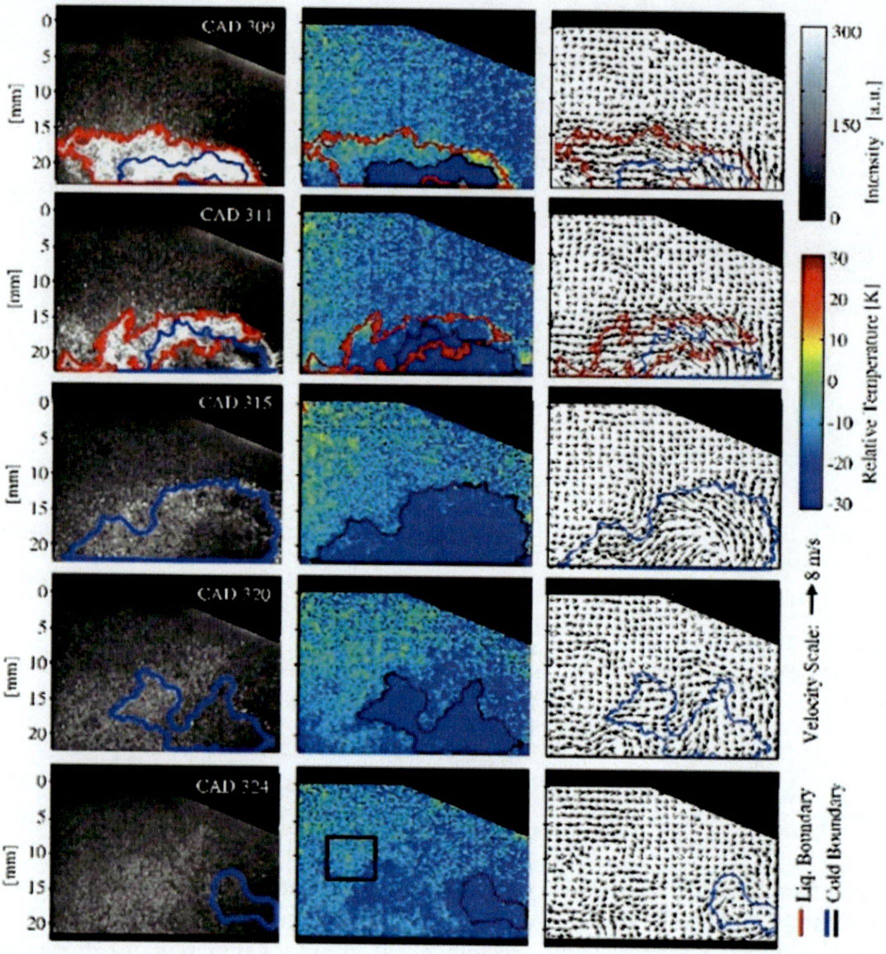

Fig. 2.12 Analysis of spray-induced thermal stratification with multiple optical diagnostics (Peterson et al. 2015)

2.4.4 A Case Study of Fuel Stratification in Reactivity Controlled Compression Ignition (RCCI) Combustion Engine by Optical Diagnostics

RCCI combustion is a promising low-temperature combustion (LTC) concept for expanding the limited operating window of LTC with simultaneously low emissions of soot and NO_x. Reactivity stratification due to low and high reactivity fuels is the key parameter to control the combustion in RCCI mode. Thus, advanced optical diagnostics of RCCI is required for simultaneous measurement of fuel and temperature stratification to analyze the effect of reactivity stratification at different engine

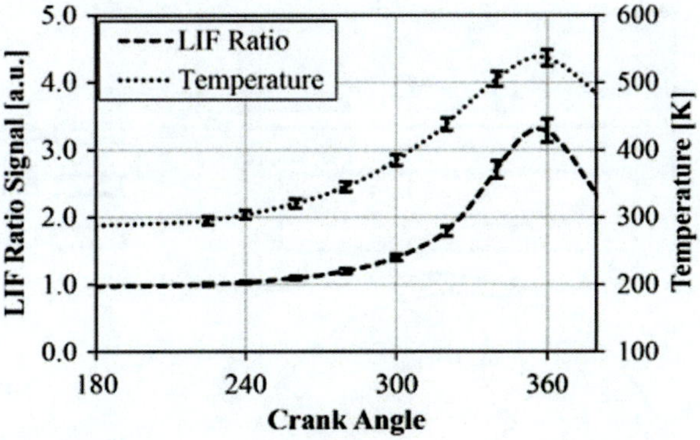

Fig. 2.13 Temperature and LIF ratio variations with crank angle (Peterson et al. 2015)

operating conditions. Tang et al. (2017) used toluene-PLIF to study fuel stratification in RCCI combustion engines. They used formaldehyde and OH-PLIF to investigate the low- and high-temperature heat releases.

A light-duty optical engine was modified to operate in RCCI mode combustion up to 7 bar IMEP. Iso-octane was introduced via port injection and n-heptane was injected late in the compression stroke via direct injection. Specifications of the engine are given in Table 2.6.

For providing optical access of the laser beam into the engine combustion chamber, a quartz rim covering 36 mm of the stroke length was installed. A 40-mm width slice was cut from the piston rim to enable the laser to pass into the piston bowl during combustion. Laser sheet was generated by a 10 Hz Nd:YAG laser. 30 mm wide and 0.8 mm thick laser sheet was adjusted to pass 10 mm below the fire-deck for successful analysis of early combustion events. A common Bowditch design was used for capturing the signals with the help of a 45° inclined mirror and an ICCD camera. The details of the piston design are given in Fig. 2.14.

Table 2.6 Engine specifications for optical diagnostics of RCCI combustion (Tang et al. 2017)

Engine type	4 stroke
Bore	92 mm
Stroke	100 mm
Displacement	0.66L
Connecting rod length	155 mm
Compression ratio	11
Combustion chamber shape	Cylindrical
Combustion chamber diameter	63 mm
DI pressure	600 bar
Holes number of DI injector	6
Spray included angle of DI injector	150°
Injector hole diameter of DI injector	0.15 mm

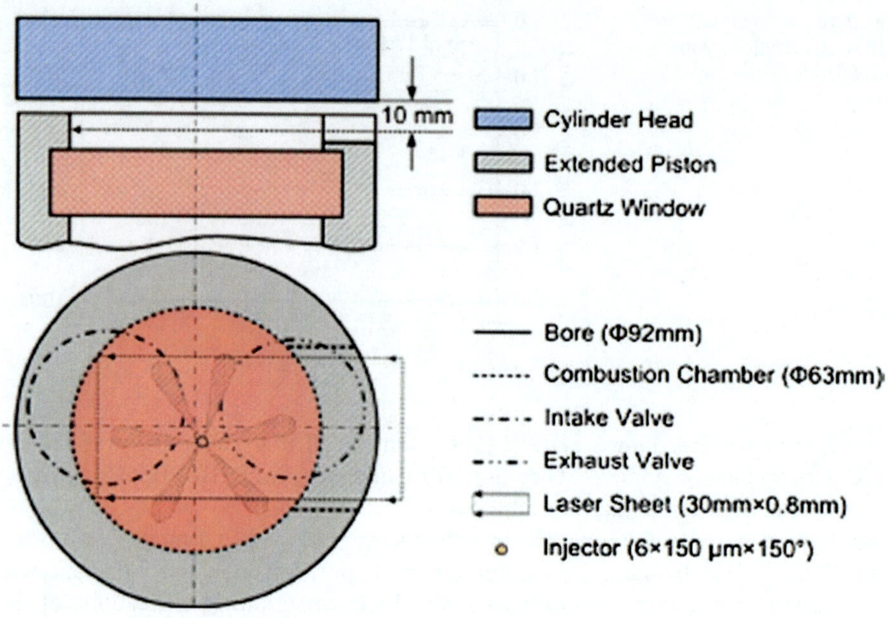

Fig. 2.14 Piston modifications for laser access in optical RCCI engine (Tang et al. 2017)

In this study, three different laser beams were used for excitation. First laser beam having 266 nm wavelength was used for toluene-PLIF under motored engine condition to study fuel stratification. Toluene (1% v/v) was used as tracer for both fuels. A 355-nm laser pulse was used for formaldehyde-PLIF to study the combustion events. An edge filter (390–480 nm) was employed for the formaldehyde signal acquisition. No PAH signals were detected, indicating low level of soot formation. 282.95 mm excitation was used for OH-PLIF with 308–325 nm edge filter. Engine operating conditions are given in Table 2.7. In Table 2.7, different RCCI combustion modes have been categorized as RCCI-xx, where xx represents the start of main injection timing.

Results from toluene-PLIF revealed that in case of RCCI-90, the fuel stratification was minimum. As ignition was retarded, the extent of stratification increased with more high reactive heptane near the combustion chamber wall. Maximum stratification was achieved by RCCI-10. The results of formaldehyde/OH PLIF are shown in Figs. 2.15 and 2.16. From Fig. 2.15, it can be observed that formaldehyde starts building-up much earlier in RCCI-90 compared to RCCI-25 and RCCI-10. This is well justified by the results of fuel stratification, which depicted the homogenous distribution of heptane. Thus, the low-temperature heat release started relatively earlier with formaldehyde formation. Comparison of Figs. 2.15 and 2.16 shows that the formaldehyde disappeared at the locations, where OH formation is started. Formaldehyde was consumed as temperature increased and OH started

Table 2.7 Engine operating conditions for various PLIF experiments (Tang et al. 2017)

		RCCI-90	RCCI-25	RCCI-10
Engine speed (rpm)		1200		
Inlet air pressure (bar)		1		
Inlet air temperature (°C)		75		
Fuel injection pressure(bar)	Port	3		
	Direct	600		
Iso-octane (% vol)		70		
N-heptane (% vol)		30		
Start of port injection(°aTDC)		−300		
Start of direct injection (°aTDC)		−90	−25	−10
Global equivalence ratio		0.77		
Gross IMEP (bar)		7.0	6.8	6.9

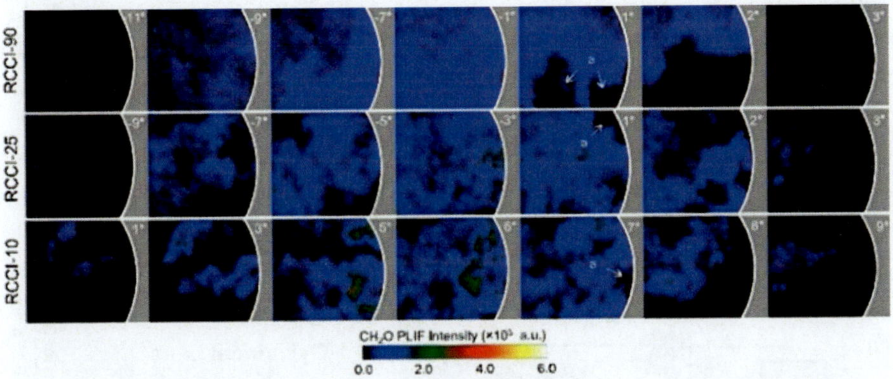

Fig. 2.15 PLIF image of formaldehyde (Tang et al. 2017)

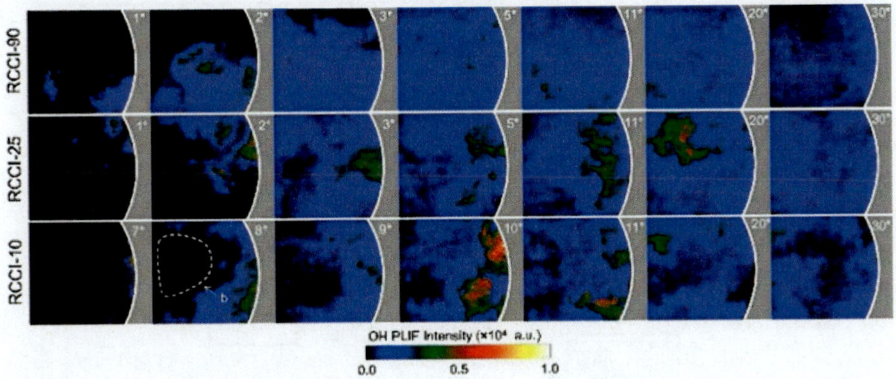

Fig. 2.16 OH PLIF images (Tang et al. 2017)

forming. Also, a higher concentration of OH can be observed near the cylinder walls for RCCI-10, which finally moved towards the centre. Combustion images also revealed that blue flames started at the periphery and auto-ignition occurred later at the canter for RCCI-10. Therefore, a high degree of fuel stratification led to more controlled combustion. Retarded ignition timing increased control over combustion and extended high-temperature combustion to the centre of the bowl. This may result in a reduction in unburned HC emissions.

2.4.5 A Case Study of Flame Evolution in Gasoline Compression Ignition (GCI) Engine by LIF Technique

GCI is a novel LTC technique, in which low octane fuels can be used for achieving higher engine performance with lower emissions. Tang et al. (2017) employed multiple optical diagnostic techniques to understand the flame evolution in a light-duty optical engine operated in GCI mode (Fig. 2.17). Optical engine specifications

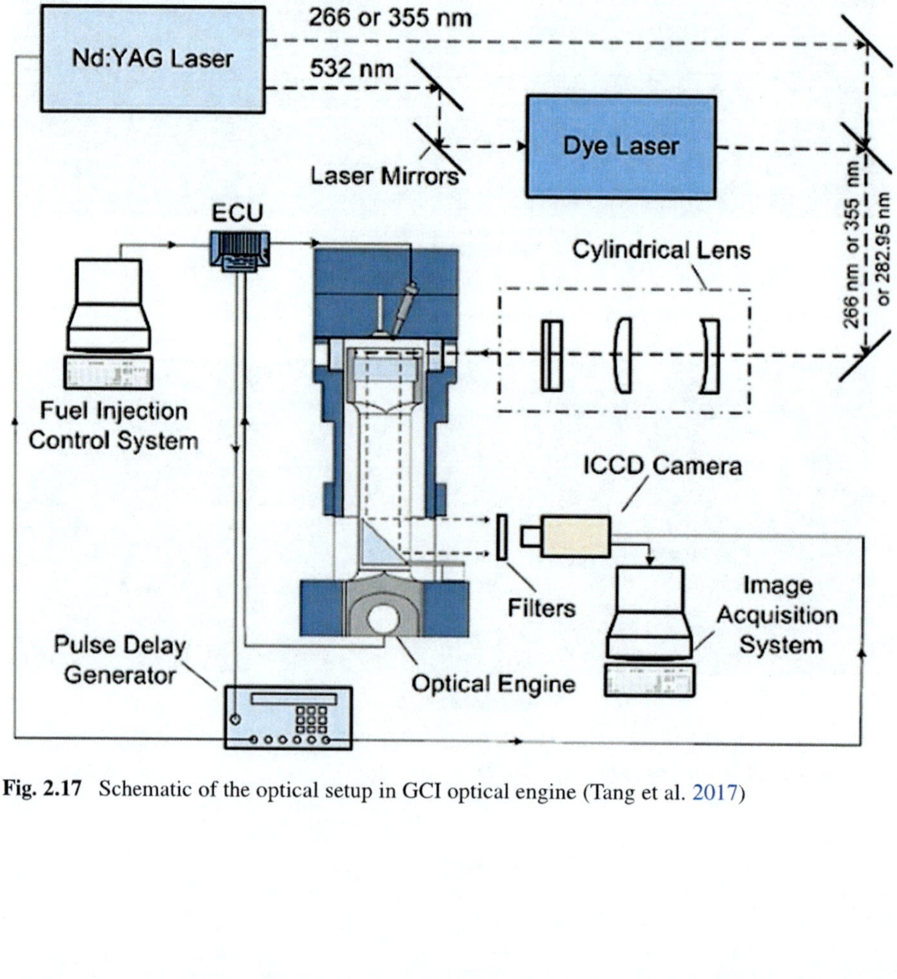

Fig. 2.17 Schematic of the optical setup in GCI optical engine (Tang et al. 2017)

Table 2.8 Optical GCI engine specifications (Tang et al. 2017)

Bore	92 mm
Stroke	100 mm
Displacement	0.664 L
Connecting rod length	155 mm
Compression ratio	11
Combustion chamber shape	Cylindrical
Combustion chamber diameter	63 mm
Common rail pressure	600 bar
Holes number for DI injector	6
Spray included angle	150°
Injector hole diameter	0.15 mm

Table 2.9 Engine operating conditions (Tang et al. 2017)

	PPC-25	PPC-90
Engine speed (rpm)	1200	–
Intake pressure (bar)	1	–
Intake temperature (°C)	125	–
DI pressure (bar)	600	–
Fuel	PRF70	–
Fuel mass (mg/cycle)	26	–
Direct injection (°ATDC)	−25	−90
Overall equivalence ratio	0.77	–
IMEP (bar)	5.99	5.90

are listed in Table 2.8. Table 2.9 depicts the details of engine operating conditions used in this study. In PPC-25 and PPC-90, the numbers indicate the main fuel injection timing. Single-injection strategy was adopted for both cases. The purpose of this investigation was to explore the fundamental combustion behaviour such as ignition, flame development, etc. in GCI combustion engines.

Investigations were carried out in a naturally aspirated, single-cylinder optical engine, in which Bowditch piston was used. An optical cylindrical combustion chamber was designed with a flat quartz window at the bottom, which provided optical access to the combustion chamber. A cut-out section of 40 mm was provided on the ring of the combustion chamber for the entry of the laser sheet into engine combustion chamber near TDC. A UV reflection mirror was mounted on the cylinder block. In this engine, modifications in the combustion chamber reduced the compression ratio to ~11 (Fig. 2.18).

A thin horizontal laser sheet (~1 mm thick, 30 mm wide) was produced by cylindrical lens group and then it was directed into the combustion chamber through the quartz ring window. The laser sheet was located in a plane 10 mm below the fire deck.

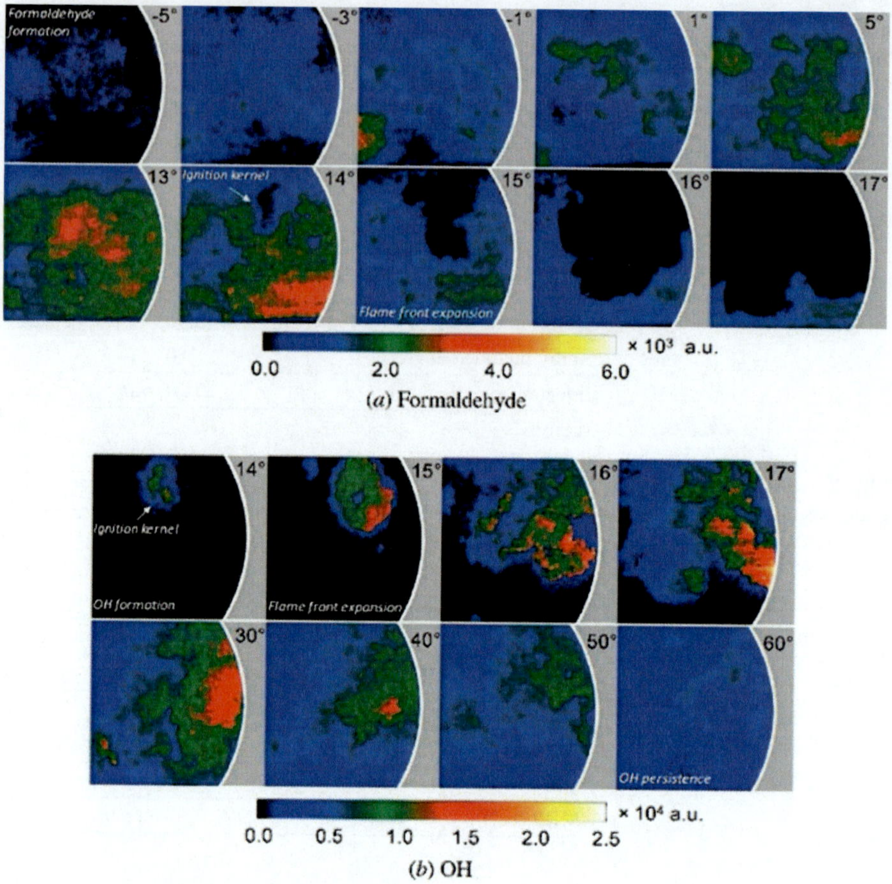

Fig. 2.18 Formaldehyde and OH PLIF for flame evolution in GCI engine (Tang et al. 2017)

For fuel tracer PLIF diagnostics, fourth harmonic Nd:YAG laser with 266 nm wavelength and 60 mJ/pulse energy was used for fuel tracer excitation. For formaldehyde and OH PLIF diagnostics, the third harmonic Nd:YAG laser with 355 nm wavelength and 70 mJ/pulse energy was used for formaldehyde excitation. For OH PLIF, the second harmonic of the Nd:YAG laser pumped dye laser, which generated a beam of 282.93 nm wavelength to excite Q_1 (6) line of the A-X (1, 0) transition of OH.

The results were divided into four stages:

i. Multiple auto-ignition kernels emerging in the fuel-rich regions.
ii. Flame front propagation of auto-ignition kernels towards fuel-lean regions.
iii. Auto-ignition in the end-gas in fuel-lean regions.
iv. A 'burnout' stage in the whole combustion chamber after the end of main heat release.

Formaldehyde LIF marks the low-temperature heat release at −5° bTDC. The distribution covers the entire combustion chamber by 13° aTDC. At 14° aTDC the flame kernel appears near the bowl periphery, much like conventional diesel combustion. OH LIF signal marks the start of high-temperature heat release, which then consumed the formaldehyde as combustion progressed. From this study, it was concluded that variations in the degree of fuel stratification caused by different fuel injection timings were responsible for this transformation in the flame development pattern in GCI engines.

2.4.6 A Case Study for Fuel Distribution in GDI Engine by LIF Technique

This study was focused on high-speed imaging to investigate the cycle-to-cycle variations and captured the rare events during HCCI combustion operated at a particular engine operating condition (Fig. 2.19).

In order to study the combustion behaviour of fuel within the cylinder, LIF technique was used in an optical gasoline direct injection (GDI) engine. In this study, hydroxyl radical LIF images were acquired in an optical square-piston engine.

For optical access (Fig. 2.20), laser A and laser B beams were directed into beam splitter, from where these laser beams were directed towards mirror M2. This beam was focused towards optical combustion chamber using various lenses, which made the combustion chamber optically accessible. In this study, crucial processes including the intake air-flow, mixture preparation, ignition, flame kernel growth,

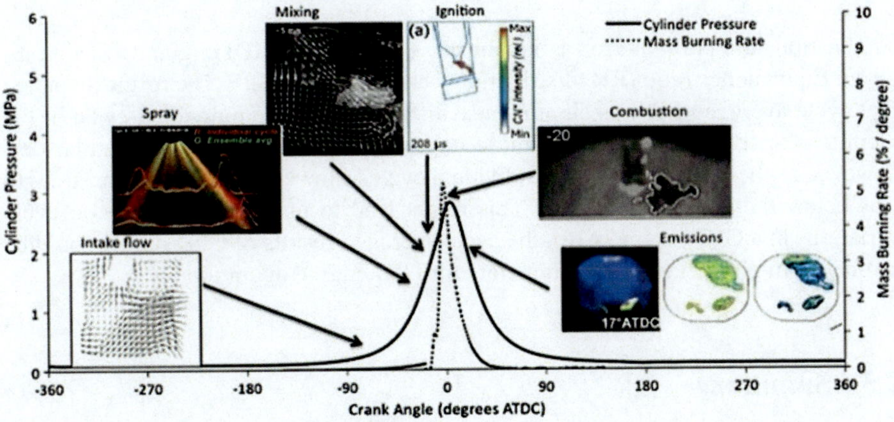

Fig. 2.19 The occurrence and timing of processes that were studied with high-speed imaging in SIDI engines are illustrated along with the in-cylinder pressure and mass-burn rate in a late injection SIDI engine experiment (Sick et al. 2010)

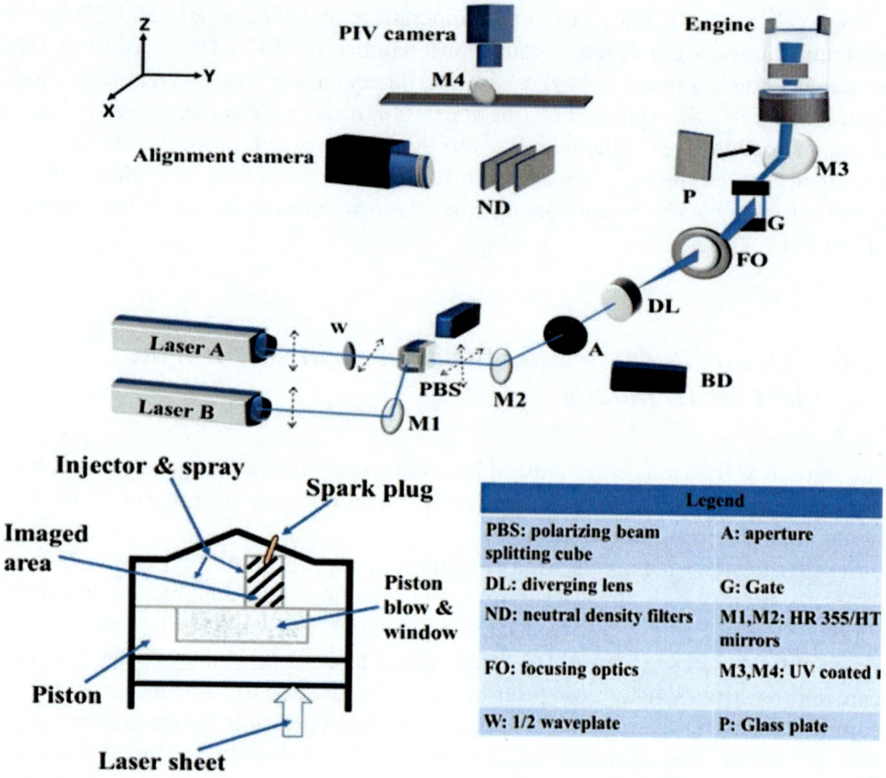

Fig. 2.20 Experimental setup

combustion, and pollutant formation during a cycle in the GDI engine were investigated. Equivalence ratio (ER) distribution was obtained by LIF. The results depicted that cycle averaged ER was significantly different from the individual cycle in the vicinity of spark plug. During the initial stage of spray, there were some similarities, however, as the cycle progressed, individual cycles showed that the mixture strength was below the combustible limit. This might lead to misfiring and cycle-to-cycle variations in a GDI engine. Also, the cycle averaged results of CFD model could be erroneous in predicting ideal parameters for GDI engine operation.

2.5 Summary

Laser-induced fluorescence (LIF) is a popular tool for state-of-the-art quantitative analysis of temperature and species distribution. Combustion researchers are putting significant efforts to make the fuel–air mixture further leaner. Under such a scenario, precise control of fuel–air mixture is important because local over-leaning may lead to combustion failure, resulting in emission spikes. LIF can be effectively used to avoid the possibility of such incidents. In this chapter, the scope of LIF application in IC engines has been discussed. This chapter provides a brief information related to different aspects of LIF such as fundamentals, components of LIF setup, methodology, and applications. A brief discussion on the design of optical access for the experimental IC engines has been provided. In IC engines, LIF has been successfully implemented to investigate in situ temperature stratification and autoignition phenomenon during combustion. Various factors affecting LIF results have been also discussed, which showed that both, the temperature and pressure affect the accuracy of quantitative measurements using LIF. Several case studies related to the application of LIF in IC engines show the relevance of LIF technique for better realization of in-cylinder processes and combustion. Development of advanced and user-friendly commercial tools for LIF would help develop effective emission control technology for IC engines and improve the fuel economy.

References

Banko AJ, Benson MJ, Gunady IE, Elkins CJ, Eaton JK (2020) An improved three-dimensional concentration measurement technique using magnetic resonance imaging. Exp Fluids 61(2):53

Bardi M, Di Lella A, Bruneaux G (2019) A novel approach for quantitative measurements of preferential evaporation of fuel by means of two-tracer laser induced fluorescence. Fuel 239:521–533

Bessler WG, Schulz C, Hartmann M, Schenk M (2001). Quantitative in-cylinder NO-LIF imaging in a direct-injected gasoline engine with exhaust gas recirculation. SAE Trans pp 1837–1846

Bessler WG, Schulz C, Shin DI, Lee T, Jeffries JB, Hanson RK (2002) Strategies for NO laser-induced fluorescence in methane/air flames at pressures between 1 and 60 bar. In: Laser applications to chemical and environmental analysis, p FB4. Optical Society of America

Brackmann C, Nygren J, Bai X, Li Z, Bladh H, Axelsson B, Denbratt I, Koopmans L, Bengtsson PE, Aldén M (2003) Laser-induced fluorescence of formaldehyde in combustion using third harmonic Nd: YAG laser excitation. Spect Acta Part A Mol Biomol Spectrosc 59(14):3347–3356

Burkert A, Triebel W, Stafast H, König J (2002) Single-shot imaging of gas temperatures in low-temperature combustion based on laser-induced fluorescence of formaldehyde. Proc Combust Inst 29(2):2645–2651

Burkert A, Paa W, Reimert M, Klinkov K, Eigenbrod C (2013) Formaldehyde LIF detection with background subtraction around single igniting GTL diesel droplets. Fuel 111:384–392

Chen B, Feng L, Wang Y, Ma T, Liu H, Geng C, Yao M (2019) Spray and flame characteristics of wall-impinging diesel fuel spray at different wall temperatures and ambient pressures in a constant volume combustion vessel. Fuel 235:416–425

Cheong J, Wigger S, Füßer HJ, Kaiser SA (2020) High-resolution LIF-Imaging of the oil film thickness in the piston-ring/cylinder-liner contact in an optical tribometer. Tribol Int 106230

Cordier M, Itani L, Bruneaux G (2020) Quantitative measurements of preferential evaporation effects of multi-component gasoline fuel sprays at ECN Spray G conditions. Int J Engine Res 21(1):185–198

Deschamps BM, Smallwood GJ, Prieur J, Snelling DR, Gülder ÖL (1996) Surface density measurements of turbulent premixed flames in a spark-ignition engine and a Bunsen-type burner using planar laser-induced fluorescence. In: Symposium (international) on combustion, vol 26, no 1, pp 427–435. Elsevier

Donkerbroek AJ, Van Vliet AP, Somers LMT, Frijters PJM, Klein-Douwel RJH, Dam NJ, Meerts WL, ter Meulen JJ (2010) Time-and space-resolved quantitative LIF measurements of formaldehyde in a heavy-duty diesel engine. Combust Flame 157(1):155–166

Fujikawa T, Fukui K, Hattori Y, Akihama K (2006) 2-D temperature measurements of unburned gas mixture in an engine by two-line excitation LIF technique. SAE technical paper 2006-01-3336

Gamba M, Miller VA, Mungal MG, Hanson RK (2015) Temperature and number density measurement in non-uniform supersonic flow fields undergoing mixing using toluene PLIF thermometry. Appl Phys B 120(2):285–304

Gessenhardt C, Schulz C, Kaiser SA (2015) Endoscopic temperature imaging in a four-cylinder IC engine via two-color toluene fluorescence. Proc Combust Inst 35(3):3697–3705

Haessler H, Bockhorn H, Pfeifer C, Kuhn D (2012) Formaldehyde-LIF of dimethyl ether during auto-ignition at elevated pressures. Flow Turbul Combust 89(2):249–259

Hildingsson L, Johansson B, Hultqvist A, Särner G, Richter M, Aldén M (2005) Simultaneous formaldehyde and fuel-tracer LIF imaging in a high-speed diesel engine with optically accessible realistic combustion chamber. SAE Technical Paper 2005-24-008

Itani LM, Bruneaux G, Di Lella A, Schulz C (2015) Two-tracer LIF imaging of preferential evaporation of multi-component gasoline fuel sprays under engine conditions. Proc Combust Inst 35(3):2915–2922

Kaiser SA, Schild M, Schulz C (2013) Thermal stratification in an internal combustion engine due to wall heat transfer measured by laser-induced fluorescence. Proc Combust Inst 34(2):2911–2919

Koban W, Koch JD, Hanson RK, Schulz C (2004) Absorption and fluorescence of toluene vapor at elevated temperatures. Phys Chem Chem Phys 6(11):2940–2945

Krämer H, Einecke S, Schulz C, Sick V, Nattrass SR, Kitching JS (1998). Simultaneous mapping of the distribution of different fuel volatility classes using tracer-LIF tomography in an IC engine. SAE Trans 1049–1060

Kranz P, Kaiser SA (2019) LIF-based imaging of preferential evaporation of a multi-component gasoline surrogate in a direct-injection engine. Proc Combust Inst 37(2):1365–1372

Lachaux T, Musculus MP (2007) In-cylinder unburned hydrocarbon visualization during low-temperature compression-ignition engine combustion using formaldehyde PLIF. Proc Combust Inst 31(2):2921–2929

Li Y, Zhao H, Ma T (2006) Flow and mixture optimization for a fuel stratification engine using PIV and PLIF techniques. J Phys Conf Ser 45(1):59. IOP Publishing

Li B, Jonsson M, Algotsson M, Bood J, Li ZS, Johansson O, Aldén M, Tunér M, Johansson B (2013) Quantitative detection of hydrogen peroxide in an HCCI engine using photo fragmentation laser-induced fluorescence. Proc Combust Inst 34(2):3573–3581

Li B, Zhang D, Yao M, Li Z (2017) Strategy for single-shot CH3 imaging in premixed methane/air flames using photofragmentation laser-induced fluorescence. Proc Combust Inst 36(3):4487–4495

Lind S, Trost J, Zigan L, Leipertz A, Will S (2015) Application of the tracer combination TEA/acetone for multi-parameter laser-induced fluorescence measurements in IC engines with exhaust gas recirculation. Proc Combust Inst 35(3):3783–3791

Lind S, Retzer U, Will S, Zigan L (2017) Investigation of mixture formation in a diesel spray by tracer-based laser-induced fluorescence using 1-methylnaphthalene. Proc Combust Inst 36(3):4497–4504

Ma X, Xu H, Jiang C, Shuai S (2014) Ultra-high speed imaging and OH-LIF study of DMF and MF combustion in a DISI optical engine. Appl Energy 122:247–260

Ma L, Lei Q, Ikeda J, Xu W, Wu Y, Carter CD (2017) Single-shot 3D flame diagnostic based on volumetric laser induced fluorescence (VLIF). Proc Combust Inst 36(3):4575–4583

Malmqvist E, Jonsson M, Larsson K, Aldén M, Bood J (2016) Two-dimensional OH-thermometry in reacting flows using photofragmentation laser-induced florescence. Combust Flame 169:297–306

Miles PC (2014) The history and evolution of optically accessible research engines and their impact on our understanding of engine combustion. In: ASME 2014 internal combustion engine division fall technical conference. American Society of Mechanical Engineers Digital Collection

Nygren J, Hult J, Richter M, Aldén M, Christensen M, Hultqvist A, Johansson B (2002) Three-dimensional laser induced fluorescence of fuel distributions in an HCCI engine. Proc Combust Inst 29(1):679–685
Omar M, Dominique T, Patrizio M, Jérôme B (2020) Investigation on the conditions leading to the micro-explosion of emulsified fuel droplet using two colors LIF method. Exper Thermal Fluid Sci 110106
Ottenwälder T, Schulz C, Raffius T, Koß HJ, Grünefeld G, Heufer KA, Pischinger S (2018) Quantitative nitrogen oxide measurements by laser-induced fluorescence in diesel-like n-heptane jets with enhanced premixing. Combust Flame 188:250–261
Peterson B, Baum E, Böhm B, Sick V, Dreizler A (2013) High-speed PIV and LIF imaging of temperature stratification in an internal combustion engine. Proc Combust Inst 34(2):3653–3660
Peterson B, Baum E, Böhm B, Sick V, Dreizler A (2014) Evaluation of toluene LIF thermometry detection strategies applied in an internal combustion engine. Appl Phys B 117(1):151–175
Peterson B, Baum E, Böhm B, Sick V, Dreizler A (2015) Spray-induced temperature stratification dynamics in a gasoline direct-injection engine. Proc Combust Inst 35(3):2923–2931
Peterson B, Baum E, Dreizler A, Böhm B (2019) An experimental study of the detailed flame transport in a SI engine using simultaneous dual-plane OH-LIF and stereoscopic PIV. Combust Flame 202:16–32
Qin S, Krohn B, Petrov V, Manera A (2020) Velocity and scalar fields of a turbulent buoyant jet in the self-similar region. Nucl Technol 206(2):307–321
Raman V, Tang Q, An Y, Shi H, Sharma P, Magnotti G, Chang J, Johansson B (2020) Impact of spray-wall interaction on the in-cylinder spatial unburned hydrocarbon distribution of a gasoline partially premixed combustion engine. Combust Flame 215:157–168
Schießl R, Maas U (2003) Analysis of end gas temperature fluctuations in an SI engine by laser-induced fluorescence. Combust Flame 133(1–2):19–27
Schmitt S, Wick M, Wouters C, Ruwe L, Graf I, Andert J, Hansen N, Pischinger S, Kohse-Höinghaus K (2020) Effects of water addition on the combustion of iso-octane investigated in laminar flames, low-temperature reactors, and an HCCI engine. Combust Flame 212:433–447
Schrewe MR, Ghandhi JB (2007) Near-wall formaldehyde planar laser-induced fluorescence measurements during HCCI combustion. Proc Combust Inst 31(2):2871–2878
Schulz C, Sick V (2005) Tracer-LIF diagnostics: quantitative measurement of fuel concentration, temperature and fuel/air ratio in practical combustion systems. Prog Energy Combust Sci 31(1):75–121
Schulz C, Yip B, Sick V, Wolfrum J (1995) A laser-induced fluorescence scheme for imaging nitric oxide in engines. Chem Phys Lett 242(3):259–264
Scott B, Willman C, Williams B, Ewart P, Stone R, Richardson D (2017) In-cylinder temperature measurements using laser induced grating spectroscopy and two-colour PLIF. SAE Int J Engines 10(4):2191–2201
Shen X, Gao X, Li Z, Li B, Zheng C, Sun Z, Ni M, Cen K, Aldén M (2012) PLIF diagnostics of NO oxidization and OH consumption in pulsed corona discharge. Fuel 102:729–736
Shi H, Tang Q, An Y, Raman V, Sim J, Chang J, Magnotti G, Johansson B (2020) Study of spray/wall interaction in transition zones from HCCI via PPC to CI combustion modes. Fuel 268:
Shibata G, Nishiuchi S, Xie P, Takai S, Ogawa H, Kobashi Y (2020) Measurements of fuel adhesion on cylinder walls and fuel wall-flow behavior with post diesel fuel injections. Int J Engine Res 21(2):352–366
Sick V, Drake MC, Fansler TD (2010) High-speed imaging for direct-injection gasoline engine research and development. Exp Fluids 49(4):937–947
Strozzi C, Claverie A, Prevost V, Sotton J, Bellenoue M (2019) HCCI and SICI combustion modes analysis with simultaneous PLIF imaging of formaldehyde and high-speed chemiluminescence in a rapid compression machine. Combust Flame 202:58–77
Tang Q, Liu H, Li M, Geng C, Yao M (2017a) Multiple optical diagnostics on effect of fuel stratification degree on reactivity controlled compression ignition. Fuel 202:688–698

Tang Q, Liu H, Li M, Yao M, Li Z (2017b) Study on ignition and flame development in gasoline partially premixed combustion using multiple optical diagnostics. Combust Flame 177:98–108
Tang Q, Liu H, Ran X, Li M, Yao M (2020) Effects of direct-injection fuel types and proportion on late-injection reactivity controlled compression ignition. Combust Flame 211:445–455
Thurber MC (1999) Acetone Laser-induced fluorescence for temperature and multi parameter imaging in gaseous flows. PhD thesis, Stanford University
Tran KH, Guibert P, Morin C, Bonnety J, Pounkin S, Legros G (2015) Temperature measurements in a rapid compression machine using anisole planar laser-induced fluorescence. Combust Flame 162(10):3960–3970
Verbiezen K, Klein-Douwel RJH, Van Vliet AP, Donkerbroek AJ, Meerts WL, Dam NJ, Ter Meulen JJ (2007) Quantitative laser-induced fluorescence measurements of nitric oxide in a heavy-duty Diesel engine. Proc Combust Inst 31(1):765–773
Williams B, Ewart P, Wang X, Stone R, Ma H, Walmsley H, Cracknell R, Stevens R, Richardson D, Fu H, Wallace S (2010) Quantitative planar laser-induced fluorescence imaging of multi-component fuel/air mixing in a firing gasoline-direct-injection engine: effects of residual exhaust gas on quantitative PLIF. Combust Flame 157(10):1866–1878
Wu Y, Xu W, Lei Q, Ma L (2015) Single-shot volumetric laser induced fluorescence (VLIF) measurements in turbulent flows seeded with iodine. Opt Express 23(26):33408–33418
Zabeti S, Drakon A, Faust S, Dreier T, Welz O, Fikri M, Schulz C (2015) Temporally and spectrally resolved UV absorption and laser-induced fluorescence measurements during the pyrolysis of toluene behind reflected shock waves. Appl Phys B 118(2):295–307
Zhang Z, Li T, Shi W (2019a) Ambient Tracer-LIF for 2-D quantitative measurement of fuel concentration in gas jets. Energy 171:372–384
Zhang Q, Su J, Du W, Yuan H, Yang F, Tan S (2019b) Experimental study on mixing phenomenon inside reactor down-comer under single-loop injection using laser induced fluorescence. Prog Nucl Energy 117:
Zhao H, Peng Z, Ma T (2004) Investigation of the HCCI/CAI combustion process by 2-D PLIF imaging of formaldehyde. SAE technical paper 2004-01-1901

Chapter 3
Challenges and Opportunities of Particle Imaging Velocimetry as a Tool for Internal Combustion Engine Diagnostics

Ashutosh Jena, Akhilendra Pratap Singh, and Avinash Kumar Agarwal

Abstract To make further progress on in situ reduction of pollutant formation in engines, the understanding of combustible mixture preparation is vital. Time-dependent development of flow structures and distribution of turbulent kinetic energy (TKE) are instrumental in charge preparation for both premixed as well as mixing controlled combustion phases. Non-linear and unsteady nature of in-cylinder air motion has remained a grey area ever since the days of initial development of internal combustion (IC) engines. The optimization of in-cylinder flow structures and development of numerical models for an in-depth understanding of the in-cylinder processes has become critical in view of the need for complying with stringent emission regulations. This can be realized by in-cylinder flow visualization and continuous tuning and validation of Computational Fluid Dynamics (CFD) models. Particle Image Velocimetry (PIV) has evolved as a pioneering tool for the investigation of intake air-flow structure, flow interaction and fluid motion. However, there are several challenges for the utilization of PIV for in-cylinder flow investigations in IC engines. While the intricate geometry of engine creates hindrance for optical access, the dynamic nature of ambient conditions complicates the selection of seeds. The work presented in this chapter summarizes these critical issues along with the possible solutions. Comprehensive literature on the evolution of PIV as a diagnostics tool for engine application has been covered. A brief review on the impact of flow structures on the combustion and pollutant formation has been also discussed. This chapter is useful for thorough understanding of PIV and its applications in IC engine and provides direction for further innovations in the field.

Keywords PIV · Optical diagnosis · Fuel–air mixing · Engines · Pollutant formation

A. Jena · A. P. Singh · A. K. Agarwal (✉)
Engine Research Laboratory, Indian Institute of Technology Kanpur, Kanpur 208016, India
e-mail: akag@iitk.ac.in

A. P. Singh and A. K. Agarwal (eds.), *Novel Internal Combustion Engine Technologies for Performance Improvement and Emission Reduction*, Energy, Environment, and Sustainability, https://doi.org/10.1007/978-981-16-1582-5_3

3.1 Introduction

Growing demand for clean and efficient power plants for automotive sector has led to innovations in advanced combustion technologies. The last two decades of research have unleashed the potential of low temperature combustion (LTC) technologies to address such challenges. However, the advancement in combustion strategy also calls for modifications in engine design to achieve these goals. LTC is pushing the equivalence ratio towards lean limits. Hence, the understanding of efficient mixing and fuel stratification is highly important to maintain combustion stability without compromising the engine performance and exhaust emissions. This is where the role of in-cylinder fluid motion and spray–flame interactions play a critical role. It is well established that in-cylinder turbulence has significant effect on combustion and emissions (Konno et al. 1992; Selamet et al. 2004). Combustion control by enhancing turbulence to mitigate pollutant formation is an active area of research since last few decades (Konno et al. 1992; Selamet et al. 2004; Jaichandar and Annamalai 2012; Zhou et al. 2017; Karthickeyan 2019). Intake valve masking, and variations in port geometry have shown promising results in reducing emissions (Udayakumar et al. 2003; Porpatham et al. 2013; Bari and Saad 2015). Different combustion geometries have been explored to investigate the effect of flow behaviour on the pollutant formation. Optimization of various design parameters affecting the in-cylinder flows also affect the formation of soot, hydrocarbons (HC) and carbon monoxide (CO). On the other hand, oxides of nitrogen (NOx) may increase due to improved combustion, leading to rise in peak in-cylinder temperature (Kaplan 2019). This shows the requirement of more involved research in design optimization to meet stringent emission norms. However, due to cost and time constraints, investigations are limited to a small number of geometries.

In the last two decades, Computational Fluid Dynamics (CFD) has emerged as a popular tool in automotive industry for optimizing the engine designs. Recent outcomes obtained by variations in port and bowl geometry designs have shown significant potential in in-cylinder flow optimization and reducing exhaust emissions (Brijesh et al. 2015; Huang et al. 2017; Wang et al. 2020; Fatehi et al. 2019). Development of CFD models requires experimental results for model validation. Therefore, in-cylinder flow-field measurements are of prime importance for the development of cleaner engines. Traditionally, Hot Wire Anemometry (HWA), paddle wheel, spark discharge, and smoke wire methods, etc. have been used to measure the fluid velocity in gas-phase studies (Kaplan 2019; Kuniyoshi et al. 1980; Cho et al. 1990). However, these methods are intrusive in nature and affect the flow-field in the area of interest (AoI), as shown in Fig. 3.1.

In recent years, laser-based diagnostics tools are extensively used due to their non-intrusive measurement capabilities. Laser Doppler Anemometry (LDA) and Laser Doppler Velocimetry (LDV) have revealed useful details of the in-cylinder flows. These techniques are excellent for single point measurements; however, Particle Image Velocimetry (PIV) provides flow-field information in a plane, which is useful for studying the flow characteristics in the engine combustion chamber and

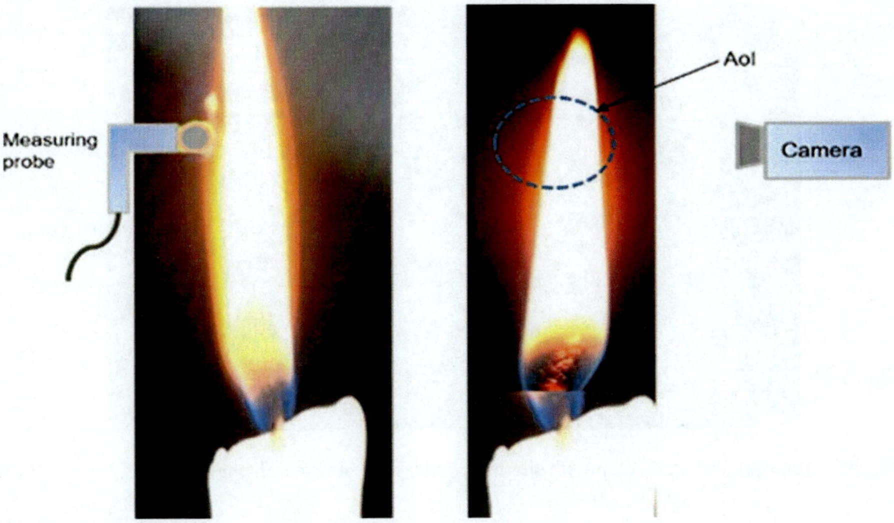

Fig. 3.1 Intrusive measurement (left) and non-intrusive measurement using imaging techniques (right)

spray–flow interactions. PIV has, therefore, evolved as a first-choice tool among different laser diagnostic techniques over time. This is a decade-old technology, which provides temporal measurement of velocity field in the flow. The basic principle of PIV is based on indirect measurement of velocity of the seed particles in the flow, where seed particles are assumed to follow the flow-field faithfully. This assumption is possible in PIV because of significantly smaller size of seed particles relative to the scale of the flow. The regime of the flow-field to be measured is illuminated by a double pulsed, high-intensity laser sheet, where two successive illuminations take place within a fraction of a second. The scattered light from the seed particles is captured by high-speed cameras, usually placed perpendicular to the plane of illumination. From these two consecutive images, the changes in location of a particular seed particle are determined. With the knowledge of time lag between the two successive illuminations and distance covered by a seed particle, the velocity of the flow-field is calculated in the form of a velocity vector. A sufficient number of particles should be present in the volume of interest to maintain precision in the measurement of the flow-field. Identifying a particular particle in the successive images, particularly in complex turbulent flow-field is difficult. This could be achieved by different correlation algorithms. Figure 3.2 shows the setup for PIV measurements in an optical research engine at ERL, IIT Kanpur.

Subsequent sections of this chapter will discuss various challenges faced in the application of PIV in real engines and the solutions to these problems. A brief idea of various engine designs has been provided for better understanding to readers. Fundamental issues with laser alignment and seeding of particles have been also

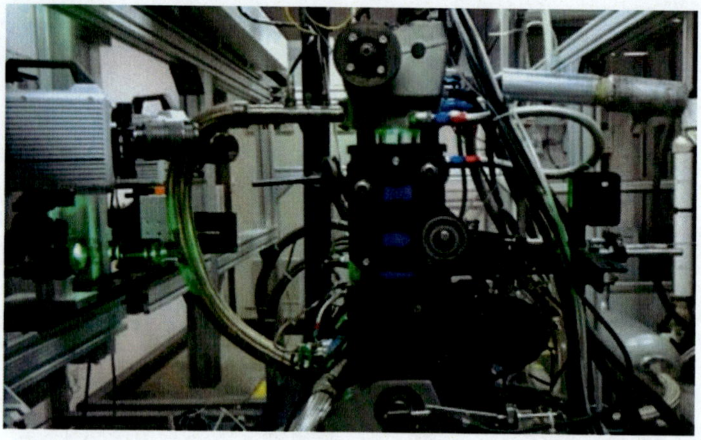

Fig. 3.2 Tomographic PIV setup with single-cylinder optical research engine

discussed in subsequent sections with comprehensive literature review. To gain a better understanding, several case studies have been provided with special attention to spray, combustion, and in-cylinder flow dynamics.

3.2 Development of PIV as a Tool for Flow Visualization

3.2.1 Engine Modifications

An optical access is a key requirement for application of any visualization technique for investigating reacting and non-reacting fluid flows. Constant volume combustion chambers (CVCC) and rapid compression machines (RCM) have relatively simpler designs and offer useful insights of some of the engine processes. Several designs have already been developed to provide optical access to the engine combustion chamber. An elaborate discussion on historical evolution of optical engines can be found in a previous study (Miles 2014). Based on the extent of optical access required, there are several variations in optical engine designs; however, most optical engines consist of Bowditch style elongated piston. The basic structure of the optical research engine is shown in Fig. 3.3. In this section, few common design variations and their limitations are discussed.

Engine with metallic liner and piston with optical window: In this design (Fig. 3.4a), a small optical window is provided in the piston rim. The bowl geometry is simple (flat), unlike realistic pistons used in the engines. Piston reciprocates in a metallic liner, similar to metallic engines. Visualization is possible from the bottom of the piston by placing a 45° inclined mirror. In such configuration, observation and analysis can be made in a swirl plane only. Such a design is useful for studying

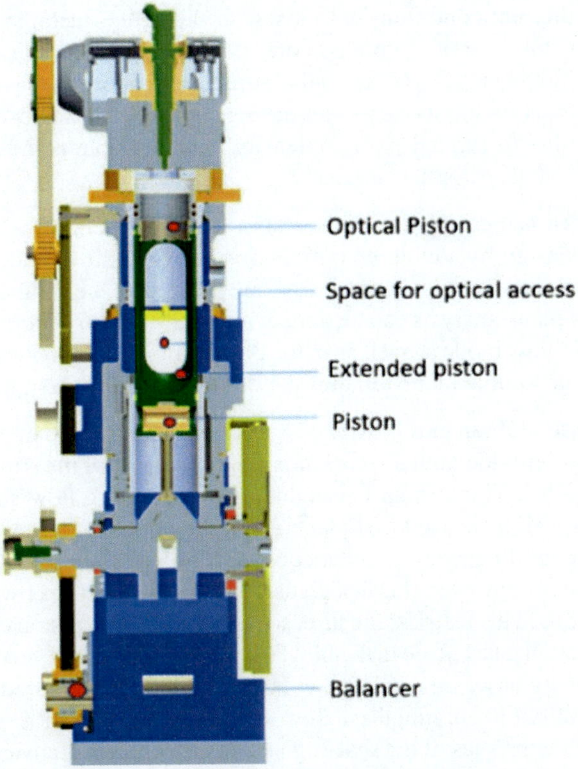

Fig. 3.3 Optical research engine with details of fundamental modifications for optical access (AVL nd)

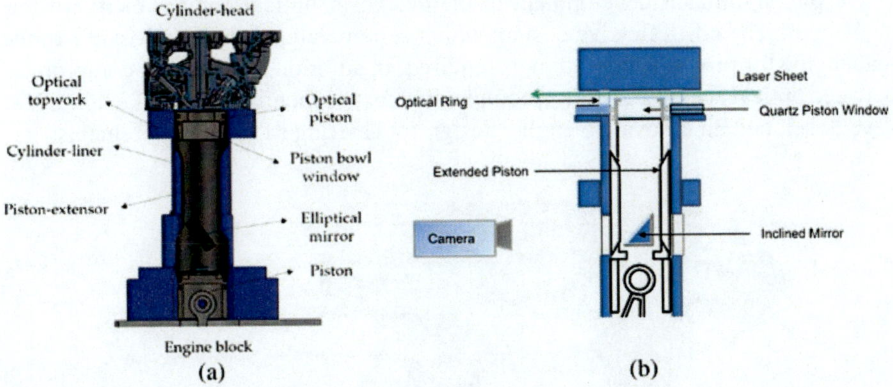

Fig. 3.4 **a** Research engine with metallic liner and optical piston (Benajes et al. 2018), **b** Research engine with optical rim for laser access

combustion as thermal conditions are closest to the actual metallic engines. Also, the run time for this type of engine is more. PIV is possible in this type of optical engine for flow-field analysis, which yields sufficient scattering. Gas-phase analysis can be done by tracking the soot particles during combustion. However, the accuracy is questionable due to limited particle density. Also, tumble plane analysis is not possible with this type of optical engines.

Optical rim with optical piston: This design (Fig. 3.4b) adopts a small variation to the previous design by adding an optical ring between the cylinder head and the liner. This optical ring provides access of the laser sheet into the combustion chamber. Therefore, swirl plane analysis can be done without the loss of accuracy. Analysis of flow-field in cold flow mode as well as in the combustion mode is possible. However, the analysis of the tumble plane remains the drawback of this design as well.

Engine with optical liner and optical piston: In this design, the liner is made fully optical in order to provide optical access for maximum part of the stroke. The design is shown in Fig. 3.2. This design is capable of investigating flow structures in the tumble plane as well as in the swirl plane. Multiple optical accesses provide scope for multiple optical diagnoses simultaneously, in order to understand complicated in-cylinder physics. However, this design uses a simplistic piston bowl shape similar to previous designs. This restricts the laser to pass through the piston cavity. In diesel engines, combustion takes place in the bowl cavity, when piston is very close to TDC. Therefore, in-cavity processes during combustion cannot be captured in this type of optical configuration. Resolving these flow structures is important to understand the heat transfer characteristics of the real engines. Hence, a further advancement to this design introduces a realistic piston bowl geometry. This is the most innovative design used till date for optical diagnostics. Figure 3.5 shows the design with realistic piston bowl geometry.

Despite significant development in the design of optical engine, there are few challenges. The cost-effective design, which can operate as close as possible to the real engine for prolonged duration, is required. In addition, a minimum compromise between optical access and design complexity must be incurred. Benajes et al. (2018) have developed an innovative engine design for catering to these requirements.

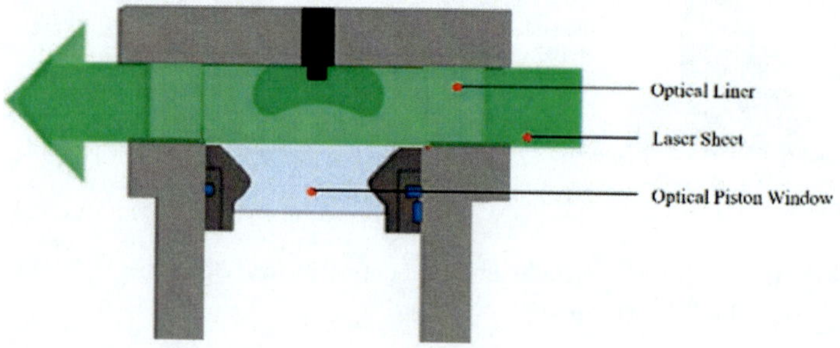

Fig. 3.5 Research engine with full optical access and realistic piston bowl geometry

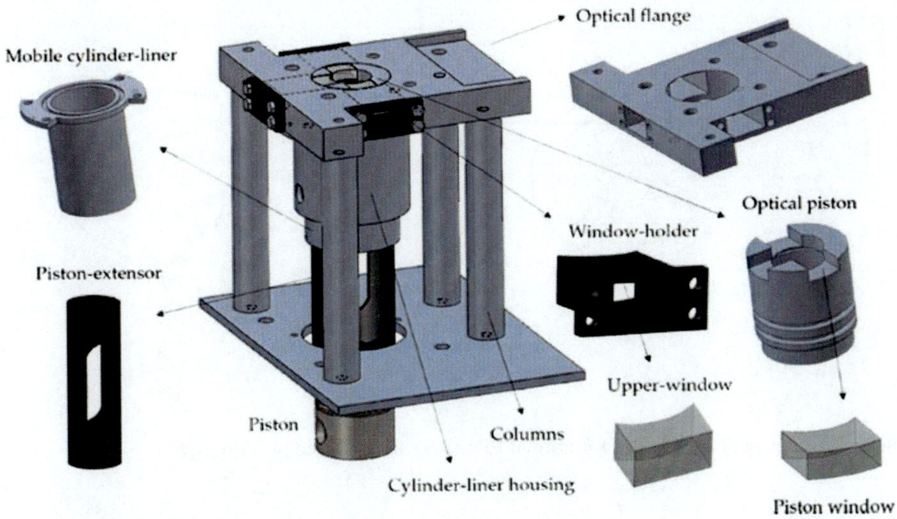

Fig. 3.6 Key modifications for minimising down-time of the optical engine (Benajes et al. 2018)

Figure 3.6 illustrates major changes in design of engine components. Optical windows during engine operation are often subjected to fouling. Quartz piston also deteriorates quickly under high-temperature conditions. Under such circumstances, cleaning and fixing take around 4–5 h. Benajes et al. (2018) proposed to replace the integral top work (Head and liner) with multiple sets of easily removable parts to tackle this issue. As shown in Fig. 3.6, the columnar design incorporates a mobile water-cooled cylinder liner, which can be detached by unscrewing four fasteners. This provides access to the piston so that the piston rings can be easily replaced in 4 min. A separate optical flange accommodates two optical windows screwed to it with the help of window holders. In case of any fouling issues, the windows can be removed, cleaned and fixed in 2 min with experience. There is also a provision to replace the optical window with metal dummies for metallic engine experiments. The piston windows are fixed and glued with high-temperature resistant adhesives. Conical shape further arrests its degree of freedom.

Engine with endoscopic PIV: Endoscopic PIV is the simplest design ever developed for in-cylinder PIV investigations. The design requires neither extended piston nor any sophisticated optical windows. Any commercial engine can be adopted and modified for endoscopic PIV study. Two holes of ~10 mm diameter are required for laser and sensor. Despite the drawback of a limited field of view, this technique has been extensively used for investigations of in-cylinder processes in real engine conditions. La-Vision's FlowMaster Endoscopic PIV system is the commercial diagnosis tool developed for this purpose. Several studies have been conducted with endoscopic PIV for high-temperature applications (Rottier et al. 2010; Nishiyama et al. 2012). Dierksheide et al. (2002) demonstrated an endoscopic PIV for in-cylinder 2D flow analysis. They used a borescope in this experiment along with suitably

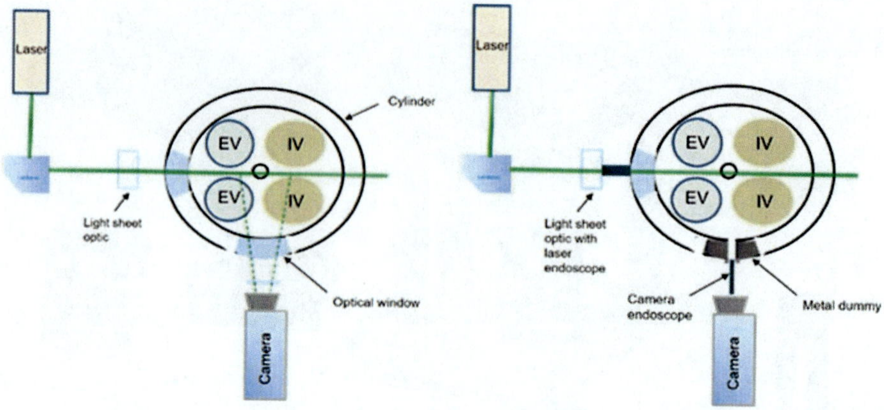

Fig. 3.7 Arrangement of optics for normal PIV (left) and endoscopic PIV (right)

placed achromatic lenses and rod lenses for transmitting the light through the guide tube. An endoscope of 8 mm tube diameter was selected. The fibre bundle provided for illumination around the lenses was removed. Therefore, 6 mm lenses could be easily accommodated in place of 4.5 mm lenses. Fibres are used for illumination, which is done by laser through an additional endoscope. This enabled the use of large 6 mm rod lenses, which boosted the brightness of the image. Larger lens diameter and smaller number of lenses should be used to increase the imaging quality. The signals were acquired through the CCD camera (FlowMaster 3S, LaVision, Gottingen, Germany). Distortion in the image was taken care by image correction tools (DaVis ver.6, LaVision, Gottingen, Germany). An 8-mm laser endoscope was used with 7-mm cylindrical lense having −3.7 mm focal length. A light sheet having 45° divergence angle was produced. The sheet thickness was in the range of 1–2 mm near the measurement location. A 200-mJ laser (PIV200, Spectra-Physics, Mountain View, California) was used for this study. The endoscope was tested for 400 mJ. Silica lenses were used instead of glass for handling such high laser power. Figure 3.7 illustrates the arrangement of the optics.

The same setup was used for normal PIV measurement. During endoscopic PIV, the large optical windows were replaced by dummies having provision for endoscopes. The field of view was 84 × 66 mm. Authors compared the results obtained by endoscopic PIV with normal PIV. Excellent agreement in the flow structures was observed among both techniques.

3.2.2 *Optics and Challenges*

The most common lasers used in PIV applications are diode-pumped Nd:YAG or Nd:YLF lasers with wavelength in the green range (510–527 nm). The application of UV laser and PIV without laser has also been explored; however, these techniques are not matured enough. A detailed review on laser selection has been discussed in previous studies (Thurow et al. 2012; Gord et al. 2008). The next important system is the image acquisition system. In PIV, either charge-coupled device (CCD) or complementary metal oxide semi-conductor (CMOS) cameras were used for high-speed image acquisition. CCD sensors have high dynamic range, low noise, high uniformity and sensitivity. On the other hand, CMOS cameras are highly responsive to high frame rate and anti-blooming, hence better suited for high-speed imaging. CCD cameras are applied to PIV investigations with a double shutter mechanism. The exposure time for the second image in PIV image pair depends upon the readout time of the first image. This creates an issue while imaging the high-intensity flames during combustion. Shutter time for the first image can be easily controlled by software; however, the second image procedure takes relatively longer time due to the readout burden of the first one, which continues till some part of the second image is acquired. This causes sufficiently high shutter opening for the second image with the CCD. Li et al. (2008) proposed a mechanical shutter mechanism to overcome this problem. Detailed discussion related to selection aspects of image acquisition devices like CCD, CMOS or intensified camera is given in previous studies (Hain et al. 2007; Weber et al. 2011).

There are several challenges associated with successful implementation of these two basics devices (laser and camera) for PIV investigations. The ability of PIV to resolve complicated structures depends on two basic aspects i.e. imaging and image processing/analysis by various cross-correlation algorithms. Quality of the image captured relies on the laser and appropriate flow seeding. Quality of PIV results is significantly affected by the capability of seeds to follow the flow pattern faithfully; however, the laser pulses should also meet the precision timing requirements. Window fouling, alignment of optics and image correction are several other critical issues. Window fouling can be easily tackled by manual procedures, and image correction can be done by various in-house and commercially developed software packages. A key step in PIV setup is a successful alignment of laser light sheets. Precision velocity mapping is possible if: (a) sheet overlap is achieved, (b) distance is calibrated properly and (c) variation in light-sheet thickness is kept to a minimum. Beam misalignment is one of the common issues, which adversely affects the PIV results. The effectiveness of cross-correlation depends heavily upon the overlap of two successive pulses (Grayson et al. 2016). Cross-correlation is applied by tracking a particular particle in the image pair. Sheet mismatch would result in the loss of a concerned particle in the second image due to slight shifting of the illumination plane from the plane of the particle motion. Figure 3.8 represents the Gaussian intensity distribution of the first laser sheet with four particles inside the sheet thickness. The sheet thickness is measured along the intensity distribution at half of the maximum intensity. In Fig. 3.8, the green curve shows the intensity distribution for the second pulse. This figure shows a relative shift in the laser sheet profile (between blue and green), which leads to escape out of the two particles. Therefore, only one of the three particles in the first image is available for analysis. Two others drift out of the second pulse laser sheet and said to be the lost particles. Clearly, similar thing can

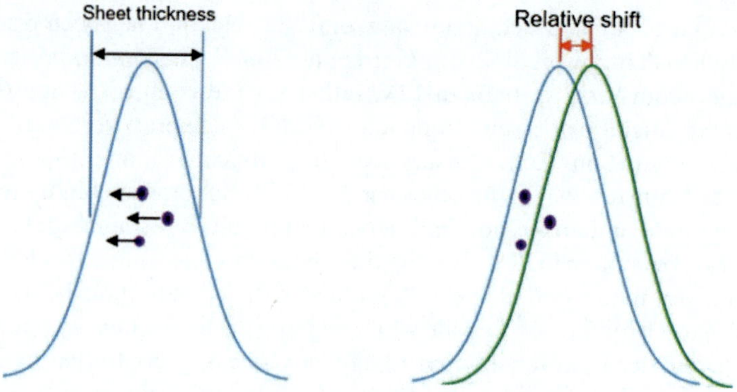

Fig. 3.8 Intensity distribution for the laser sheet. First pulse (left) and second pulse with the relative shift from the first pulse (right)

happen due to out of plane velocity, if the particle moves out of the sheet thickness owing to its normal velocity component.

This is similar to out-of-plane velocity condition in three-dimensional flow. Intentionally misaligned laser pulse technique has been explored to tackle the severe case of out-of-plane velocity components. Several studies have been carried out to understand this issue (Nobach 2011; Scharnowski and Kähler 2016). Increasing particle losses would result in low-quality image acquisition with lesser number of vectors, which may not properly represent the actual flow structures. Thus, detecting possible misalignment prior to the experiment will ensure a high-quality image acquisition for ensuring reliable results. Pfadler et al. (2009) used an alignment control system for proper alignment. This system felicitated the measurement of thickness, separation and parallelism between the consecutive light sheets. Online monitoring of these parameters during the experiment is also important. This is done by directing a fraction of sheets to the monitoring unit. The monitoring system consists of a CCD camera, neutral density filter for signal attenuation and a movable rail to enable online alignment by adjustment along the optical axis.

Grayson et al. (2016) developed a low-cost laser profiling device for laser beam profiling and sheet alignment. This device consists of mirror-less interchangeable lens camera. A Canon EOS-M camera body, 18 Megapixels/230pixel/mm was selected for imaging. 22.3 × 14.9 mm^2 APS-C sensor of the device was used for imaging the laser beams similar to PIV experiments. The camera was triggered by a wired remote coupled to a Wi-Fi-enabled SD card, in order to make the real-time monitoring with remote access possible. To protect the sensor and electronics in the camera from damage by high power laser pulse, attenuation of the signal was required. Neutral density filters were used for this purpose. The filter strength was flexible to work with range of laser powers, which made the device versatile for many applications. Such filters have been successfully implemented by other researchers as well. These filters come with threaded mount in order to adjust the strength by stacking multiple filters. Generally, laser-grade density filters are recommended; however, photographic neutral density filters are economical and easily available. Considering the laser power and cost, the filters should be selected in an optimized manner. Authors preferred to go with a mix of both. For low power laser pulse, photographic neutral density filter of different strengths was employed. A laser grade filter can be threaded on top, in order to handle the high laser power of the order of 400 mJ.

There are many other methods such as traversing knife method, and paper-burn method, which have been implemented by researchers (Mullin and Dahm 2005; Ganapathisubramani et al. 2005). However, development and implementation of such techniques require numerous trials and experience. Commercial laser beam profiler (Ophir, FX-50) with tailor-made software packages was suitable for this purpose. Naka et al. (2016) used this commercial tool for precise alignment. Figure 3.8 shows the measured intensity profiles. A perfect overlap can be seen for all four laser pulses with respective wavelengths and polarization (Fig. 3.9).

Reflection of the laser from the engine boundary creates a problem in imaging the particles and their cross-correlations. In addition, high refection intensity may damage the sensor of the imaging system. This may happen when the laser is not allowed to pass through the volume of integration completely. For example, during the analysis of the tumble plane, illuminating laser enters the cylinder from the bottom, hence cannot escape the head. Such high intensity regimes can be avoided by shielding them physically, when the area of interest is far from the boundaries. However, this remains a compromise because near-wall fluid dynamics is of particular interest in many cases. For instance, spark plug is located close to the boundary in gasoline engines and near spark turbulence has a significant impact on the nucleation of flames. Similarly, flows near the intake valves and in the piston cavity take part in fuel–air mixing and heat transfer processes. Thus, researchers opted for alternative solutions. Fundamentally, the sensors can be protected either by absorbing the incident laser beam or by directing the reflection away from the camera. This is possible by coating the reflecting surface by applying black paint, anodized aluminium, or fluorescent paint, or by surpassing diffuse reflection with the help of surface polishing

Fig. 3.9 Intensity distribution of laser pulses after alignment (Naka et al. 2016)

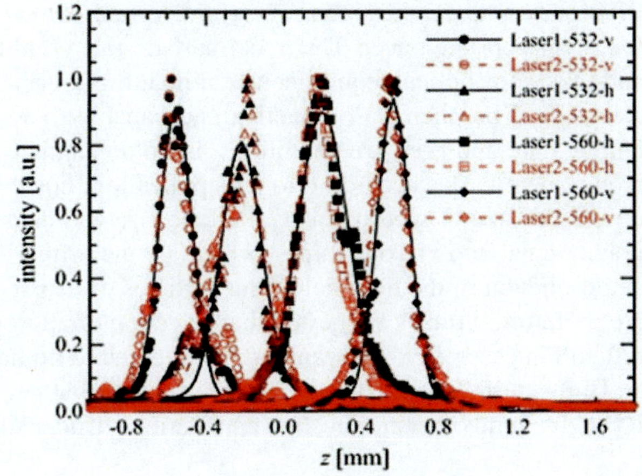

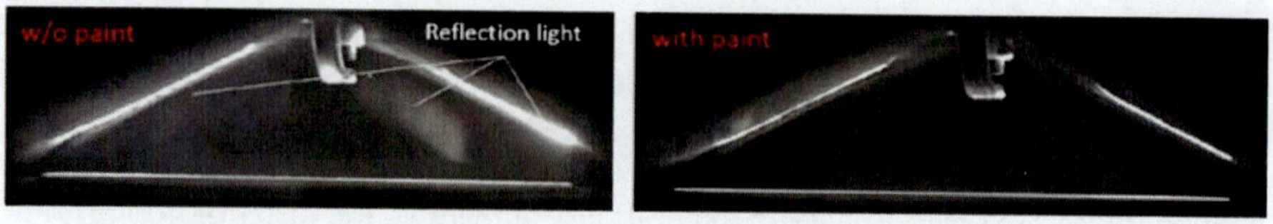

Fig. 3.10 Reflection of laser sheet from the surfaces (Nishiyama et al. 2019)

(Watanabe et al. 2004; Rostamy et al. 2012; Velte et al. 2008; Lee 2011). Nishamiya et al. (2019) applied commercially available black ink in order to absorb the incident laser. The interrogation regime was maximized by restricting the blanking zone to 1 mm thickness near the fire-deck. Figure 3.10 shows the difference between the coated and uncoated cases.

Paterna et al. (2013) carried out experiments with different materials in order to investigate the capability of mitigating the reflection issue. They reported that fluorescent paint on the woods and polished steel can be used to achieve 7–30 times reduction in reflection. Additionally, application of bandpass filter further improves these results. Bisel et al. (2017) found that fluorescent rhodamine 6G coating along with bandpass filter successfully suppressed the trailing edge reflection. Electro-polished steel can be viable solution for engine applications because it poses no health risks and can sustain high temperatures as well. There are several other techniques like proper orthogonal decomposition, and ghost particle elimination, which mitigate the reflection without any physical change in the setup (Mendez et al. 2017; Elsinga and Tokgoz 2014). These techniques are mostly applied during post-processing of the acquired images.

3.2.3 Seeding Considerations

This section is based on the selection of suitable tracer material for application in PIV for various applications. There are several pivotal properties that are required for an appropriate tracer. The tracer particle must faithfully follow the flow pattern and satisfy the optical requirements for imaging and analysis. In addition, seed particles should be chemically inactive and capable to withstand extreme in-cylinder temperature and pressure conditions. In cold chamber analysis, the flow following requirement varies according to the application; however, this is critical because it affects the size of seed particles. Particle size and density both decide the effect of gravitational and viscous forces, which are the primary sources of error in velocity determination in the flow-field. The particles of density matching with that of fluid are preferred. Studies suggested that selection of appropriate seed particles is more difficult in case of gas flow analysis, compared to liquid flow analysis.

Highly turbulent gas dynamics is the general characteristics of in-cylinder flows in IC engines and other combustion applications. Under such cases, the flow-following accuracy can be characterized by matching amplitudes of flow oscillations and particles at a given angular frequency. The particles must be able to resolve the least time-scale of the flow or else as per the requirement of the study. In a study carried out by Haghooie et al. (1986), it was reported that aluminium particles of 1 μm diameter having density 4 g/cm^3 satisfactorily (90% amplitude matching) matched 10 kHz fluctuations, therefore they could be used for PIV investigations in IC engines. Generally, IC engines have a flow frequency of the order of few kHz.

Smaller particles tend to follow the fluctuations more efficiently. Larger particles are more prone to suffer velocity lag due to higher inertia. The difference between the flow velocity (V) and the particle velocity (V_p) is called the velocity slip (V_s) (Fig. 3.11). V_s is a function of Stokes number, which varies with second power of the particle diameter. Stokes number determines the characteristic time scale of the particle motion. Lower Stokes number makes the particle more sensitive to high frequency velocity fluctuations in a turbulent flow-field. However, the scattering intensity is a function of particle size. In Mie scattering, the intensity is proportional to the square of the particle diameter. Light scattered by μm-sized seeds is often overshadowed by intense scattering from larger particles. Thus, particles of smaller mean diameter require relatively higher laser power and more sensitive camera, which are not economical. In many PIV studies, it has been shown that 1 μm alumina particles are capable to follow 10 kHz velocity fluctuations with 90% amplitude accuracy. However, particles of larger size can match the same level of accuracy only up to 2 kHz frequency or less. One solution to this has been seen in the application of liquid bubbles filled with lighter gases. Helium-filled soap bubbles have been investigated as a possible contender. Soap bubbles are neutrally buoyant and of the order of few mm size. A more detailed information can be found in previous studies (Macháček 2002), but bubbles are not suitable for boundary layer investigations, which are critical in heat transfer applications as well as the understanding of Spray wall impingement phenomenon. PIV investigations of sprays have been performed without the use of any tracer particles. Sufficient number density of liquid droplets acted as tracer particles and yielded satisfactory results (Zama et al. 2012).

A majority of studies have used additional tracer materials to obtain required control over imaging requirements. As mentioned earlier, seeding of liquid flow-field can be easily achieved by using suspension of solid particles. However, for gaseous flows (as in IC engines), window fouling is a major challenge. Further, abrasive nature of tracer particles may compromise the life of expensive optical parts. Hence, liquid seed materials have emerged as an attractive alternative for motored cold chamber

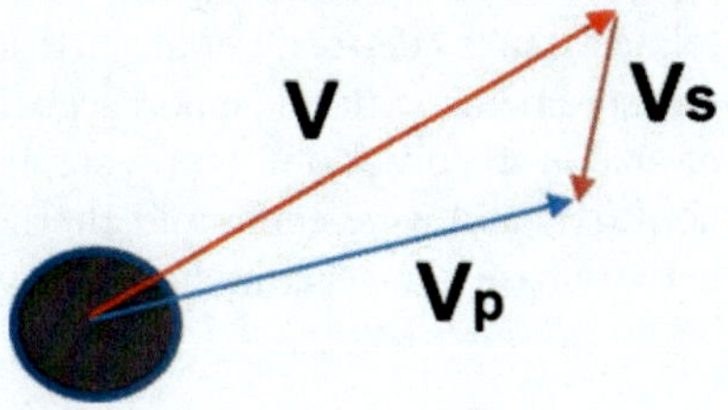

Fig. 3.11 Velocity triangle for tracer particles in fluid flow

PIV analysis. For liquid seed materials, their vapour pressure must be adequate to avoid vaporization. It is worthwhile to note here that toxic seeding particles may raise significant health concerns to the experimentalist human resources upon inhalation. Vegetable oils are preferred as seed material because of their negligible toxicity (Tsiogkas nd; da Costa et al. 2017). Aleiferis et al. (Aleiferis and Behringer 2017) seeded olive oil droplets having a density 920 kg/m^3 ~150 mm upstream of the intake valve to achieve homogenous seed distribution. Velocity resolution up to 0.1 CAD was achieved with droplets in the range of 1–2 μm.

The next challenge encountered in PIV seeding is handling the two-phase flow. Spray–flow interaction is critical to understand the effect of injection timing on the combustion and cycle-to-cycle variations in the engines. In such case, velocity field of the spray (liquid phase) and the in-cylinder air (gas phase) needs to be resolved. The droplet size distribution of the spray may range from few μm to few hundred μm depending upon the injection timing, fuel properties, and injection parameters. Thus, larger spray droplets would scatter more light, which would hinder the imaging of smaller seeds. However, for early injection (port fuel injection (PFI) and gasoline direct injection (GDI) engines in homogenous mode), the fuel droplets are vaporized to reduced diameters of the order of few μm, which is similar to the size of seed particles. The number density of such droplets is also very less. Aleiferis and Behringer (2017) confirmed that there are no significant variations in images acquired by running the engine in fuelled and non-fuelled seeded modes. The conflict between the spray droplets and the seeds arises in late-cycle injection, common to diesel engines and GDI engines in stratified mode. In another study, scattering from larger droplets was masked by providing a specific threshold. Scattering intensity increased six times if the seeds of size ~0.5 μm were discarded from the acquisition. Thus, droplets larger than 1 μm were excluded to confirm that the results represent gas-phase velocity only. Authors reported that the droplet sizes above the applied threshold were present in the spray vortex only. The intensity decayed with time and was similar to seed intensity 2° CA after the end of injection. PIV investigations were performed on the image data acquired at 2° CA after the end of injection to discard any bias (Stiehl et al. 2016). Another way to resolve this issue was the colour-based method (Towers et al. 1999; Driscoll et al. 2003). This method introduced μm-sized fluorescence seeds to overcome Mie scattering from larger droplets. The seeds were doped with laser dye. The gas-phase velocity was obtained by analysing the images of these fluorescence seeds. However, the toxicity and thermal quenching nature of these dyed particles remained major concerns. Cost-effective doped particle generation posed another challenge for their application. KR620-doped particles were reported to be viable alternative due to their low toxicity, however with signal yield-related issues (Petrosky et al. 2015). Fan et al. (Fan et al. 2016) demonstrated PIV measurements with phosphorescence signals using the ZnO particles. The particles of size of the order of 2 μm were excited by UV laser (355 nm). ZnO particles are non-toxic and were economic alternative to fluorescent particles for the application in two-phase non-reacting flows. However, despite having a strong signal yield at low temperature, signal strength dropped as the temperature shot above 500 K. Mie scattering technique along with algorithms based on signal intensity is another option. However, these algorithms are highly complicated.

Application of PIV in high-temperature investigations is a upcoming area of research. Flame flow interactions play a vital role in determining combustion quality, charge utilization and pollutant formation. High in-cylinder temperature requires the tracer particles to be resilient enough to sustain the intense reacting environment of the combustion chamber. Flame luminosity poses an additional challenge in image processing. Thus, liquid droplets are no longer an option for combustion PIV. Barring the effect of abrasion, refractory solid particles are suitable for such harsh conditions. Witze and Baritaud (1986) performed exhaustive experiments in order to characterize solid particles for PIV applications. They found that TiO_2 and Al_2O_3 particles of sub-μm size perform very well under cold-flow conditions; however, they exhibit poor optical characteristics in high-temperature conditions because they are engulfed by the flames. Silica particles have been extensively used as seed particles in many combustion PIV studies. In order to obtain adequate intensity and successfully imaging of both the flame fronts and the flow-fields, seeding density needs to be adjusted carefully. Nishamiya et al. (2019) investigated the flow-field in a combustion environment. Controlling the air-flow rate, ideal seeding density was achieved. They reported that 1 mm^3 of seed particles in bulk volume per cycle yielded satisfactory results. This was confirmed by analyzing the images for different seeding densities. Figure 3.12 illustrates the effect of over-seeding and under-seeding on the PIV images. Density may vary depending upon a particular application, particle size, and optics characteristics.

It is also important to realize that in engine applications of PIV, the cylinder volume shrinks as the piston approach TDC. This further increases the seeding density. Not only particle size but also high seed density also produces intensely blooming images due to multiple scattering effects. Hence the seeding rate must be optimized for the entire range of motion for the intended CA interval for investigations. Figure 3.13 shows the variations in intensity with the CA. Optimized seeding density enabled the author to keep the intensity variations within limits.

Titanium dioxide has also been implemented for high-temperature PIV applications (Najafabadi et al. 2017; Tanov et al. 2018). TiO_2 particles of size 2–3 μm were able to follow the flow with a time lag of 30 μs for the thermodynamic condition of 35 bar and 900 K. This is of the order of Kolmogorov time-scale for typical turbulent flows in IC engine applications (Tanov et al. 2018). Thus, the particles are expected to faithfully follow the flow at high-temperature engine environment. To get rid of particle agglomeration, the particles are baked for 24 h before the experiment.

Recent studies have reported a novel laser-induced incandescence-based PIV. Detailed discussion on LII was done by Michelsen et al. (2015). The incandescence signals from the particles can be distinguished from others. Fan et al. (2018) investigated LII-based PIV methods for two phases non-reacting flow environment. After exhaustive search, tungsten carbide (WC) particles (0.2 μm) were selected as tracer particles. The particles had low Strokes number and high melting point (2800 °C).

Particle images	Analyze	
	Velocity (Ordinary PIV)	Flame detection (Combustion PIV)
	-	-
	-	-
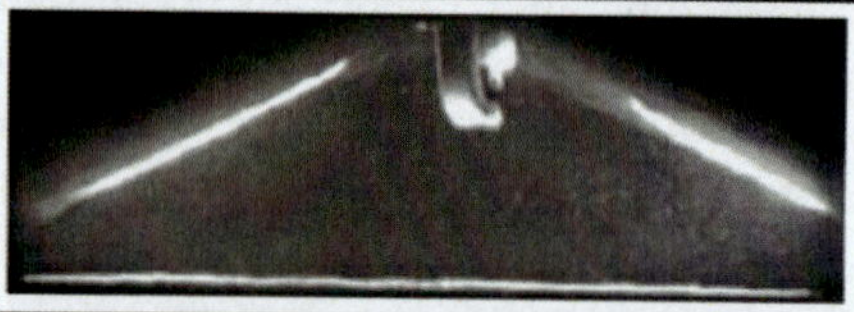	✓	-
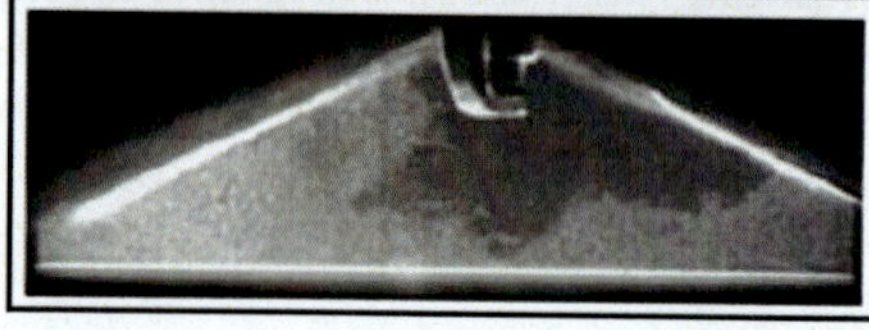	✓	✓

Fig. 3.12 Effect of seed concentration on image quality from analysis point of view (Nishiyama et al. 2019)

These particles can follow the flow faithfully even at high temperatures (Fan et al. 2018). WC particles are non-toxic and less expensive, which makes them a suitable candidate for LII-based PIV measurements.

3.3 Application of PIV

3.3.1 PIV Studies in Flow Bench and Constant Volume Chambers

3.3.1.1 Application in Flow-Field Analysis of High-Pressure Spray

To meet the emission norms, boost pressure in the DI diesel engines has to be increased. Despite this fact, PIV investigations of spray under boosted conditions still remain a grey area of research. Under the condition of high gas density, behaviour

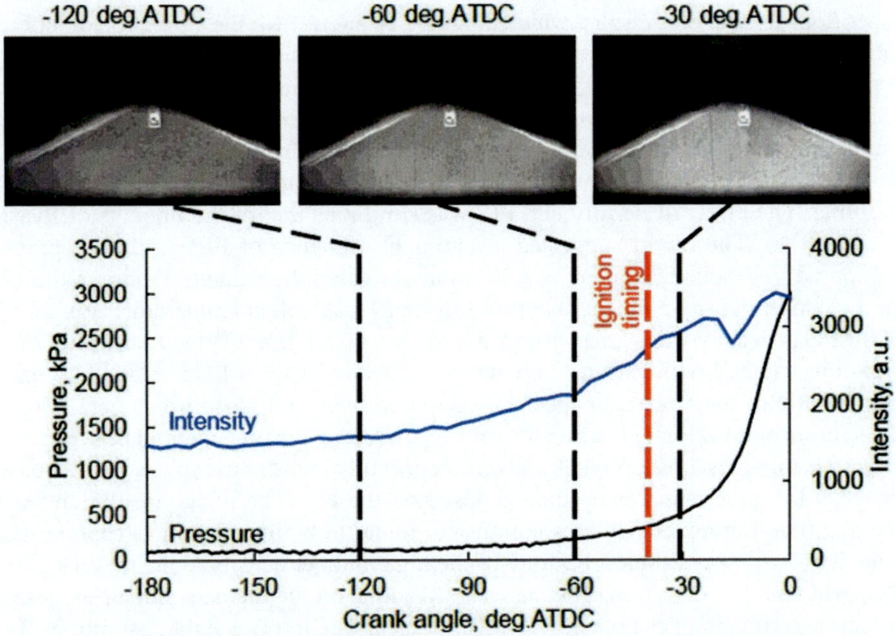

Fig. 3.13 Intensity variations with piston movement during the compression stroke (Nishiyama et al. 2019)

of the spray needs to be studied to understand the mixture formation. Typical experimental setup for a spray study in constant volume combustion chamber (CVCC) by PIV is shown in the Fig. 3.12. Pyari et al. (2016) investigated the flow-field in spray by using a diesel-like single hole injector. Schematic of the experimental setup is shown in Fig. 3.14. They reported non-linear relationship between the flow-field of spray and the ambient density. However, fuel injection pressure (FIP) and velocity field exhibited a linear relation. Spray tip vortex formation took place due to velocity gradient along the spray radius. Turbulent behaviour resulted in relatively higher shot-to-shot variations. Authors recommended ensemble averaging of minimum 25

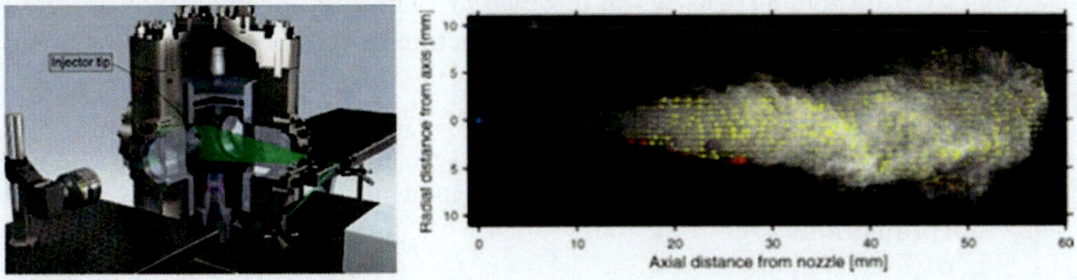

Fig. 3.14 Schematic of the experimental setup (left) and Velocity field in the spray (right)

repetitions to get good results, which could be of interest for the validation of simulation models. The study also compared these results with 1-D spray models and found the variations to be within 7%. A recent study investigated gasoline spray characteristics using high-pressure diesel injector in high-density conditions using multiple optical diagnostics techniques including PIV (Thongchai and Lim 2020). In this study, 5% biodiesel was blended with gasoline to increase the lubricity of gasoline. The effect of density and FIP was similar as that of the findings of Pyari et al. (2016). The report suggested a dominant influence of FIP on diesel sprays compared to gasoline. Zama et al. (2017) investigated the velocity field in a diesel spray in high-density ambient conditions using PIV. Ambient temperature was set at 300 K. Gas density was varied from 5.8 kg/m^3 to 46.5 kg/m^3. Velocity distributions were investigated with various ambient gas densities. The AoI was around 40 mm from the tip of the nozzle. Images were acquired using a high-speed camera. Since the spray droplet velocity is a function of FIP, therefore, the time interval between the successive images must be synchronized accordingly in order to capture the velocity flow-field. Frame rate was optimized based on the FIP. The image resolution was fixed at 0.08 mm/pixel. This resolution was found to be insufficient to capture the velocity of all spray droplets because of the high number density in the diesel spray. The grid size was 0.92 mm. The analysis was done on the incident side of the laser because the image of the other side was not clear due to prior light absorption. By analyzing the velocity in the high-density environment, re-entertainment of spray droplets at the periphery was observed. Vortex formation near the periphery was also encountered. The effect of post impingement can also be analyzed with the help of PIV. In the work reported by Zama et al. (2017), the effect of post impingement velocity and ambient gas density was investigated. A common rail direct injection (CRDI) system was used in the CVCC. A continuous-wave laser was used with cylindrical lens to form a 1 mm thick light sheet. No seeding was done for visualization and diesel spray droplets illuminated themselves. A digital high-speed video camera (Vision Research, PhantomV2511) was used for image acquisition, whose frame rate was set at 270,833 fps and 140,086 fps.

Impingement distance was set constant at 40 mm. To avoid masking of the rolled-up portion of the impinging spray, a slender bar was used as an impingement plate. By analyzing the flow-field, a position correlation between the velocity of impinging spray and peak velocity along the wall was developed. Inverse relationship of the peak flow velocity with ambient density was discovered. Flame impingement may decrease the flame temperature due to heat transfer. Reduction in velocity after impingement was also reported.

3.3.1.2 Application in Flow Bench Studies Relevant to IC Engine Applications

Design and optimization of intake port and valve lift profile are important in the generation of swirl and tumble motion in IC engines. Fluid motion during the intake stroke has significant effect on turbulence intensity and flow structure during combustion.

Cycle-to-cycle variations in SI engines can be controlled in a better way by understanding and optimizing the flow characteristics. Heat transfer and fuel–air mixing during conventional diffusion-controlled CI combustion and LTC depend on the in-cylinder flow. However, optimization of geometry in the engine requires significant modifications in the head design. Thus, the number of modifications for optimization is not economical and rather quite time-consuming. Flow bench study can be a practical solution to this problem. Simple design and lighter materials, which can be machined easily, provide flexibility to test multiple designs with the variations in geometric parameters. The flow rates can be easily controlled to achieve the Reynolds number similar to real engine operating conditions. The valve lift can be adjusted to generate multiple data sets required for optimization study and CFD model validation. This approach can save significant amount of time and money. Several recent studies have been conducted in a steady-state flow bench to get better insights and understanding of in-cylinder flows. El Adawy et al. (2018) investigated the effect of boost pressure and valve lift variations on flow structures of a GDI engine. For this purpose, pentroof design of a GDI engine head was mounted on the flow bench. A centrifugal compressor was used to maintain the pressure differences across the suction valve. The valve lift was also varied for each applied pressure difference. A Plexiglass cylinder with a flat piston was fixed to the head for providing optical access. The suction side of the compressor was connected to the cylinder on the flow bench with a stepper motor bypass system for pressure control. The study revealed that the in-cylinder flow structure during suction was independent of boost pressure and relied heavily on the valve lift. The vortices preserved their shape but got strengthened as the pressure difference increased for constant valve lift. The turbulent kinetic energy (TKE) exhibited a close relationship with applied pressure difference or the boost pressure. A similar study on the flow bench reported an increase in vorticity and TKE with increasing valve lift (El-Adawy et al. 2017). Vester et al. (2019) compared the flow characteristics of two different intake ports by employing a unique water analogue model. The flow during suction can be assumed to be incompressible. Therefore, incompressible fluid like water can be used, given that the geometrical characteristics are retained by the model. Thus, the air was replaced by water. Another feature unique to the design was that the intake ports could be rotated manually to perform tumble plane analysis at different angular orientations in the cylinder. Schematic of the setup can be seen in Fig. 3.15.

The test was conducted on the heavy-duty engine cylinder head models of Scania CV AB. One had twin inlets into the head and the other one had a single inlet, which was further split inside the head. The results exhibited relatively lesser cycle-to-cycle variations at fixed valve lift. This was mainly due to coherent flow structures. Out of the three different fixed valve lifts investigated, fully opened valve resulted in large-scale structures. The flow was more organized and stable. On the other hand, smaller opening resulted in higher conversion of TKE into small-scale turbulence. Rebault et al. (2016) also investigated the effect of valve lift in the swirl and tumble flows. To reconstruct the flow-field, 3D3C concept was applied using stereoscopic PIV measurements. In this method, the laser and sensors were kept stationary and the cylinder head was rotated about the fixed axis (Fig. 3.16). This helped in the

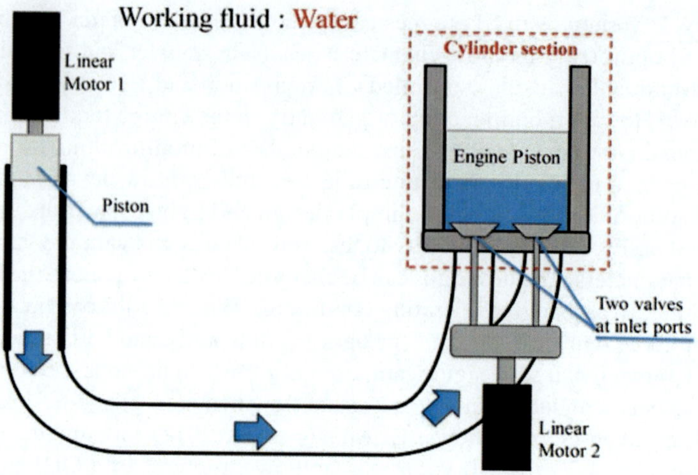

Fig. 3.15 Schematic of the water analogue model (Vester et al. 2019)

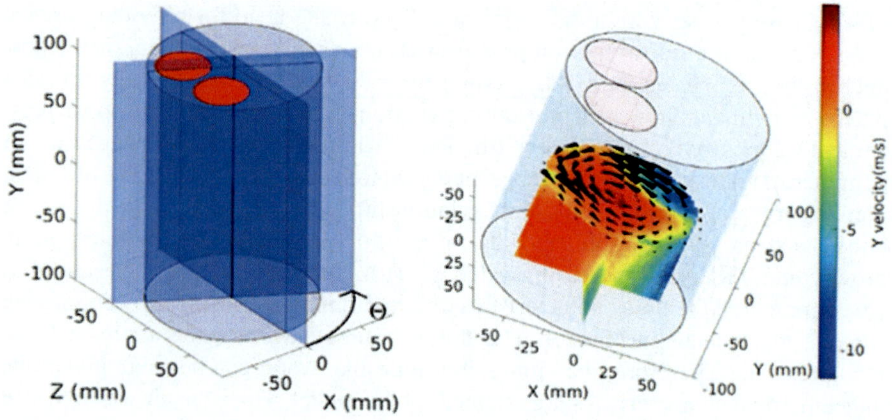

Fig. 3.16 Rotation of the cylinder for 3D3C reconstruction (left) and reconstructed flow-field (right) (Rabault et al. 2016)

construction of the 3D flow structures in the tumble plane. Images were acquired at 10° angular intervals and the cylinder sweep was achieved by 18 images. The findings outlined similar results as other studies related to the valve profile. Author also carried out instantaneous swirl centre analysis. The finding suggested that higher flow stability was achieved by high swirl cylinder head. These results are important in the understanding of cycle-to-cycle variations.

3.3.2 *PIV for In-Cylinder Investigations*

3.3.2.1 Application to Flow Structure Analysis During Motoring

PIV studies in motoring condition also called for cold chamber analysis, revealing useful information about the flow structures developed during intake and early compression stroke and its dissipation during late compression stroke. The intake port geometry, piston speed, valve lift, etc. played an important role in determining the in-cylinder flow characteristics. In motoring conditions, the flow analysis is done in a single phase; hence it does not involve complications of multiphase flows or elevated temperatures during reactions. These analyses can be done by using oil seeding, which eliminates the issue of window abrasion. This is relatively simpler yet a powerful method to understand the in-cylinder flow behaviour. Thus, the majority of research in IC engines with PIV was carried out under cold chamber conditions. Tsiogkas et al. (2019) conducted cold-flow analysis in a single-cylinder optical GDI engine. The study was carried out at fixed throttle (25%) and at two different engine speeds (1000 and 1500 rpm). Results showed a counter-clockwise tumble flow at both engine speeds during the intake stroke. Similar flow structures during suction were also reported by da Costa et al. (2017) and Stansfield et al. (2007). da Costa et al. (2017) found a large cycle-to-cycle variations and strong relation between volumetric efficiency and flow structures. Stansfeild et al. (2007) observed a shift in tumble ratio and flow structures during the transition from lower to higher engine speeds. Volumetric efficiency, engine speed and throttle position are closely related in SI engines. Alam et al. (2020) investigated the effect of throttle opening on cycle-to-cycle variations. Histogram of vorticity was taken as the basis to quantify these variations. They found a similar level of variations for 50% and wide-open throttle conditions. Singh et al. (2015) carried out tomographic PIV analysis in a different plane to study the effect of engine speed on flow behaviour. The study reported that swirling intensity increased with increasing engine speed. Higher vorticity concentration was observed near the intake valve. Large-scale flow structures were observed during the intake stroke, which broke down to small-scale turbulence during compression stroke. A study carried out by Agarwal et al. (2018) revealed that highly disordered air motion during intake stroke became homogenous during the compression stroke. A higher velocity magnitude was observed near the injector during compression. Sun et al. (2019a) reported significantly higher TKE generated during the intake stroke, which decayed to 1.6–3.3% of initial value during the compression stroke. Higher TKE was observed near the swirl centre. Large-scale structures dissipate into small-scale structures with loss of TKE. Small-scale structures are of particular interest to understand mixing and oxidation during combustion. Small-scale structures in boundary layers near surface play an important role in heat transfer. Alharbi and Sick (2010) investigated the effect of these small-scale structures during compression and expansion strokes. A combination of micro PIV and particle tracking velocimetry (PTV) was used in this investigation. They found sub-millimetre sized vertical structures near the cylinder walls. Flow intensity decayed during the end of compression and continued up to 40° CAD aTDC. At this point, flow reversal also takes place. Bucker et al. (2013) conducted experiments to understand the relationship between flow structures during suction and compression strokes. Two different engines having different intake geometries (tumble port and filling) were used in this study. A strong vortex structure was observed for the engine having tumble port, which conserved the kinetic energy during later part of the intake stroke. The study revealed a strong relationship between flow during suction and compression strokes. The effect of port geometry was investigated by Agarwal et al. (2017) by using port deactivation in a single-cylinder optical research engine. The tomographic PIV method was applied to this study. They found that flow originated from the swirl port was dominant compared to the tangential port, when both ports were active. A comparison between the flow from individual ports by deactivating one of them at a time revealed that the flow energy got dissipated at a higher rate for the tangential port deactivation. Combustion analysis suggested that the swirl port open (SPO) mode exhibited superior combustion characteristics. Zha et al. (2015) also investigated the effect of port geometry on flow structures during compression by using a realistic piston bowl shape. They found that the swirl axis tilting was independent of the port geometry. The piston bowl geometry could be possibly responsible for this. However, the swirl centre motion was dependent upon the port geometry. To understand the effect of squish on the swirl centre, Sun et al. (2019b) investigated the flow-field using different bowl geometries. Their study revealed that large squish flow can move the swirl centre towards the cylinder axis. Larger squish flow also increased the radial component of velocity.

3.3.2.2 Application to Multiphase and Reacting Flows

Multiphase flow is of particular interest in IC engine applications. The spray–fluid interactions play a vital role in spray break-up, spray atomization, vaporization and fuel–air mixing. In high-pressure direct injection sprays, induced turbulence is critical in determining the level of TKE during the combustion process. In-cylinder geometry interacts with the flow structures. Variation in these flow structures interacts with spray and vice versa. This complex interaction between the air-flow and spray is key to understand cyclic variations. Critical aspects of seeding particles in multiphase flow investigations through PIV have been discussed in Sect. 2.3. Despite the complexity of the phenomenon, several in-cylinder studies have been conducted to understand the physics. Zama et al. (2019) investigated the location of glow-plug on fluid structures and their interactions with the spray. Ma et al. (2017) investigated the near-wall boundary layer by using a high-speed micro-PIV. The study contributed to the understanding of wall heat transfer model. They reported that the non-equilibrium wall function model was more accurate compared to equilibrium wall function model. Tanov et al. (2017) investigated the effect of injection strategy on the heat transfer characteristics during combustion. Convective heat transfer is a function of flow-field in the cylinder. The study applied PIV in reacting multiphase flow condition to analyze the influence of different injection strategies on the flow-field. The velocity field obtained by PIV was further processed to determine tangential velocity and TKE. The investigations were performed in the tumble plane. Thus, to obtain details of swirl flow, additional computational modelling was used. The 1-D model used for the calculations was validated with CFD (CONVERGE software) results and experimental results from two different engines. The angular velocity of the swirl was assumed to be uniform for the entire gas volume. Applying these methods, zone dependent heat transfer coefficients were analyzed for single and multiple injections. They reported that single injection strategy exhibited higher heat loss with the highest overall heat transfer coefficient, followed by triple and double injection strategies. The zone dependent analysis revealed that the central zone contributed to the highest heat loss for all injection strategies. Their study is an important contribution in applying PIV to understand engine heat loss behaviour. As an extension of this study, Tanov et al. (2018) investigated the distribution of TKE inside piston bowl and its effect on the wall heat transfer characteristics for multiple injection strategies. A similar approach was used by Tanov et al. (2017) for swirl evaluation.

They observed that injection events contributed significantly to the TKE level. The peak was observed near the bowl periphery for all three injection strategies (single, double and triple). Combustion-driven turbulence contributed more to the bowl centre. Figure 3.17 shows the variations of TKE and heat transfer coefficient distribution over the bowl w.r.t. CAD. White dashed lines represent crank angle position corresponding to 10% mass fraction burned (MFB) (CA_{10}) and crank angle position corresponding to 90% (CA_{90}). The black dotted lines represent the difference in the heat transfer coefficient with the motored case. It can be observed that the single injection case resulted in relatively higher heat transfer coefficient compared to double and triple injection cases (Fig. 3.17). For the double injection case, bowl centre attained the lowest TKE. The distribution of TKE for triple injection was moderate and relatively uniform. This satisfactorily explained the heat transfer characteristics and was in general agreement with previous studies. Significantly higher rise in TKE during single injection event also signified the relationship between spray momentum and spray induced turbulence. Since ambient density affected the spray flow-field behaviour, therefore, the effect of thermodynamic conditions on spray-fluid interaction was explored by Najafabadi et al. (2017). They investigated the spray-induced turbulence by conducting injection sweeps of the main injection. The effect of a number of nozzles on the in-cylinder turbulence generation was also investigated. Results were correlated with the heat release rate for each injection strategy. The findings revealed that advancing injection timings reduced the turbulence during injection. 5-hole injector yielded slightly more turbulence compared to 7-hole injector due to higher injected mass per hole, which induced more turbulence. However, the heat release rate and combustion driven turbulence were higher for 7-hole injector due to superior atomization and combustion efficiency.

Spray–fluid interaction is also of importance in the SI engines. For direct injection SI (DISI) engine, the spray structure was affected by the fluid motion. Geschwindner et al. (2020) investigated this phenomenon in a GDI engine under Spray-G conditions. The results revealed that weak tumble motion was unable to retain their

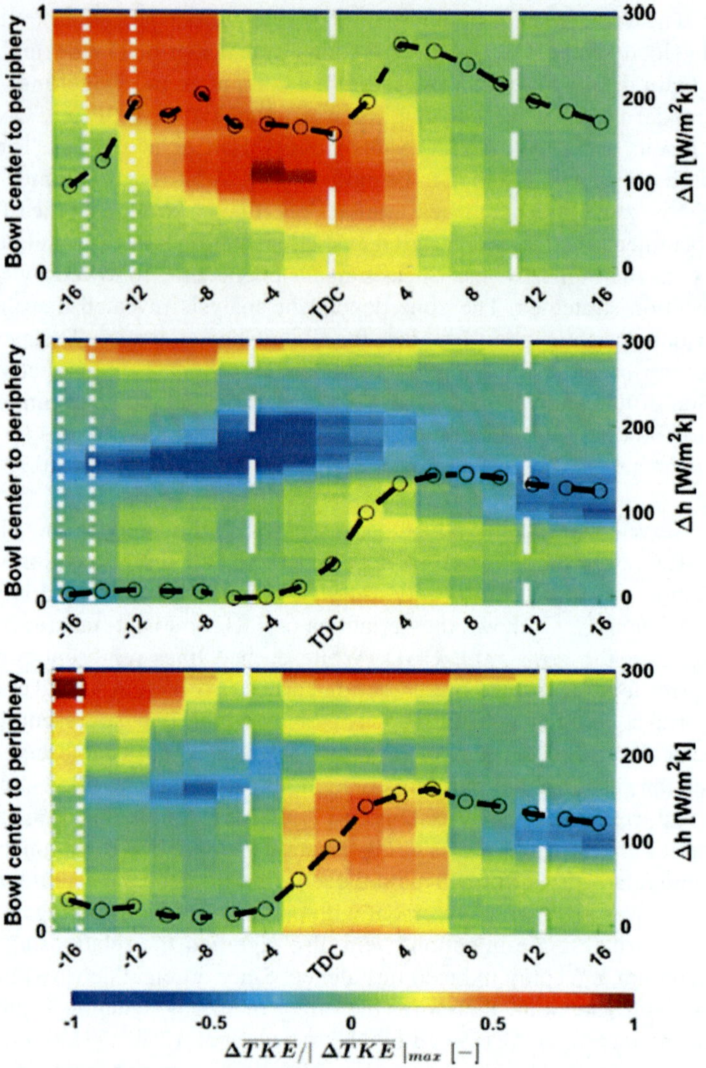

Fig. 3.17 Distribution of TKE and heat transfer coefficient (Tanov et al. 2018)

momentum during the spray. The structure broke into counter-rotating vortices. However, stronger tumble was not that much affected. Increased engine speed showed the tendency to minimize the spray collapse. Spray penetration was hardly affected by the in-cylinder flows. A comparison with constant volume chamber spray analysis showed that there was no significant difference. However, when the piston approached, spray-wall impingement occurred. Stiehl et al. (2016) conducted a similar investigation in a GDI engine with multiple injections. They found that the spray structure of the second injection during compression was affected by large-scale tumble vortices during the intake. This large-scale vortex showed cycle-to-cycle fluctuation due to injection during suction. Late-cycle injection controlled and governed the combustion. Therefore, the first injection made significant contributions to the cycle-to-cycle variations in multiple injection GDI engines. Authors also found that changes in intake port geometry had no significant effect on the stability of the second injection. Nishamiya et al. (2019) investigated the flow structures and flame interactions for a SI engine with port fuel injection. Such analysis was challenging due to the fact that high temperatures affected the seed particle structures and hence their optical characteristics. Also, combustion luminosity posed additional challenges. Authors used the gradient in seed particle density in the burnt and unburnt region to identify the flame front. The study found that the global vortex location significantly affected flame growth in a particular direction. The flames also interacted with the flow to enhance the intensity in the unburned region.

3.4 Air Motion and Pollutant Formation

Incomplete combustion of fuel is the main reason for pollutant formation in IC engines. This is greatly affected by the fuel–air mixing and local temperature. Both fuel–air mixing and heat transfer are function of local flow-fields. Flame temperature and turbulent mixing are two important factors affecting the soot formation as well as oxidation. Several investigations have been carried out to correlate the flow dynamics with transport and oxidation of intermediate species. Razak et al. (2019) investigated the effect of turbulent Schimidt number (Sc_t) on pollutant formation using numerical analysis. Simulations were carried out for different values of Sc_t under Spray A condition (a case of low-temperature combustion condition under 15% oxygen described by Engine Combustion Network (ECN)). The velocity field was validated using experimental data in ECN library obtained through PIV. Results depicted that C_2H_2 increased with increasing Sc_t, however, the soot prediction was not accurate. Authors concluded that soot formation was significantly influenced by Sc_t, hence required further investigations to understand the correlation between soot formation and Sc_t. Taylor et al. (2020) investigated the effect of ethanol blending and velocity field on sooting characteristics of a swirling jet. They reported that ethanol evaporation affected the flow structures downstream of the jet. Figure 3.18 showed the time averaged value of soot volume fraction (SVF, proportional to size of the black dots) and particulate number-size distribution. Results depicted that the flow-field variation had marginal effect on the particle size.

In diesel engines, spray-induced turbulence and spray–swirl interactions have crucial role in soot oxidation and late-cycle pollutant formation. However, this area has not been experimentally explored much. Grüneberger et al. (2019) carried out in-cylinder PIV experiments along with flame transport analysis to understand the movement of soot cloud. They reported that initial evolvement of the flames was similar at low and high engine loads. Backflow of soot clouds into the central combustion chamber was also observed at all engine loads. However, the turbulent whirling

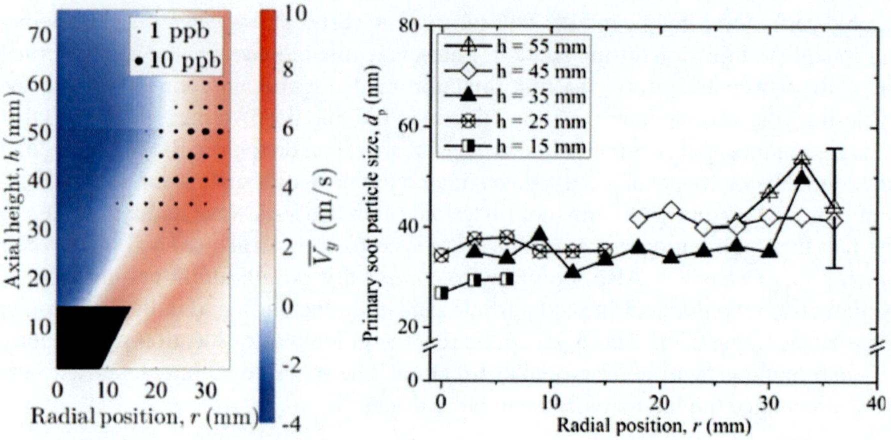

Fig. 3.18 SVF (left) and soot particle size distribution (Rault et al. 2020)

motion was observed at 20° aTDC (Fig. 3.19), which collapsed after the end of the injection. This was mainly due to strong momentum of the spray, which was present for longer duration at higher engine loads. They also examined the effect of FIP on spray–swirl interactions and reported that higher FIP led to higher velocity during spray collapse. A strong whirling motion was developed at the bowl centre post injection. The velocity of the whirl increased with increasing FIP. This type of flow is important for late-cycle oxidation of the pollutants. Similar structures have been also reported in other investigations (Dembinski 2014; Zha et al. 2018).

Dembinski (2014) carried out experiments to study the effect of swirl and tumble on soot formation. Two different cylinder heads were used and port-deactivation method was adopted to achieve different values of swirl and tumble. Findings revealed that soot decreased with increasing swirl and increased with increasing tumble. For a constant swirl number (SN) of 2.5, higher value of tumble resulted in 50% higher soot. The swirl–tumble interaction could have played a key role here. They reported that the swirl centre got disturbed by tumble, which led to asymmetric charge distribution in the combustion chamber and higher soot was formed in the regions of weak flow. Higher SN also led to higher angular momentum in the bowl centre, which resulted in increased rate of soot oxidation during post-combustion phase. At higher SN, the soot emissions decreased from 0.19 FSN to 0.05 FSN. The authors also investigated the effect with two different bowl geometries i.e. flat piston and shaped piston on the flow behaviour. They reported relatively higher peripheral velocity for shaped piston. The shaped piston had higher bowl volume distribution towards the peripheri, which led to higher momentum and velocities.

The peripheral velocity affects the flame impingement and flame heat transfer. Higher FIP led to shorter residence time of flames near the bowl wall. This could be determining factor in soot formation, soot deposit on piston surface, and also on the heat loss (Osada et al. 2015). Agarwal et al. (2017) investigated the effect of three different port configurations namely single port open (SPO), tangential port

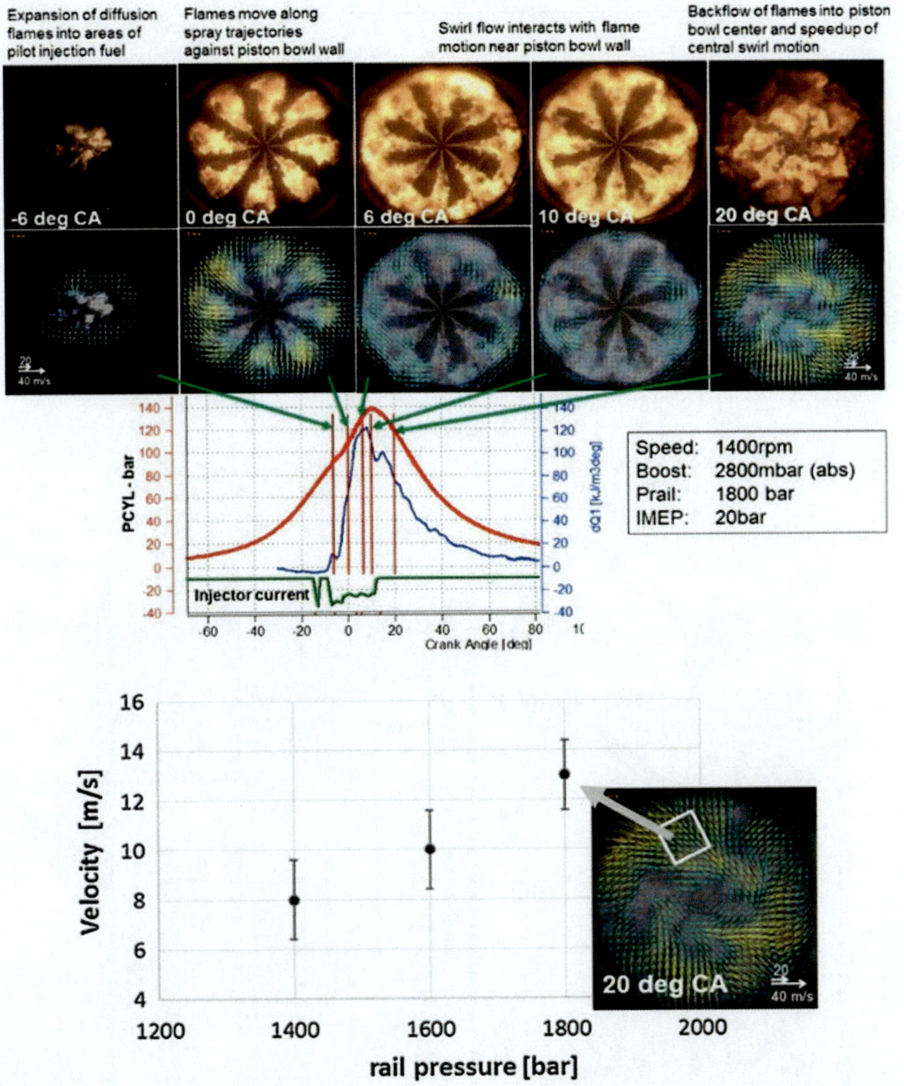

Fig. 3.19 Flame velocity distribution with CAD (top) and effect of fuel rail pressure on velocity after the end of injection (right) (Grüneberger et al. 2019)

open (TPO) and both port open (BPO) configurations on the number size distribution of particulate emissions. Flow-field investigations were carried out using tomographic PIV to study the effect of fluid motion on the particulate emission characteristics. Figure 3.20 shows the TKE variations for different port configurations along with particulate number size distributions. The figure depicts that TPO configuration resulted in the highest number of particulate emissions. This could have been due

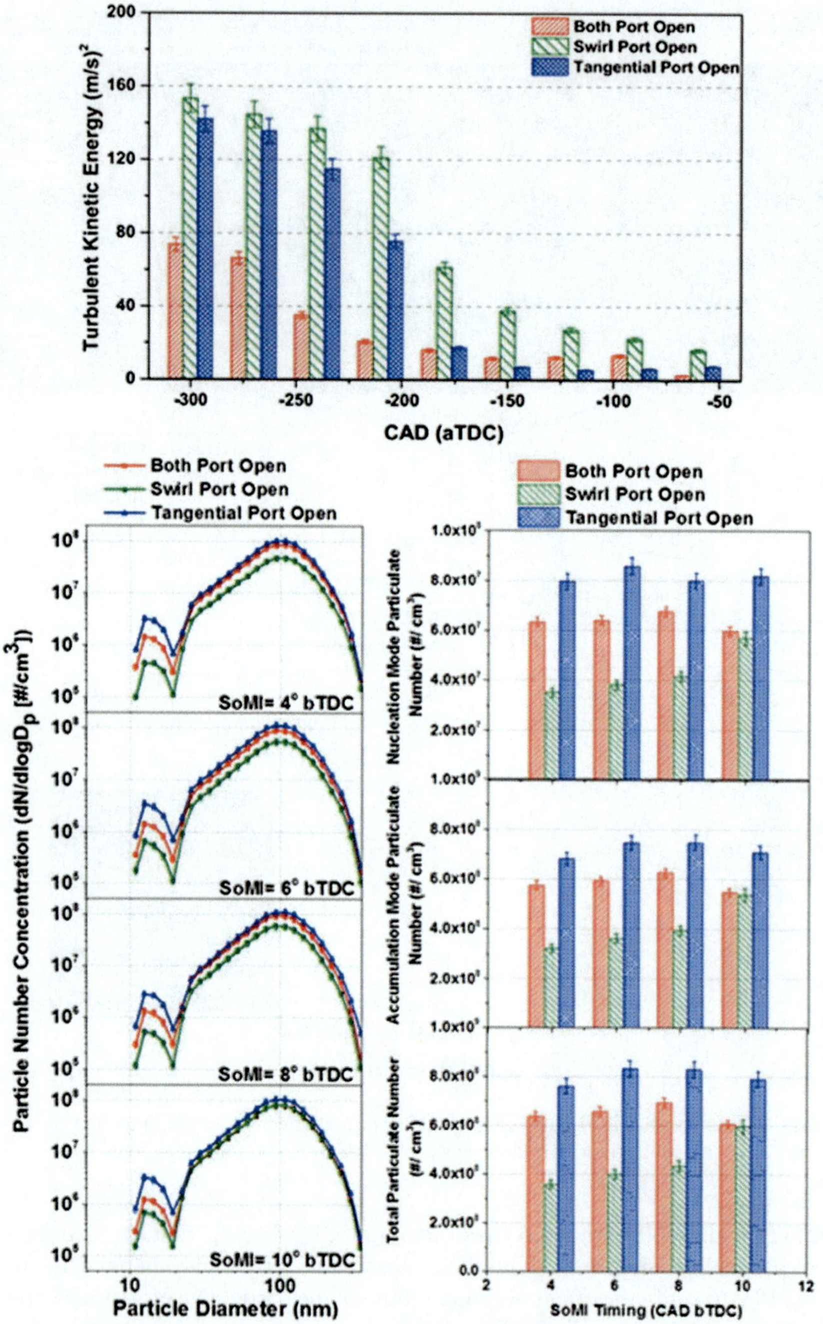

Fig. 3.20 Variations in TKE with CAD (top) and particulate size distributions (bottom) for different port configurations (Agarwal et al. 2017)

to higher tumble motion owing to deactivated swirl port. However, the nature of particle number size distribution was similar in all cases. From Fig. 3.20b, it can be observed that the total particulate number (TPN) is more influenced by the SoMI timings in SPO configuration. At advanced SoMI timings, TPN of SPO configuration approached a value similar to TPN of BPO configuration. A retarded SoMI timing could have promoted higher spray–swirl interaction, resulting in late-cycle soot oxidation. This phenomenon was discussed earlier in this section. This quantitative analysis of the particulate emissions further confirmed the dominating effect of spray-momentum on the spary–swirl interactions and soot formation.

3.5 Way Forward

Extensive research in laser diagnostics has helped develop PIV technique as a front-line tool for real-time flow diagnosis in real geometry IC engines. PIV has been extensively used in IC engines to understand the flow structures and cycle-to-cycle variations. Several studies have investigated the physics of high temperature reacting flows along with spray–gas interactions. However, the effects of bulk gas motion on the combustion and emissions are yet to be understood completely. It has been found that combustion chamber design, intake port design and configuration and piston geometry have a significant effect on the development of vortices inside the cylinder, however, no definite understanding, or reliable tools are available to optimize the design for these components. Hence, it is a challenge for the researchers to develop correlations between design parameters and flow parameters along with optimization tools to enhance the flow structure interactions for improving the in-cylinder processes. Complexity of optical engines puts limitations on design variations, which need to be resolved. This can be resolved by splitting the main problem into a number of sub-problems. Flow bench study can significantly help in this regard. Comparative studies between existing optical engine geometries and its kinematically similar models on flow benches will help in the understanding of sources, which create differences. Flow bench models provide flexibility to change the geometry economically. Experimenting with a range of geometries can help bridge the gap between qualitative ideas to quantitative predictive tools. This also provides an excellent platform for validation of CFD-based model. CFD models are validated by comparing the pressure vs. crank angle curves and heat release rate curves. Few recent studies have explored using PIV results for simulation studies (da Costa et al. 2017; Willman et al. 2020; Van Dam et al. 2017; Van Dam and Rutland 2016). Such studies should be conducted extensively in the future.

The difference between thermodynamic conditions of an optical engine and a metallic engine is another gap, which should be minimized. Matching the heat transfer characteristics is a major issue of this problem. This issue can be addressed by changing ambient conditions for the optical engines like intake air temperature, coolant temperature, lubricating oil temperature, etc. Heat release rate can be a benchmark for the comparison of two designs. Such studies may help develop fundamental insights. Multiple optical diagnosis tools should be the next step for effective use of PIV in engine development. To understand the effect of fluid motion on combustion, spray–flame visualization should be carried out with flow structure analysis. Variations in piston geometry affect the flow structures. By developing knowledge about the flow–flame interactions will help design efficient combustion chambers.

Innovations in developing PIV as an engine diagnostics tool must go on. Endoscopic PIV is the least explored technique, owing to its drawback of small interrogation volume and alignment issues. However, a combination of endoscopic PIV and normal PIV can be explored and developed further. For example, with realistic optical piston design, image quality has been a major concern. This is one of the reasons why such designs have not been explored extensively. A probable solution could be the use of optical piston with metallic cylinder liners. Laser sheet can be admitted through the piston window to analyze the tumble plane. For imaging, an endoscopic camera can be used. To enlarge the field of view, one of the exhaust valves of the engine can be replaced by an interrogation window. Such designs have been implemented for combustion visualization and flow investigations in the engines (Kashdan and Thirouard 2011). At low-speed operations, this does not create any issues in engine operation. A similar design can also be explored for all metallic operation. Laser alignment creates some issues with endoscopic PIV. Therefore, the laser optics can be accommodated in the exhaust port, which will alleviate this issue to some extent. The camera can be placed at a suitable proximity to analyse a particular location in the combustion chamber. In large-bore engines, this technique can be used more effectively compared to small engines. The same interrogation volume can be investigated in an optical engine. This would help in understanding the effect of differences in design and operation parameters of optical and metallic engines.

3.6 Summary

This chapter is focused to understand the state-of-the-art of PIV as a diagnostics tool for engine applications. PIV is a decade old technology; however, the technology is continuously evolving for adaptation in different applications related to fluid flow. In last few years, PIV has gained popularity in IC engine applications for understanding complex in-cylinder processes, which have been a grey area of research thus far. However, it is yet to be explored to its full potential to develop complete understanding of the complex in-cylinder processes. There are several challenges. These challenged and their solutions have been discussed in this chapter comprehensively. Benchmark investigations have been summarized to support deep learnings. A critical review of the current state-of-the-art has also been presented to provide the updates related to PIV technique. Different verticals of applications have been discussed to provide the overview of the research status. Few case studies along with results have also been included to improve our understanding.

References

Agarwal AK, Gadekar S, Singh AP (2017) In-cylinder air-flow characteristics of different intake port geometries using tomographic PIV. Phys Fluids 29(9):

Alam A, Mittal M, Lakshminarasimhan V (2020) Analysis of in-cylinder flow and cycle-to-cycle flow variations in a small spark-ignition engine at different throttle openings. SAE Technical Paper 2020-01-0793

Aleiferis PG, Behringer MK (2017) Modulation of integral length scales of turbulence in an optical SI engine by direct injection of gasoline, iso-octane, ethanol and butanol fuels. Fuel 189:238–259

Alharbi AY, Sick V (2010) Investigation of boundary layers in internal combustion engines using a hybrid algorithm of high speed micro-PIV and PTV. Exp Fluids 49(4):949–959

Bari S, Saad I (2015) Performance and emissions of a compression ignition (CI) engine run with biodiesel using guide vanes at varied vane angles. Fuel 143:217–228

Benajes J, Pastor JV, García A, Monsalve-Serrano J (2018) Redesign and characterization of a single-cylinder optical research engine to allow full optical access and fast cleaning during combustion studies. Exp Techn 42(1):55–68

Bisel TT, Dahlberg JL, Martin TR, Owen SS, Keanini RG, Tkacik PT, Narayan N, Goudarzi N (2017) A comparison of flat white aerosol and rhodamine (r6g) fluorescent paints and their effect on the results of tomographic piv measurements. In: ASME 2017 international mechanical engineering congress and exposition. American Society of Mechanical Engineers Digital Collection

Brijesh P, Abhishek S, Sreedhara S (2015) Numerical investigation of effect of bowl profiles on performance and emission characteristics of a diesel engine. SAE Technical Paper 2015-01-0402

Bücker I, Karhoff D, Dannemann J, Pielhop K, Klaas M, Schröder W (2013) Comparison of PIV measured flow structures in two four-valve piston engines. In: New results in numerical and experimental fluid mechanics VIII, pp 633–640. Springer, Berlin, Heidelberg

Cho IY, Fujimoto H, Kuniyoshi H, Ha JY, Tanabe H, Sato GT (1990) Similarity law of entrainment into diesel spray and steady spray. SAE Technical Paper 900447

da Costa RBR, Braga RM, Júnior CAG, Valle RM, Huebner R (2017) PIV measurements and numerical analysis of in-cylinder tumble flow in a motored engine. J Brazilian Soc Mech Sci Eng 39(10):3931–3945

Dembinski HW (2014) The effects of injection pressure and swirl on in-cylinder flow pattern and combustion in a compression–ignition engine. Int J Engine Res 15(4):444–459

Dierksheide U, Meyer P, Hovestadt T, Hentschel W (2002) Endoscopic 2D particle image velocimetry (PIV) flow-field measurements in IC engines. Exp Fluids 33(6):794–800

Driscoll KD, Sick V, Gray C (2003) Simultaneous air/fuel-phase PIV measurements in a dense fuel spray. Exp Fluids 35(1):112–115

El Adawy M, Heikal MR, Aziz ARA, Munir S, Siddiqui MI (2018) Effect of boost pressure on the in-cylinder tumble-motion of GDI engine under steady-state conditions using Stereoscopic-PIV. J Appl Fluid Mech 11(3):733–742

El-Adawy M, Heikal MR, Aziz ARA, Siddiqui MI, Wahhab HAA (2017) Experimental study on an IC engine in-cylinder flow using different steady-state flow benches. Alexandria Eng J 56(4):727–736

Elsinga GE, Tokgoz S (2014) Ghost hunting—an assessment of ghost particle detection and removal methods for tomographic-PIV. Measur Sci Technol 25(8):

Fan L, Gao Y, Hayakawa A, Hochgreb S (2016) Temperature and velocity measurements by thermographic particle image velocimetry with ZnO tracers. Exp Fluids 58:1–34

Fan L, McGrath D, Chong CT, Jaafar MNM, Zhong H, Hochgreb S (2018) Laser-induced incandescence particle image velocimetry (LII-PIV) for two-phase flow velocity measurement. Exp Fluids 59(10):156

Fatehi H, Lucchini T, D'errico G, Karlsson A, Bai XS, Andersson Ö (2019) Effect of in-cylinder flow structures on late cycle soot oxidation in a quiescent heavy-duty diesel engine. Combus Sci Technol, 1–21

Ganapathisubramani B, Longmire EK, Marusic I, Pothos S (2005) Dual-plane PIV technique to determine the complete velocity gradient tensor in a turbulent boundary layer. Exp Fluids 39(2):222–231

Geschwindner C, Kranz P, Welch C, Schmidt M, Böhm B, Kaiser SA, De la Morena J (2020) Analysis of the interaction of Spray G and in-cylinder flow in two optical engines for late gasoline direct injection. Int J Engine Res 21(1):169–184

Gord JR, Meyer TR, Roy S (2008) Applications of ultrafast lasers for optical measurements in combusting flows

Grayson K, de Silva CM, Hutchins N, Marusic I (2016) Laser light sheet profile and alignment effects on PIV performance. In: 18th international symposium on applications of laser and imaging techniques to fluid mechanics, Lisbon, Portugal (pp 4–7)

Grüneberger P, Hengge E, Winklhofer E (2019) Flame transport in a heavy-duty diesel engine. Autom Engine Technol 4(1–2):1–8

Haghgooie M, Kent JC, Tabaczynski RJ (1986) Verification of LDA and seed generator performance. Exp Fluids 4(1):27–32

Hain R, Kähler CJ, Tropea C (2007) Comparison of CCD, CMOS and intensified cameras. Expe Fluids 42(3):403–411

Huang S, Zhao C, Zhu Y (2017) Numerical investigation on effects of combustion chamber structure and intake air humidification on combustion and emission characteristics of marine diesel engine. SAE Technical Paper 2017-01-2254

Jaichandar S, Annamalai K (2012) Effects of open combustion chamber geometries on the performance of pongamia biodiesel in a DI diesel engine. Fuel 98:272–279

Kaplan M (2019) Influence of swirl, tumble and squish flows on combustion characteristics and emissions in internal combustion engine-review. Int J Autom Eng Technol 8(2):83–102

Karthickeyan V (2019) Effect of combustion chamber bowl geometry modification on engine performance, combustion and emission characteristics of biodiesel fuelled diesel engine with its energy and exergy analysis. Energy 176:830–852

Kashdan J, Thirouard B (2011) Optical engines as representative tools in the development of new combustion engine concepts. Oil & Gas Sci Technol-Revue d'IFP Energies Nouvelles 66(5):759–777

Konno M, Chikahisa T, Murayama T (1992) Reduction of smoke and NOx by strong turbulence generated during the combustion process in DI diesel engines. SAE Transactions, pp 811–819

Kumar Agarwal A, Gadekar S, Pratap SinghA (2018) In-cylinder flow evolution using tomographic particle imaging velocimetry in an internal combustion engine. J Energy Resour Technol 140(1)

Kuniyoshi H, Tanabe H, Sato GT, Fujimoto H (1980) Investigation on the characteristics of diesel fuel spray. SAE Transactions, pp. 2998–3014

Lee T (2011) PIV study of near-field tip vortex behind perforated Gurney flaps. Exp Fluids 50(2):351–361

Li YH, Wu CY, Chen BC, Chao YC (2008) Measurements of a high-luminosity flame structure by a shuttered PIV system. Measur Sci Technol 19(4):

Ma PC, Ewan T, Jainski C, Lu L, Dreizler A, Sick V, Ihme M (2017) Development and analysis of wall models for internal combustion engine simulations using high-speed micro-PIV measurements. Flow, Turbulence Combus 98(1):283–309

Macháček M (2002) A quantitative visualization tool for large wind tunnel experiments (Doctoral dissertation, ETH Zurich)

Mendez MA, Raiola M, Masullo A, Discetti S, Ianiro A, Theunissen R, Buchlin JM (2017) POD-based background removal for particle image velocimetry. Exp Thermal Fluid Sci 80:181–192

Michelsen HA, Schulz C, Smallwood GJ, Will S (2015) Laser-induced incandescence: Particulate diagnostics for combustion, atmospheric, and industrial applications. Progress Energy Combus Sci 51:2–48

Miles PC (2014) The history and evolution of optically accessible research engines and their impact on our understanding of engine combustion. In ASME 2014 internal combustion engine division fall technical conference. American Society of Mechanical Engineers Digital Collection

Mullin JA, Dahm WJ (2005) Dual-plane stereo particle image velocimetry (DSPIV) for measuring velocity gradient fields at intermediate and small scales of turbulent flows. Exp Fluids 38(2):185–196

Najafabadi MI, Tanov S, Wang H, Somers B, Johansson B, Dam N (2017) Effects of injection timing on fluid flow characteristics of partially premixed combustion based on high-speed particle image velocimetry. SAE Int J Eng 10(4):1443–1453

Naka Y, Tomita K, Shimura M, Fukushima N, Tanahashi M, Miyauchi T (2016) Quad-plane stereoscopic PIV for fine-scale structure measurements in turbulence. Exp Fluids 57(5):63

Nishiyama A, Jeong H, Ikeda Y, Sawada R (2012) Application of Endoscopic Stereo PIV to 3-D exhaust gas flow measurements in a practical si engine. In: 16th international symposium on applications of laser techniques to fluid mechanics, Lisbon, Portugal, July (pp. 9–12)

Nishiyama A, Le MK, Furui T, Ikeda Y (2019) Simultaneous in-cylinder flow measurement and flame imaging in a realistic operating engine environment using high-speed PIV. Appl Sci 9(13):2678

Nobach H (2011) Influence of individual variations of particle image intensities on high-resolution PIV. Exp Fluids 50(4):919–927

Osada H, Uchida N, Zama Y (2015) An analysis on heat loss of a heavy-duty diesel engine by wall-impinged spray flame observation. SAE Technical Paper 2015-01-1832

Paterna E, Moonen P, Dorer V, Carmeliet J (2013) Mitigation of surface reflection in PIV measurements. Measur Sci Technol 24(5):

Payri R, Viera JP, Wang H, Malbec LM (2016) Velocity field analysis of the high density, high pressure diesel spray. Int J Multiphase Flow 80:69–78

Petrosky BJ, Lowe KT, Danehy PM, Wohl CJ, Tiemsin PI (2015) Improvements in laser flare removal for particle image velocimetry using fluorescent dye-doped particles. Measur Sci Technol 26(11):

Pfadler S, Dinkelacker F, Beyrau F, Leipertz A (2009) High resolution dual-plane stereo-PIV for validation of subgrid scale models in large-eddy simulations of turbulent premixed flames. Combus Flame 156(8):1552–1564

Porpatham E, Ramesh A, Nagalingam BJEC (2013) Effect of swirl on the performance and combustion of a biogas fuelled spark ignition engine. Energy Conver Manag 76:463–471

Rabault J, Vernet JA, Lindgren B, Alfredsson PH (2016) A study using PIV of the intake flow in a diesel engine cylinder. Int J Heat Fluid Flow 62:56–67

Rault TM, Vishwanath RB, Gülder ÖL (2020) Impact of ethanol blending on soot in turbulent swirl-stabilized Jet A-1 spray flames in a model gas turbine combustor. Proceedings of the Combustion Institute

Razak MF, Salehi F, Chishty MA (2019) An analysis of turbulent mixing effects on the soot formation in high pressure n-dodecane sprays. Flow, Turbul Combus 103(3):605–624

Rostamy N, Sumner D, Bergstrom DJ, Bugg JD (2012) Local flow-field of a surface-mounted finite circular cylinder. J Fluids Struct 34:105–122

Rottier C, Godard G, Corbin F, Boukhalfa AM, Honoré D (2010) An endoscopic particle image velocimetry system for high-temperature furnaces. Measur Scie Technol 21(11):

Scharnowski S, Kähler CJ (2016) Estimation and optimization of loss-of-pair uncertainties based on PIV correlation functions. Exp Fluids 57(2):23

Selamet A, Rupal S, He Y, Keller PS (2004) An experimental study on the effect of intake primary runner blockages on combustion and emissions in SI engines under part-load conditions. SAE Technical Paper 2004-01-2973

Singh AP, Gupta A, Agarwal AK (2015) Tomographic particle image velocimetry for flow analysis in a single cylinder optical engine. SAE Int J Mater Manuf 8(2):472–481

Stansfield P, Wigley G, Justham T, Catto J, Pitcher G (2007) PIV analysis of in-cylinder flow structures over a range of realistic engine speeds. Exp Fluids 43(1):135–146

Stiehl R, Bode J, Schorr J, Krüger C, Dreizler A, Böhm B (2016) Influence of intake geometry variations on in-cylinder flow and flow–spray interactions in a stratified direct-injection spark-ignition engine captured by time-resolved particle image velocimetry. Int J Engine Res 17(9):983–997

Sun Y, Che Z, Sun K, Wang T, Li Y, Bai H (2019a) Measurements of turbulence sources in a swirl-supported diesel engine. Int J Engine Res, p 1468087419870415
Sun Y, Sun K, Wang T, Li Y, Lu Z (2019b) Effects of squish flow on tangential flow and turbulence in a diesel engine. J Eng Gas Turbines Power 141(5)
Tanov S, Salvador-Iborra J, Andersson Ö, Olmeda P, García A (2017) Influence of the number of injections on piston heat rejection under low temperature combustion conditions in an optical compression-ignition engine. Energy Conv Manag 153:335–345
Tanov S, Pachano L, Andersson Ö, Wang Z, Richter M, Pastor JV, García-Oliver JM, García A (2018) Influence of spatial and temporal distribution of Turbulent Kinetic Energy on heat transfer coefficient in a light duty CI engine operating with Partially Premixed Combustion. Appl Thermal Eng 129:31–40
Thongchai S, Lim O (2020) Macroscopic spray behavior of a single-hole common rail diesel injector using gasoline-blended 5% biodiesel. Energies 13(9):2276
Thurow B, Jiang N, Lempert W (2012) Review of ultra-high repetition rate laser diagnostics for fluid dynamic measurements. Measur Sci Technol 24(1):
Towers DP, Towers CE, Buckberry CH, Reeves M (1999) A colour PIV system employing fluorescent particles for two-phase flow measurements. Measur Sci Technol 10(9):824
Tsiogkas VD, Chraniotis A, Kolokotronis D, Tourlidakis A (2019) In-cylinder flow measurements in a transparent spark ignition engine. SAE Technical Paper 2019-24-0099
Udayakumar R, Arasu PV, Sriram S (2003) Experimental investigation on emission control in CI engines using shrouded inlet valve. SAE Technical Paper 2003-01-0350
Van Dam, N. and Rutland, C., 2016. Understanding in-cylinder flow variability using large-eddy simulations. Journal of Engineering for Gas Turbines and Power, 138(10)
Van Dam N, ZengW, Sjöberg M, Som S (2017) Parallel multi-cycle LES of an optical pent-roof DISI engine under motored operating conditions. In: ASME 2017 internal combustion engine division fall technical conference. American Society of Mechanical Engineers Digital Collection
Velte CM, Hansen MOL, Cavar D (2008) Flow analysis of vortex generators on wing sections by stereoscopic particle image velocimetry measurements. Environ Res Lett 3(1):
Vester AK, Nishio Y, Alfredsson PH (2019) Investigating swirl and tumble using two prototype inlet port designs by means of multi-planar PIV. Int J Heat Fluid Flow 75:61–76
Wang G, Yu W, Li X, Yang R (2020) Influence of fuel injection and intake port on combustion characteristics of controllable intake swirl diesel engine. Fuel 262:
Watanabe S, Kato H, Kwak DY, Shirotake M, Rinoie K (2004) Stereoscopic PIV measurements of leading edge separation vortices on a cranked arrow wing. Measur Sci Technol 15(6):1079
Weber V, Brübach J, Gordon RL, Dreizler A (2011) Pixel-based characterisation of CMOS high-speed camera systems. Appl Phys B 103(2):421–433
Willman C, Scott B, Stone R, Richardson D (2020) Quantitative metrics for comparison of in-cylinder velocity fields using particle image velocimetry. Exp Fluids 61(2):62
Witze PO, Baritaud TA (1986) Particle seeding for mie scattering measurements in combusting flows. In: Proceedings of the third international symposium on applications of laser anemometry to fluid mechanics, Lisbon, Vol 5
User's guide optical version for engine type 5402.070, AVL
Tsiogkas VD, Chraniotis A, Kolokotronis D, Tourlidakis A, Cold flow measurement in optical internal combustion engine using PIV
Zama Y, Ochiai W, Furuhata T, Arai M (2012) Velocity measurement inside a diesel spray by using time-resolved PIV under high ambient density condition. In: Proceedings of 12th triennial international conference on liquid atomization and spray systems, pp 1–8
Zama Y, Odawara Y, Furuhata T (2017) Experimental investigation on velocity inside a diesel spray after impingement on a wall. Fuel 203:757–763
Zama Y, Tajima T, Yatabe W, Furutaha T, Hayakawa T, Okinaka M (2019) Experimental investigation on behavior of diesel spray after interference with a circular cylinder (No. 2019-01-2285). SAE Technical Paper

Zha K, Busch S, Miles PC, Wijeyakulasuriya S, Mitra S, Senecal PK (2015) Characterization of flow asymmetry during the compression stroke using swirl-plane PIV in a light-duty optical diesel engine with the re-entrant piston bowl geometry. SAE Int J Eng 8(4):1837–1855
Zha K, Busch S, Warey A, Peterson RC, Kurtz E (2018) A study of piston geometry effects on late-stage combustion in a light-duty optical diesel engine using combustion image velocimetry. SAE Int J Engines 11(6):783–804
Zhou F, Fu J, Ke W, Liu J, Yuan Z, Luo B (2017) Effects of lean combustion coupling with intake tumble on economy and emission performance of gasoline engine. Energy 133:366–379

Chapter 4
Dimethyl Ether Spray Characteristics for Compression Ignition Engines

Akash Rai, Dhananjay Kumar, Utkarsha Sonawane, and Avinash Kumar Agarwal

Abstract In the current scenario, alternative fuels are being explored, which can fulfil the global energy demand and meet stringent emission norms simultaneously. Dimethyl ether (DME) is one such alternative fuel that satisfies both these requirements. It has a high cetane number, which makes it suitable for heavy-duty engine applications. However, before the implementation of these new fuels in existing engines, various parameters need to be investigated and optimised. In compression ignition (CI) engines, one of the most important parameters is fuel–air ratio, which has a tremendous effect on engine combustion, performance, and emissions. Spray characteristics have a dominating effect on fuel–air mixture preparation. It depends on various parameters such as fuel injection pressure (FIP), injector geometry, fuel properties, and ambient conditions. Selection of fuel injection strategies and injector nozzle dimensions are primarily governed by spray characteristics of the fuel, injection timings, FIP, injection angle, and injector nozzle dimensions. This chapter summarises the production and application of DME in CI engines. The focus of this chapter is to analyse spray characteristics of DME under different ambient temperature and pressure conditions. The effect of spray characteristics on the in-cylinder mixture formation and engine combustion has also been discussed.

Keywords DME · Spray investigations · Alternative fuels · Macroscopic spray characteristics · Microscopic spray characteristics

Abbreviations

SI	Spark Ignition
CI	Compression Ignition
ICE	Internal Combustion Engines

A. Rai · D. Kumar · U. Sonawane · A. K. Agarwal (✉)
Department of Mechanical Engineering, Engine Research Laboratory, Indian Institute of Technology Kanpur, Kanpur 208016, India
e-mail: akag@iitk.ac.in

A. P. Singh and A. K. Agarwal (eds.), *Novel Internal Combustion Engine Technologies for Performance Improvement and Emission Reduction*, Energy, Environment, and Sustainability, https://doi.org/10.1007/978-981-16-1582-5_4

CO	Carbon Monoxide
HC	Hydrocarbon
NOx	Nitrogen Oxide
mtoe	one million tonne of oil equivalent
DME	Dimethyl ether
bcm	billion cubic meters
TBM	Thousand million barrels
R/P	Reserve to production ratio
SMD	Sauter mean diameters
TDC	Top dead centre

4.1 Introduction

Global energy consumption is increasing continuously due to economic growth, primarily in developing countries. Energy demand cannot be fulfilled by fossil fuels due to their limited reserves. Most nations are concerned for energy security pertaining to conventional petroleum-based fuels in future. Researchers all over the world have started exploring alternative fuels that can meet the galloping energy demand and comply with the stringent emission norms simultaneously. The most widely used alternative fuels are biodiesel, natural gas, methanol, and dimethyl ether (DME). Methanol has emerged as a potential alternative fuel for spark-ignition (SI) engines. However, for compression ignition (CI) engines, DME is the most appropriate alternative fuel due to its high cetane number. Development and utilisation of DME would lead to indigenous energy production and distribution across the country and the world. However, such growth and large-scale adaptation of DME would require enormous investment and creation of new DME infrastructure.

4.1.1 Energy Scenario Worldwide

According to the international energy agency (IEA) 1999, United States, China, and Russia together produced ~31% of global fossil fuels and consumed ~41% of global energy (International Energy Agency 1999). Fossil fuels are the primary source of energy in coming years. However, energy demand is swelling drastically in the foreseeable future (Singh and Singh). According to statistical review of the world energy report (2019), growth rate in global primary energy consumption was ~2.9% in 2018, which was also the fastest growth since 2010 and almost double the average growth rate of the past 10 years (BP Statistical Review of World Energy 2019). Natural and renewable energy resources are the reason for such drastic growth. Natural gas contributes to more than 40% of the total growth, and renewable energy accounted for the second-largest energy increase. According to a report, USA, China, and India accounted for more than two-thirds of the total energy consumption growth rate. Global coal consumption grew up by ~1.4%, which is double the average growth rate of the past decade. Global coal consumption growth rate was mainly contributed by India (36 mtoe) and China (16 mtoe). Contributions of coal in primary energy have fallen to ~27.2%, which is lowest in the last 15 years. While global coal production is increased by 4.3% (162 mtoe), renewable energy is growing sharply, and primary fuel sources are declining. In the year 2018, renewable energy grew up by 14.5% (71 mtoe), which is much higher than the growth of primary energy resources. This increasement of 71 mtoe includes an increase of 32 mtoe in wind energy and 30 mtoe in solar energy.

China has the largest contribution of 32 mtoe in the growth of renewable energy production. Global nuclear energy generation grew up by 2.4% with major contribution by China and Japan (BP Statistical Review of World Energy 2019). The primary energy production and consumption majorly contributed to carbon emissions, which grew by 2% in the last 7 years. This is an alarming and fast growth rate (BP Statistical Review of World Energy 2019; Masson-Delmotte et al. 2018).

Petroleum, coal, and natural gas are the main primary energy sources to cater to global energy demand. Reservoirs of the fossil fuels are limited and might get exhausted in near future. According to the statistical review of the world energy report 2018, reserve to production ratio (R/P) is the ratio of reserve remaining at the end of any year to that of total production in that year. The numerical value of R/P simply represents the duration of time, for which remaining reserves would last, if the production rate remains the same. A comprehensive region-wise data of oil reserves of the world with R/P ratio have been listed in Table 4.1.

Table 4.1 Oil reserves worldwide (BP Statistical Review of World Energy 2019)

Region	At end 1998 (TBM)	At end 2008 (TBM)	At end 2017 (TBM)	At end 2018 (TBM)	At end 2018 (TMT)	At end 2018 (%)	R/P
North America	100	216.6	237.8	236.7	35.4	13.7	28.7
South America	95.6	196	324	325.1	51.1	18.8	136.2
Europe	21.4	14.2	13.7	14.3	1.9	0.8	11.1
CIS	121.1	144.8	144.7	144.7	19.6	8.4	11.1
Middle East	685.2	753.7	834.3	836.1	113.2	48.3	72.1
Africa	77.2	120.4	125.3	125.3	16.6	7.2	41.9
Asia	40.8	48	47.7	47.6	6.3	2.8	17.1
World	1141.2	1493.8	1727.5	1729.5	224.1	100	50

TBM: Thousand million barrels; TMT: Thousand million tonnes

4.2 Production of Dimethyl Ether (DME)

Various feedstocks can produce DME in two distinctive ways. The primary five feedstocks for DME production are methanol, natural gas, biomass, organic waste, and coal (Ateah; Xu et al. 1997; Tomatis et al. 2019; Farooqui et al. 2019; Inayat et al. 2017; Parvez et al. 2018). These five feedstocks finally converged to the same set of reactions. The type of catalyst used in the reactions mainly affects the production rate of DME. Two distinctive processes involved in DME production have been illustrated in Fig. 4.1.

4.2.1 Production of DME from Methanol

The process of methanol production from the syngas is known for many decades. Therefore, the involved reaction mechanisms, active catalysts, active reagents, and the role of each ingredient are very well known. However, the production of DME from dehydration of methanol is not much explored. The production rate of DME from dehydration of methanol is mainly dependent on the type of catalyst used. The stability temperatures of different catalysts are mentioned in Table 4.2. All these potential catalysts have been discussed in more detail in the following paragraphs.

(a) **Gamma Al_2O_3**

Reactivity of gamma Al_2O_3 catalyst greatly depends on the operating temperature, and hence the conversion rate of methanol to DME. Conversion rate of approximately 6% was achieved, when the temperature was below 150 °C and it increased

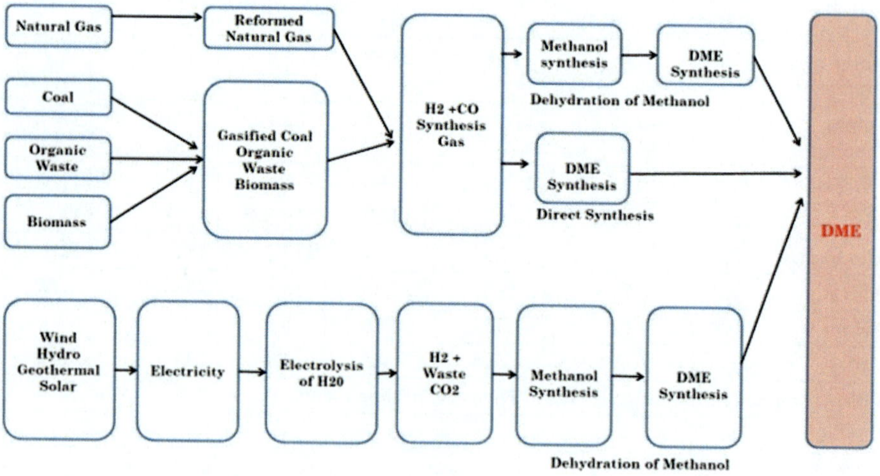

Fig. 4.1 Schematic of DME production from different feedstocks

Table 4.2 Catalytic stability as determined by methanol conversion (Xu et al. 1997)

Catalyst	Temperature (°C)
Gamma-Al_2O_3	205
TiO_2/ZrO_2	190
TIO_2/Al2O_3(Na^+)	335
TIO_2/Al2O_3(NH_4^+)	205
HZNS-5	190

very rapidly as the temperature increased from 150 to 225 °C. DME conversion reaction is an exothermic reaction and increase in the temperature above the optimum temperature can reduce the rate of reaction hence, a further increase in temperature to 400 °C reduced conversion rate to 0.8%. Water and DME are two main products of methanol dehydration. Addition of excess water reduces DME formation or conversion rate. For achieving the same conversion rate with excess water, temperature of reactants needs to be increased. This may be due to an increase in activation energy of reaction on the addition of water (Xu et al. 1997).

(b) **TiO_2/ZrO_2**

Zirconium and titanium oxide are less active than Al_2O_3. The same conversion rate can be achieved by ZrO_2 and Al_2O_3, but it requires higher operating temperature for ZrO_2. The conversion rate of 41% requires a temperature of 325 °C in case of TiO_2 catalyst. Whereas gamma Al_2O_3 achieves the same conversion rate with a lower temperature of ~190 °C (Xu et al. 1997; Bonura et al. 2013).

(c) **TiO_2/Al_2O_3**

NH_4 + TiO_2/Al_2O_3 and Na + TiO_2/Al_2O_3 are the main catalysts generated from TiO_2/Al_2O_3. Experimental results showed that NH_4 + TiO_2/Al_2O_3 can achieve conversion rate of 42% at the reaction temperature of 190 °C. At the same time, the same level of methanol conversion could only be obtained at 300 °C using Na + TiO_2/Al_2O_3 catalyst. This shows that the NH_4 + TiO_2/Al_2O_3 catalyst is much more reactive than that of Na + TiO_2/Al_2O_3 catalyst. NH_4 + TiO_2/Al_2O_3 catalyst shows comparable results with gamma Al_2O_3 (Xu et al. 1997).

(d) **HZNS-5**

HZNS-5 is a better catalyst than gamma Al_2O_3 for dehydration of methanol. HZNS-5 has both Lewis acid and Bronsted acid side, while gamma Al_2O_3 has only Lewis acid side to attack (Xu et al. 1997; Bonura et al. 2013). Addition of water harms the catalytic activity of both catalysts, but the effect is more on gamma Al_2O_3 than HZNS-5. Water is strongly absorbed by Lewis acid side of gamma Al_2O_3, which prohibits the formation of DME, while the effect of water is much less significant on H-ZSM5 due to hydrophobic nature of HZSM-5 (Xu et al. 1997; Bonura et al. 2013).

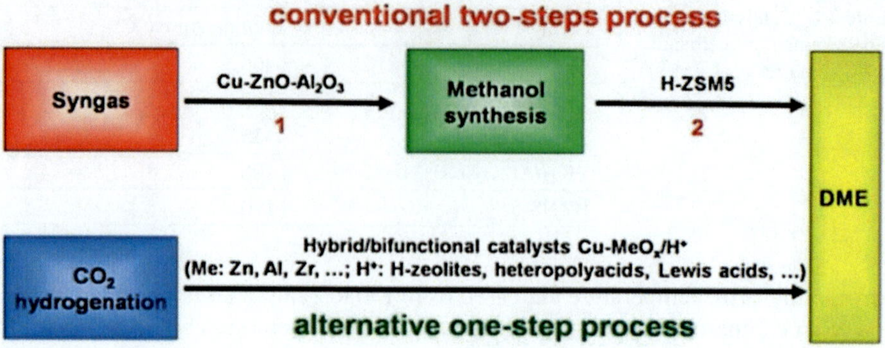

Fig. 4.2 Conversion of DME from atmospheric CO_2 (Bonura et al. 2013)

4.2.2 Conversion of DME from CO_2

There are two methods for converting CO_2 or syngas to DME. A schematic illustration has been shown in Fig. 4.2. In the first method, CO_2 is converted to methanol by a single catalyst, which is further dehydrated to DME by the second catalyst (as explained in the earlier section). In the second method, CO_2 is directly converted to DME by a single bi-functional catalyst, which performs the two processes simultaneously; the first is the formation of methanol and the second is dehydration of methanol (Bonura et al. 2013). Since it uses atmospheric CO_2 to produce DME, it reduces global warming and produces energy. Therefore, it is a sustainable source of energy.

4.2.3 Wind Power to DME via Methanol

The CO_2 concentration in the atmosphere is increasing continuously and is a serious threat to the environment. Major part of CO_2 emission is contributed by conventional vehicle, which is the backbone of the economy. In this method, CO_2 from ambient air combines with water to produce methanol using the electricity generated by wind energy. Electricity is required for electrolysis of water. Hydrogen comes out as a product of electrolysis of water, which can be achieved by solid oxide electrolysis, alkaline electrolysis, and proton exchange method. Among these methods, solid oxide electrolysis is more efficient and economical because it operates at a high temperature of ~700–900 °C. High-temperature requirement lowers the activation energy of the process, hence lowers the electricity demand. CO_2 for this method is directly taken from ambient air, as shown in Fig. 4.3 (Bos et al. 2020; Azarabadi and Lackner 2019; Veselovskaya et al. 2013). Hydrogen combines with CO_2 to form methanol or direct DME (Bos et al. 2020; Naik et al. 2011).

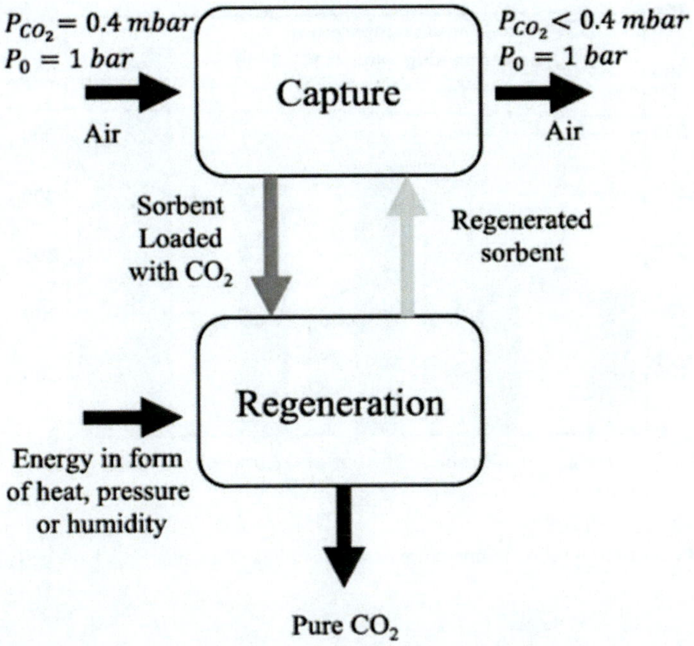

Fig. 4.3 Schematic for CO_2 capture from ambient air (Azarabadi and Lackner 2019)

4.3 DME as an Alternate Fuel in CI Engine

From the perspectives of emission reduction, DME is one of the most promising alternative fuels for CI engines. For application of DME in CI engine, its chemical and physical properties must be analysed. This section reviews different properties of DME and their comparison with diesel.

(a) **Combustion and Thermodynamic Properties**: One of the advantages of DME is that it has lower auto-ignition temperature than conventional diesel under pressurised conditions (Huang et al. 1999). Under atmospheric conditions, DME (350 °C) has higher auto-ignition temperature than diesel (250 °C). The autoignition temperature of DME depends on the ambient pressure and it decreases with an increasing pressure.
Autoignition temperature of DME is in the range of 200–250 °C in high-pressure conditions during the compression stroke. Hence, DME is having lower auto-ignition temperature than diesel under real engine conditions (Teng et al. 2004, 2003). Figure 4.4 shows the auto-ignition and lower heating value (LHV) of various fuels such as methanol, ethanol, CNG, DME, and diesel.

(b) **Cetane Number**: Higher cetane number signifies lower ignition delay, and it is defined as the period between start of injection and a noticeable rise in pressure due to combustion (Teng et al. 2003). One of the advantages of DME

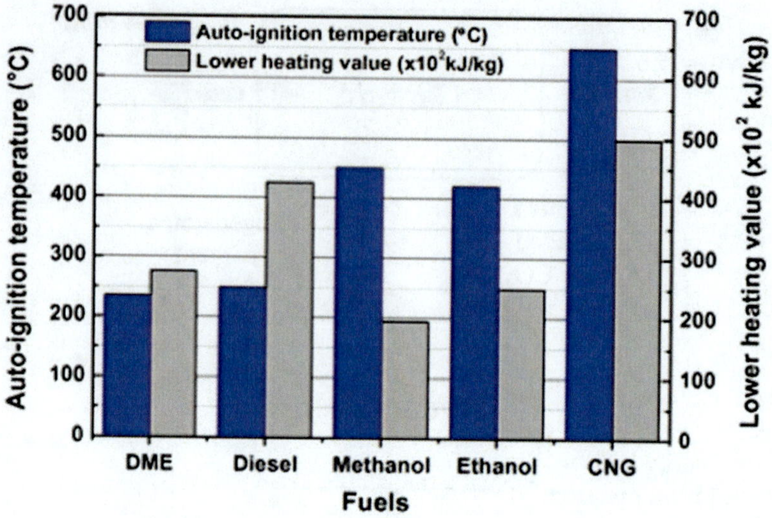

Fig. 4.4 LHV and autoignition temperature of various fuels (Teng et al. 2004, 2003; Park and Lee 2014)

is that it has a higher cetane number (55–60) than diesel (45–55), which makes it appropriate for CI engines (Teng et al. 2003). Higher cetane number shortens the ignition delay and hence premixed combustion duration will be lower. This leads to lesser noise, lesser knock, smoother combustion, and lesser emissions.

(c) **Molecular Structure**: Chemical structure of DME is $H_3C–O–CH_3$. There is no carbon–carbon bond in it, which contributes to sootless or clean combustion and lower emissions. Oxygen content of DME is 34.8% w/w, which is higher than diesel and other alternative fuels. Therefore, DME fuelled engine produces lower soot/particulate matter (PM) emissions. It plays a vital role in promoting clean combustion even in rich mixtures (Park and Lee 2014). High cetane number and lower ignition delay period of DME are due to the presence of C–O bond, which has lower binding energy (359.0 kJ/mol) compared to C–H (410.4 kJ/mol) bond in conventional diesel. Therefore, C–O bond breaks more quickly as compared to C–H bond due to lower auto-ignition temperature of DME compared to diesel (Teng et al. 2003; Arcoumanis et al. 2008).

(d) **Viscosity and Lubricity**: Viscosity and lubricity determine the applicability of any given fuel in CI engines. Main problems with DME are its lower viscosity and lubricity compared to diesel. Kinematic viscosity range of DME is 0.185–0.23 mm^2/s (cSt) and depends upon temperature and pressure (Teng et al. 2002). Low lubricity and viscosity may lead to surface wear and fuel leakage in the engine, plunger, feed pump, main pump, and fuel delivery system. To increase the viscosity and lubricity of DME, different additives are used. Many researchers have suggested use of additives with a certain composition for DME such as Hitec, Lubrizol, and Castor oil (Youn et al. 2011; Tsuchiya and Sato

2006; Sugiyama et al. 2004; Kapus and Ofner 1995; Goto et al. 2005). Also, it is advantageous to reduce clearance between two mating objects in the new injector design.

(e) **Boiling Point and Latent Heat**: Latent heat of vaporisation and boiling point play a vital role in spray atomisation process. Latent heat of vaporisation of diesel is 250 kJ/Kg at normal temperature and pressure. For DME, it is 460 kJ/Kg at temperature of 20 °C (Sezer 2011). Vaporisation of DME takes place by absorption of heat from combustion chamber. This reduces overall temperature of combustion chamber. Low boiling point of DME helps in quick evaporation, which leads to quick drop in combustion chamber temperature. Hence overall combustion chamber temperature for DME is less than conventional diesel. The formation of NOx takes place under high-temperature condition. Overall, quick reduction in combustion chamber temperature reduces NOx formation and promotes clean combustion (Park and Lee 2014).

(f) **Lower Heating Value (LHV)**: Lower heating value is one of the key factors for utilisation of any alternative fuel in the engine. DME has lower LHV as compared to diesel. LHV for DME and diesel is 28.8 MJ/Kg and 42.5 MJ/Kg, respectively (Sorenson et al. 1998). LHV of DME is approximately two-thirds of diesel. Therefore, for achieving the same power output, mass of DME injected is ~1.5 times more than the corresponding diesel. This can be done by increasing the nozzle diameter, injection duration, or fuel injection pressure (Sorenson et al. 1998; Han et al. 2020).

(g) **Compressibility of DME**: Compressibility plays a major role in the storage of any fuel. DME has low bulk modulus and hence is a highly compressible fuel. The bulk modulus of DME is lower than many hydrocarbons such as benzene, heptane, dodecane, and diesel (Sorenson et al. 1998). The work required for compression of DME by using a high-pressure injection pump is much higher than conventional diesel. It is because of high compressibility and lower density of DME compared to diesel (Sorenson et al. 1998).

4.4 Fuel Spray Characterisation

Engine performance and emissions can be optimised by analysis of fuel spray characteristics. The spray characterisation is required for detailed analysis of physical and chemical properties of DME such as boiling point, kinematic viscosity, lubricity, latent heat of vaporisation, compressibility, volatility, surface tension, ambient density, and vapour pressure and their effect on charge preparation. LHV of DME is ~65% of diesel. Hence to obtain the same engine power output, DME must be injected in a larger amount compared to diesel in same injection duration. Consequently, it is required to increase nozzle hole diameter of diesel injectors (Han et al. 2020). Vapour pressure of DME is very high, hence a detailed study on possibility of cavitation must be done before designing an injector nozzle for DME. Experiments for spray characterisation are normally performed in a constant volume spray chamber (CVSC), which may be either reacting or non-reacting spray. CVSCs are designed to maintain real combustion conditions (Microscopic and Macroscopic Spray Characterization of Biodiesel). Ambient pressure in the CVSC and fuel injection pressure can be varied to simulate engine like conditions. Spray atomisation of DME is superior to diesel due to its lower viscosity, surface tension, and boiling point (Park and Lee 2014). Spray characteristics are mainly divided into two subcategories namely macroscopic and microscopic spray characteristics. These characteristic measurements provide a detailed understanding of spray morphology and spray atomisation quality.

4.4.1 Macroscopic Spray Characteristics

Macroscopic spray characteristics are referred for spray geometrical analysis. It is mainly done by geometrical interpretation of spray images and their intensity captured by a high-speed camera. This largely includes three parameters namely spray tip penetration, spray cone angle, and spray area.

Spray penetration length: It is defined as the maximum length between the nozzle tip and to the farthest point on the spray plume, where the spray actually diminishes. Figure 4.5 shows the top view of a six-hole injector, whereas Fig. 4.6 shows the frontal view of single spray plume with horizontal and vertical spray tip penetrations (Kim et al. 2011a). Any fuel spray must have an optimum value of spray penetration length. Longer spray penetration length could lead to spray plume hitting the cylinder walls. This may result in breakup the lubricating oil layer present on the cylinder liner, leading to higher engine wear, higher soot formation, and deteriorated

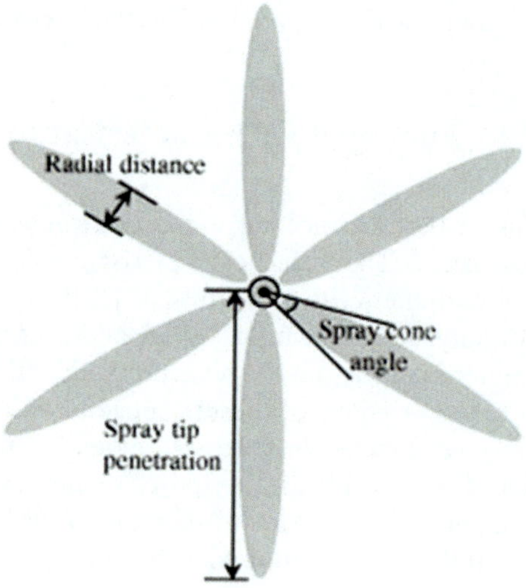

Fig. 4.5 Top view of spray plumes emerging from a six-hole injector (Park et al. 2010)

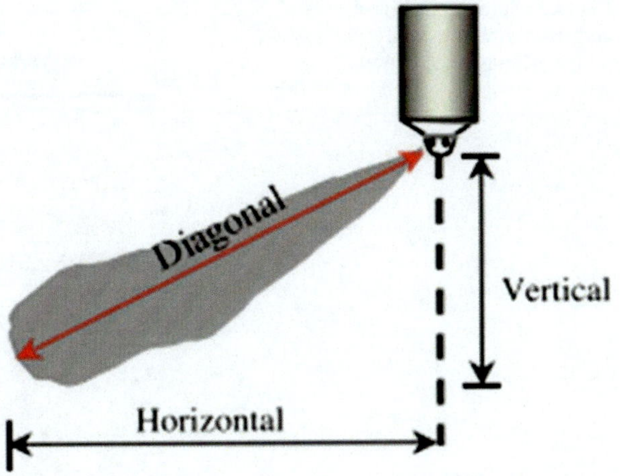

Fig. 4.6 Frontal view of a spray plume with horizontal and vertical spray tip penetration (Kim et al. 2011a)

engine performance. Shorter spray penetration length results in an inefficient air–fuel mixture formation. This may lead to poor engine combustion performance as well. Overall, spray tip penetration is one of the essential factors to obtain optimum engine performance and emissions (Characterization and of Diesel and Biodiesel Spray).

Spray cone angle: Spray cone angle is a measure of the extent of dispersion of spray cone (Fig. 4.7). A highly dispersive fuel like DME has a larger cone angle than less dispersive fuel like biodiesel. Any fuel spray should have an optimum value of cone angle for maximum utilisation of air to form efficient fuel–air mixture. A narrow spray cone angle may reduce the fuel–air mixing, whereas a large spray cone angle may lead to cylinder wetting (Microscopic and Macroscopic Spray Characterization of Biodiese; Characterization and of Diesel and Biodiesel Spray).

Spray area: It is calculated by post-processing the spray image captured by a high-speed camera. It is defined as the area covered by spray droplets within the spray regime, as shown in Fig. 4.7.

4.4.2 *Microscopic Spray Characteristics*

Microscopic spray characteristics mainly describe the spray droplet size, droplet size distribution, three components of axial and radial velocities, time for arriving of the droplet, and droplet number density. Droplet is assumed to be spherical and is measured by various diameters such as arithmetic mean diameter (D_{10}), volume mean diameter (D_{30}), and Sauter mean diameter (D_{32}).

Arithmetic Mean Diameter: It represents an average of diameters of all the droplets in a particular spray plume.

Volume Mean Diameter: It denotes the diameter of a droplet that gives the volume of a complete sample, if multiplied by the total number of droplets in that specific volume.

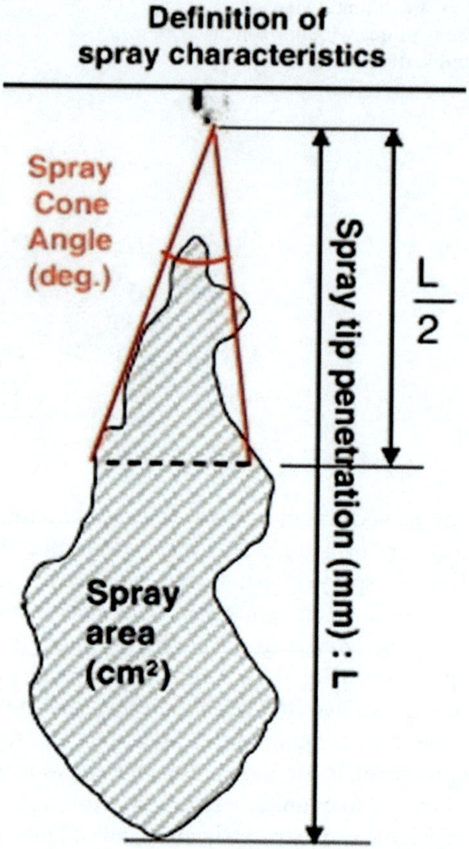

Fig. 4.7 Spray cone angle and spray area of a spray plume (Kim et al. 2010)

Sauter Mean Diameter: It is a theoretical ratio of volume to the surface area of the complete spray sample. It specifies the diameter of droplet, which is exactly same as the ratio of volume to the surface area of the complete spray sample.

Spray Droplet Velocity: Spray droplet velocity is one of the most essential factors to analyse. It must be high enough to penetrate even into a high ambient pressure chamber to form an efficient fuel–air mixture. It directly depends on the difference between chamber and fuel injection pressure. Spray velocity first increases up to a maximum value, and then reduces smoothly due to aerodynamic drag offered by the in-cylinder pressurised gas (Characterization and of Diesel and Biodiesel Spray).

For assessment of microscopic spray characteristics, Laser Doppler Velocimetry (LDV) or Laser Doppler Anemometry (LDA) techniques were used. They measure the velocity of droplets but not measure the droplet size and droplet size distribution. Another technique known as Phase Doppler Interferometry (PDI) or Phase Doppler Particle Analyzer (PDPA), which works on the same principle as of LDV, was developed by Dr. William Bachalo and Mike Houser in the year 1984. PDI technique can measure the droplet size and velocity of each droplet of inconsistent shapes, sizes, and

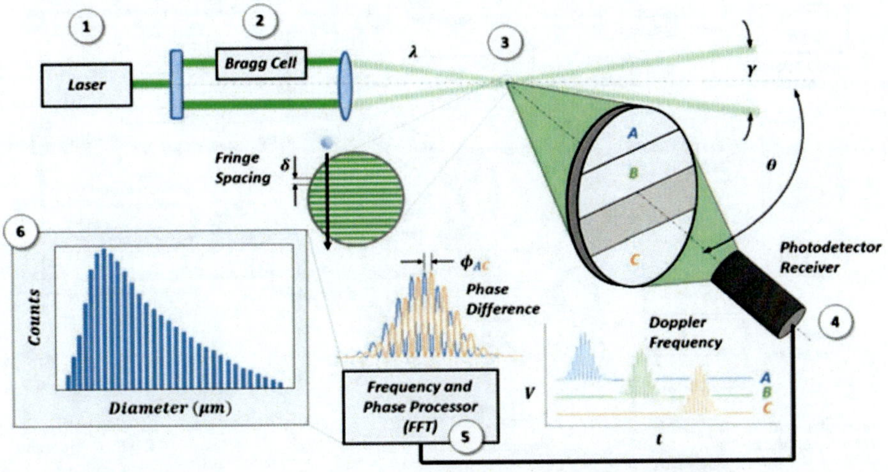

Fig. 4.8 Schematic of Phase Doppler Interferometry (PDI) or Phase Doppler Particle Analyser (PDPA) Technique for Microscopic Spray Characterisation (Fulk et al. 2017)

non-uniform distribution. Interferometry and light scattering are two basic principles behind this technique. Measurement of data depends upon the wavelength of the light source, focal length of transmitters, and receiver lenses, and separation between the detectors. Intensity of light does not have any effect on the droplet size measurement data (Microscopic and Macroscopic Spray Characterization of Biodiesel). Figure 4.8 shows the schematic of the PDI setup.

PDI system consists of a beam splitter, which divides the laser beam into two equal intensity beams. A Bragg's cell is used to provide shift in the frequency of laser beams to enhance the sharpness in size measurements. These two laser beams interact with each other at a point, which is referred as 'probe volume'. This probe volume is focused by transmitter having lens of a certain focal length. Each spray droplet scatters the light, when passed through the probe volume. The photodetector inside the receiver detects the scattered light (refraction/reflection) through the lens. Then voltage signals are sent to the analysers for measurement of droplet size distributions and droplet velocities.

Kim et al. (2010) investigated the spray characteristics of DME and biodiesel using image processing. Fuel was injected through a common rail direct injection (CRDI) system. The spray macroscopic characteristics such as the spray tip penetration, spray cone angle, and spray area were calculated. Local Sauter mean diameters (SMD) were measured by PDPA, as shown in Fig. 4.9. Two metal halide lamps illuminated the spray plume and a high-speed CCD camera (Photron, Fastcam-APX RS) captured the spray images. Detailed specifications of the instruments used in the experimental setup can be found in our previous publication (Kim et al. 2010).

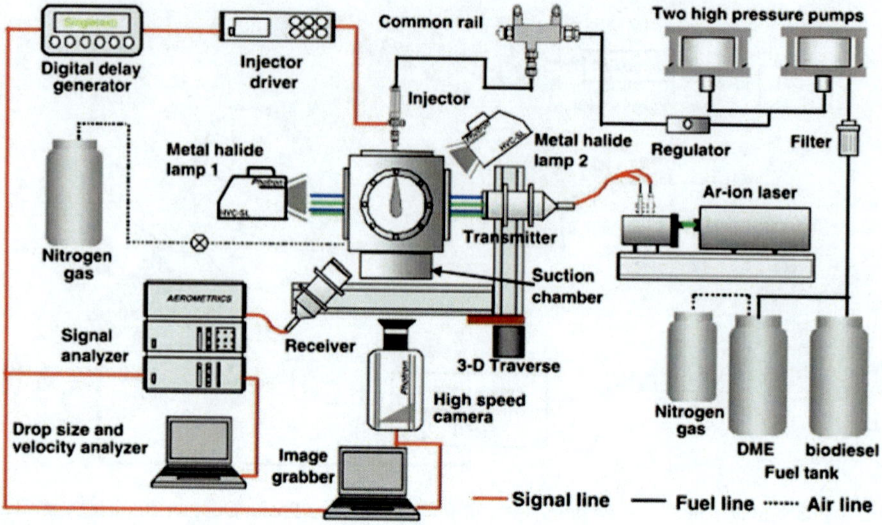

Fig. 4.9 A typical experimental layout for microscopic fuel spray characterisation (Kim et al. 2010)

4.5 Parameters Affecting Spray Characteristics

4.5.1 *Effect of Fuel Injection Pressure*

Fuel injection pressure (FIP) is one of the most important factors affecting the spray characteristics of the fuel. Higher FIP may possibly lead to wall wetting, which might result in deteriorated engine performance and higher emissions. In contrast, lower FIP leads to lesser spray penetration and coarser droplets, which form less efficient and inhomogeneous fuel–air mixture. Therefore, FIP should have an optimum value, which can provide sufficient droplet velocity to penetrate even in high-pressure combustion chamber environment, without any wall wetting. DME does not require higher FIP than diesel due to its relatively much lower viscosity and lubricity. DME fuelled modern engines have the maximum FIP of 60 MPa. DME is in the gaseous state under ambient conditions. Hence it must be pressurised to convert it into liquid state before injection in a CI engine without the requirement of any major modification in the fuel injector. To maintain DME in liquid state, pressure of 0.5 MPa is required.

Spray tip penetration is increased with an increase in FIP due to an increasing the inertia of fuel droplets. According to Levich spray breakup formula, spray breakup length is directly proportional to orifice diameter and square root of density of injected fluid, as shown in Eq. 4.1. Density of DME increases with an increasing FIP, which results in an increasing spray breakup length.

$$L_{breakup} = C_b d_{noz} \left(\rho_f / \rho_\alpha\right)^{1/2} \tag{4.1}$$

where, $L_{breakup}$ is the breakup length, C_b is constant, d_{noz} is nozzle diameter, ρ_f and ρ_a is the density of fuel and air, respectively.

Researchers mostly used a CRDI system to investigate spray atomisation of DME. Cipolat and Valentim (2013) compared the spray characteristics of diesel and DME theoretically and experimentally. They investigated spray penetration length of both fuels at a FIP of 40 and 60 MPa. They reported that diesel has a higher penetration length than DME. This may be due to higher density and viscosity of diesel compared to DME. Spray penetration length increased with increasing FIP. The possible reason may be increased momentum or inertia of fuel spray droplets (Cipolat and Valentim 2013). Kim et al. (2009) reported that SMD in case of DME was lower than diesel, mainly due to relatively lower viscosity of DME. Droplet size of DME further decreased with an increasing FIP, which leads to faster evaporation. This was due to higher droplet velocity, which resulted in a more destabilising force at higher FIP. Optimisation of FIP for adequate atomisation was required, as higher FIP increased the cost of the vehicle. Pfeifer et al. (2013) studied spray characteristics of DME at FIP of 6 and 8 MPa into a quiescent atmosphere maintained at 2 MPa. Droplet velocity and size distributions were measured at various axial and radial positions w.r.t. the nozzle. Figure 4.10 showed various measurement locations in the spray area, where (0,0) represented the injector nozzle having diameter of 2 mm. Figure 4.11 represents spatially resolved variations in the mean diameter of droplets at two FIPs of 6 and 8 MPa. To classify the circle quantitively, a circle of 50 mm mean diameter was drawn. From the reference circle of 50 mm, the comparison of the droplet size at two different locations can be done. It was observed that mean diameter decreased with an increasing FIP at constant ambient pressure. At FIP of 6 MPa, droplet diameter showed decreasing and increasing behaviours along the length. This was mainly due to instant evaporation at the beginning, which resulted into a reduction in the droplet size. After some time, smaller droplets completely evaporated and droplets having larger diameter remained in suspension. It was found

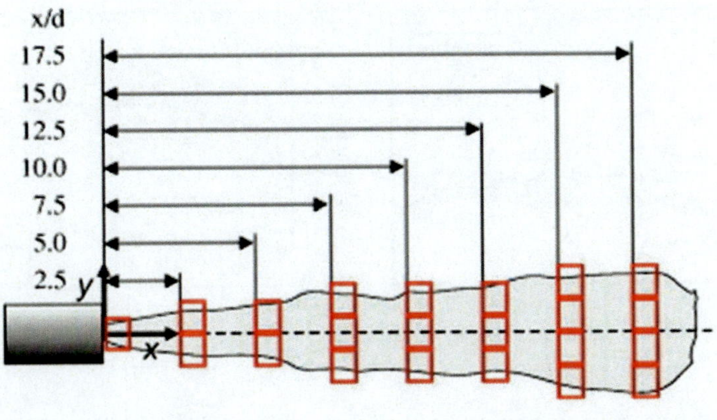

Fig. 4.10 Measurement points in the spray plume. Red rectangles represent the domain for image acquisition (Pfeifer et al. 2013)

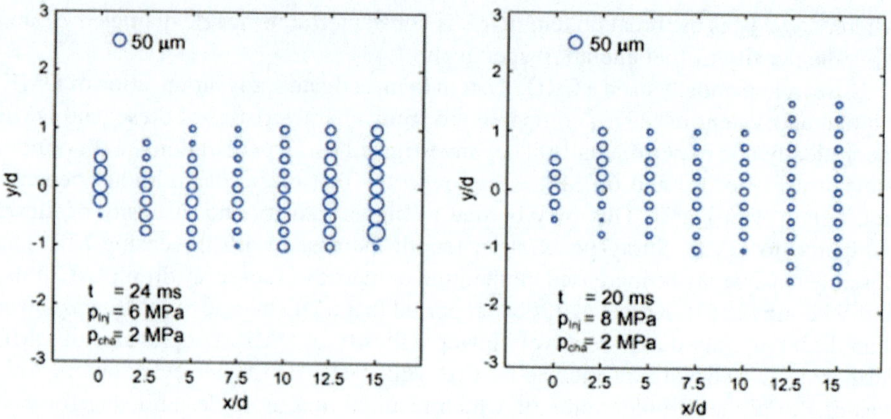

Fig. 4.11 Radial and axial variations in mean droplet diameter at $P_{inj} = 6$ MPa and 8 MPa, and $P_{cha} = 2$ MPa (Pfeifer et al. 2013)

that spread of spatial distribution was wider with increasing axial distance at FIP of 8 MPa. Wider spray spread showed efficient fuel–air mixing, which led to more complete combustion and reduction in soot and unburned hydrocarbon emissions (Pfeifer et al. 2013).

FIP has a direct correlation with the fuel droplet velocity. Fuel droplet velocity increased with an increasing FIP. This was due to an increase in driving force or momentum of fuel spray droplets. Droplet velocity first increased to a maximum due to momentum, then started decreasing due to evaporation and destabilising force in the high-pressure chamber. DME exhibited lower droplet velocity than diesel, due to less momentum of smaller droplets of DME (Cipolat and Valentim 2013). Figure 4.12 showed the mean axial velocity of DME droplets at FIP of 6 and 8 MPa

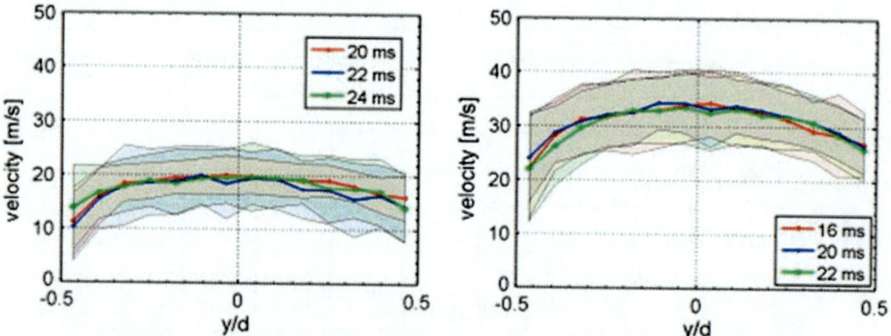

(a) $p_{inj} = 6\,MPa$ (b) $p_{inj} = 8\,MPa$

Fig. 4.12 Velocity profile at nozzle outlet (0, y) at two different FIPs **a** 6 MPa, and **b** 8 MPa with ambient pressure of 2 MPa at three different time stages, 1 mm above and below the nozzle (Pfeifer et al. 2013)

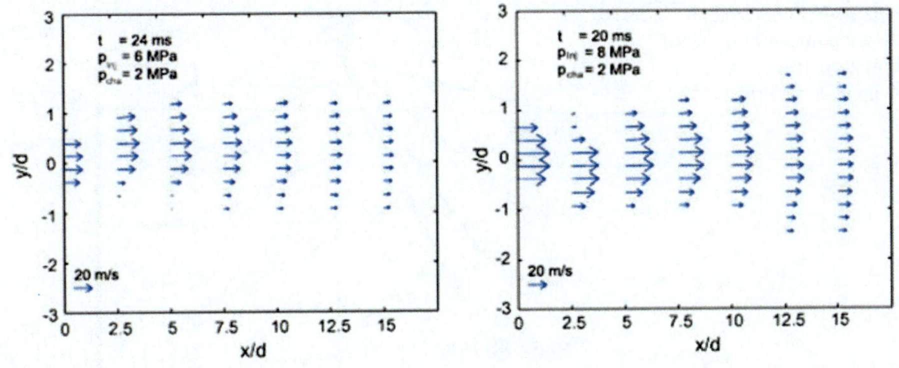

Fig. 4.13 Mean droplet velocity variations at a different spatial position at FIP of 6 and 8 MPa (Pfeifer et al. 2013)

at constant ambient pressure/temperature of 2 MPa/298 K respectively. Mean axial velocity at FIP of 6 MPa was lower than 8 MPa. This was due to relatively lower pressure difference or driving force in case of 6 MPa FIP. Similar observation was made in Fig. 4.12.

Figure 4.13 shows that droplet velocity was higher for FIP of 8 MPa (Pfeifer et al. 2013). It was found that higher velocity leads to longer penetration length, which resulted in higher spray area. Adequate spray penetration length without cylinder wall impingement leads to an efficient air–fuel mixture formation. This further reduces the possibility of rich mixture formation, soot formation, and unburned hydrocarbon emissions.

4.5.2 *Effect of Injection Timing and Injection Angle*

Injection angle is defined as the angle made by fuel spray with the nozzle axis, as shown in Fig. 4.14. Injection timing represents time, at which fuel is injected into the engine. Generally, it is referred in terms of before or after top dead centre (bTDC/aTDC). Kim et al. (2011a) experimentally investigated engine combustion, emissions and spray characteristics using two different injectors having injection angles of 60° and 156°. They numerically investigated for injection angles of 60°, 70°, 90°, 120°, and 150° and injection timings were varied from 20° to 80° bTDC. Spray characteristics were analysed using oxidation and chemical kinetics models of DME. Early injection in conventional diesel led to piston wall wetting. This was due to low temperature and ambient air density, which were not sufficient to vaporise diesel droplets. However, DME droplets formed homogenous air–fuel mixture in early injection due to quicker evaporation. This was possible due to lower boiling point of DME. Results showed that spray tip penetration and spray cone angles were higher for injection angle of 60° compared to 156°. It was also observed that

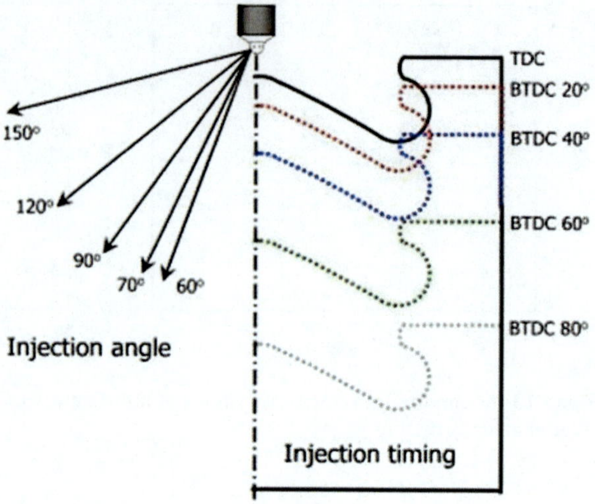

Fig. 4.14 Schematic showing different injection angles and injection timings (Kim et al. 2011a)

spray with narrow-injection angle of 70° was wetting the bottom portion of cylinder wall at injection timing of 60° bTDC. This was due to higher vertical penetration in case of narrower injection angles. Fuel got accumulated in the piston bowl at an injection angle between 60° and 90°. This resulted in higher combustion pressure and temperature due to more complete combustion. However, fuel got concentrated outside of piston bowl and into the crevices for higher injection angles of greater than 90°. This led to lower combustion pressure and temperature due to incomplete combustion and non-availability of oxygen in the crevices. NOx emissions decreased significantly with fuel injection timing advanced to 30° bTDC in HCCI combustion. This was due to increased homogeneity of air–fuel mixture, which reduced premixed combustion. It was found that peak temperature was below 2000 K at SoI timing before 70° BTDC, however, CO and HC emissions remained high. This was due to incomplete combustion of fuel accumulated in the piston bowl and crevices. NOx and PM trade-off were present at the SoI timing of 70° BTDC. IMEP and ISFC showed increasing and decreasing trends upon retarding the SoI timing due to increased work done as combustion pressure and temperature increased (Kim et al. 2011b).

4.5.3 Effect of Ambient Chamber Pressure

Ambient chamber pressure is also referred as back pressure. The spray characteristics and atomisation greatly depend on the ambient air pressure. Kim et al. (2010) compared the macroscopic and microscopic spray characteristics of DME and biodiesel at different ambient pressures. They injected fuel sprays using a common rail system at various chamber pressures. The chamber was designed to withstand pressure up to 4 MPa. It was observed that spray tip penetration of DME was shorter than biodiesel at identical ambient conditions. This was due to low boiling point and low viscosity of DME. Spray tip penetration and droplet velocity reduced with an increasing ambient pressure. This was due to an increase in resisting force with increased ambient pressure. At ambient conditions, DME showed narrower cone angle compared to biodiesel due to lower boiling point and higher vapour pressure of DME. Spray cone angle increased with an increasing ambient pressure. Results showed that spray area of DME was smaller than biodiesel. SMD of DME spray droplets was smaller than biodiesel at all radial and axial positions. Overall, DME showed superior spray atomisation than biodiesel under all ambient pressures (Kim et al. 2010). Suh and Lee (2008) compared the spray tip penetration of diesel and DME at different ambient pressures of 0.1 and 2 MPa at FIP of 60 MPa and an ambient temperature of 293 K. Results showed that DME exhibited narrower spray cone angle and lower spray penetration length than diesel. This was mainly because of the quicker evaporation of DME, particularly in the outer spray region. Experimental study showed that spray development process was slower, spray tip penetration was shorter, and spray cone angle was wider for both fuels at higher ambient pressures. Spray evaporation in outer region in the atmospheric conditions was observed, when ambient pressure was further increased (Suh and Lee 2008).

Fine spray droplet atomisation certainly leads to lower pollutant formation, wider combustion range, higher air–fuel ratio, and higher volumetric heat release rate. SMD qualitatively represents the size of an average fuel droplet. Suh et al. (2009) conducted a study to optimise the fuel spray atomisation for superior combustion under varying ambient conditions. DME spray droplet atomisation variations with respect to the ambient pressure were reported in terms of SMD. Spray arrival time increased with increasing ambient pressure, as shown in Fig. 4.15. Increase in spray arrival time indicated a reduction in fuel droplet velocity with an increasing ambient pressure. This was primarily due to increase in aerodynamic drag due to increasing ambient pressure, which reduced the spray droplet velocity. Reduction in relative velocity

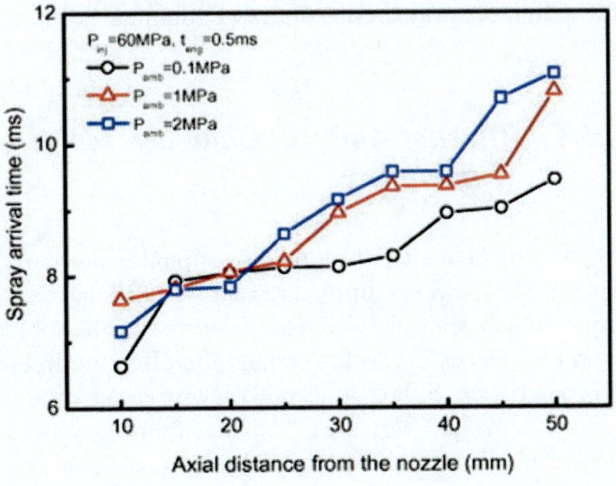

Fig. 4.15 DME spray droplets arrival time at each axial distance from the nozzle exit point at a fixed fuel injection pressure of 60 MPa under varying ambient pressures, at fixed ambient temperature of 293 K and solenoid energising duration of 0.5 ms (Suh et al. 2009)

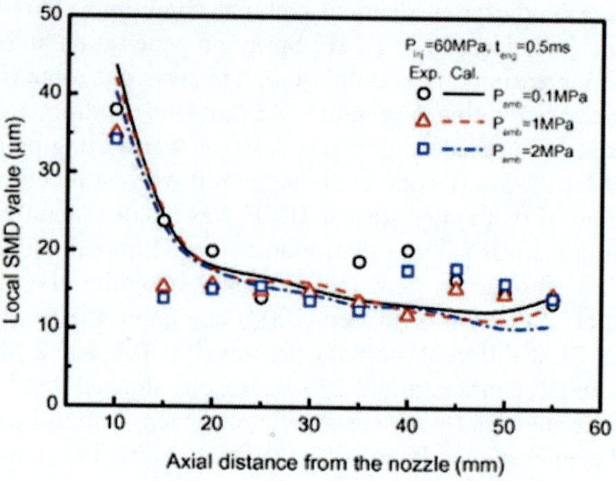

Fig. 4.16 SMD distribution of DME under various ambient pressure conditions ($P_{inj} = 60$ MPa, $T_{amb} = 293$ K) (Suh et al. 2009)

between the ambient gas and fuel spray droplets lead to an increase in velocity of gas-phase fuel and reduction in velocity of liquid-phase fuel due to momentum transfer. Therefore, DME spray under high-pressure conditions reached each measuring point rather slowly. This was due to high momentum transfer under high ambient pressure. Large fluctuations in the initial stage were observed in low ambient pressures of 0.1 MPa because lower spray droplets velocities cause instability in the spray formation process.

Figure 4.16 shows the variation of SMD of DME spray droplets with changing ambient pressure at varying axial distance from the nozzle exit. It was observed that spray droplet size decreased rapidly after exiting from the nozzle tip. The smallest droplet sizes were observed in the highest ambient pressure conditions of 2 MPa because of enhanced atomisation of DME. There was an increase in the SMD with increasing ambient pressure, which was contrastory to the general trend. This was because of higher number of droplet collisions and coalescence, which led to the formation of spray droplets of larger diameters.

4.5.4 *Effect of Ambient Chamber Temperature and Fuel Temperature*

Most spray characteristic study investigations were carried out in cold conditions in the CVSC. However, limited research studies have also explored engine-like conditions, which are quite different from the results obtained in cold conditions. It is, therefore, necessary to understand the effect of ambient temperature on fuel spray characteristics. Suh et al. (2009) investigated spray behaviour at varying ambient temperatures and pressures. They used nitrogen as ambient gas and an electric coil was used to vary the ambient temperature. They reported that SMD of fuel spray

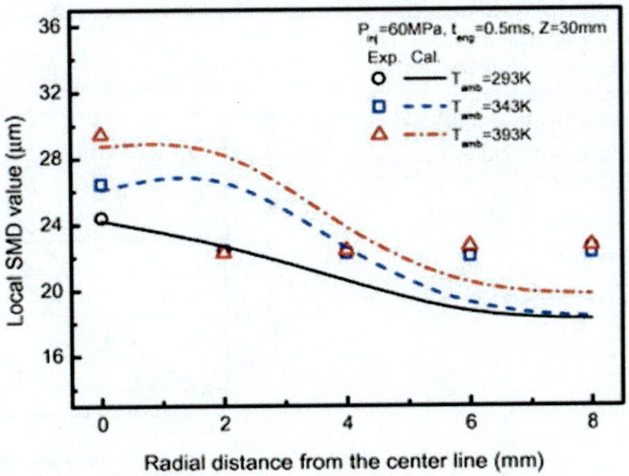

Fig. 4.17 Variation of SMD with respect to radial distance from the centre line (Suh et al. 2009)

droplets increased with increasing ambient temperature. This was mainly due to faster evaporation of smaller size spray droplets in high temperature ambient environment. As a result, only the droplets of larger diameter were detected in the chamber. Figure 4.17 shows the variations of local SMD with respect to the radial distance at different ambient temperatures. Experimental results were in good agreement with the simulation studies for SMD variations. SMD decreased along radial distance from the centre line because of evaporation of droplets in the outer region of the spray.

Similar study by Park et al. (2010) investigated experimentally and numerically the effect of fuel temperature and ambient temperature on the fuel spray atomisation and spray morphology. In this experiment, a diesel injector having six holes with a diameter of 0.126 mm, equipped with an injector controller was used. A heat exchanger boiler with thermocouple and a relief controlled the fuel temperature. Liquid DME was injected using a CRDI system having a parallel linked high-pressure pump. K-type thermocouple was installed before the injector to measure the temperature of DME. DME temperature was varied from 293 to 370 K, whereas ambient temperature was varied from 293 to 400 K. Helmholtz and Rayleigh–Taylor (KH-RT) hybrid breakup model was used for numerical analysis using KIVA-3D software.

Figure 4.18 shows the simulation results of variations in spray tip penetration with respect to the time after SoI under atmospheric conditions (P_{amb} = 0.1 MPa, T_{amb} = 293 K). DME spray exhibited lesser penetration length than diesel at all fuel temperatures. This may be due to faster evaporation and rapid momentum loss of DME droplets. Under atmospheric conditions, spray cone angle of diesel was smaller than DME, however, at higher ambient pressure conditions (P_{amb} = 3.0 MPa, T_{fuel} = 293 K), spray tip penetration and cone angle of both fuels were almost identical (Park et al. 2010).

When DME is depressurised to a pressure lower than its saturation vapour pressure, it cannot remain in liquid state, and rapid boiling takes place. This phenomenon is known as flashing. When DME is injected at a FIP of 60 MPa under atmospheric

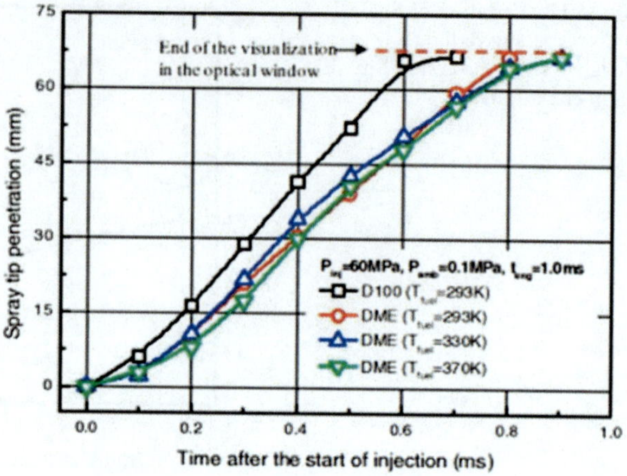

Fig. 4.18 Variations in spray tip penetration of DME and diesel at different fuel temperatures under atmospheric conditions (P_{amb} = 0.1 MPa, T_{amb} = 293 K) (Park et al. 2010)

conditions, flashing takes place. This flashing enhances the air–fuel mixture and fuel atomisation and reduces the wall wetting. DME has a wider spray cone angle compared to diesel due to flashing in the initial stages of injection. Experimental results showed that spray tip velocity first increases up to a peak value, and then it starts decreasing under atmospheric conditions (P_{amb} = 0.1 MPa, T_{amb} = 293 K), while spray tip velocity continuously decreases under high ambient pressure conditions (P_{amb} = 3.0 MPa, T_{amb} = 293 K). Spray tip velocity was higher under atmospheric conditions compared to high pressure ambient conditions. This was mainly due to high aerodynamic drag offered by high-pressure ambient air. Spray tip velocity is calculated as a change in spray tip position in 0.1 ms. Fuel temperature does not have any noticeable effect on the spray tip velocity (Park et al. 2010).

Figure 4.19 shows a comparison of the experimental spray images and spray image obtained by KIVA-3D code simulations at different fuel temperatures and ambient temperatures. Ambient pressure was fixed at 1.5 MPa. The calculated spray tip penetration, spray development process, and spray cone angle were in good agreement with the experimental results. However, in Fig. 4.19, it is shown that the calculated spray was wider than that of experimental spray results. This was because spray droplets at the tip were actively atomised and then spread in radial direction.

4.6 Summary

This chapter primarily discusses about the potential of DME as an alternative fuel for CI engines. It examines different chemical and physical properties of DME and compared them with diesel. High cetane number of DME makes it a suitable fuel for CI engines. Various production aspects of DME from different feedstocks have been discussed. DME could be produced from methanol, which is produced from syngas utilising different catalysts and reaction mechanism. We have briefly

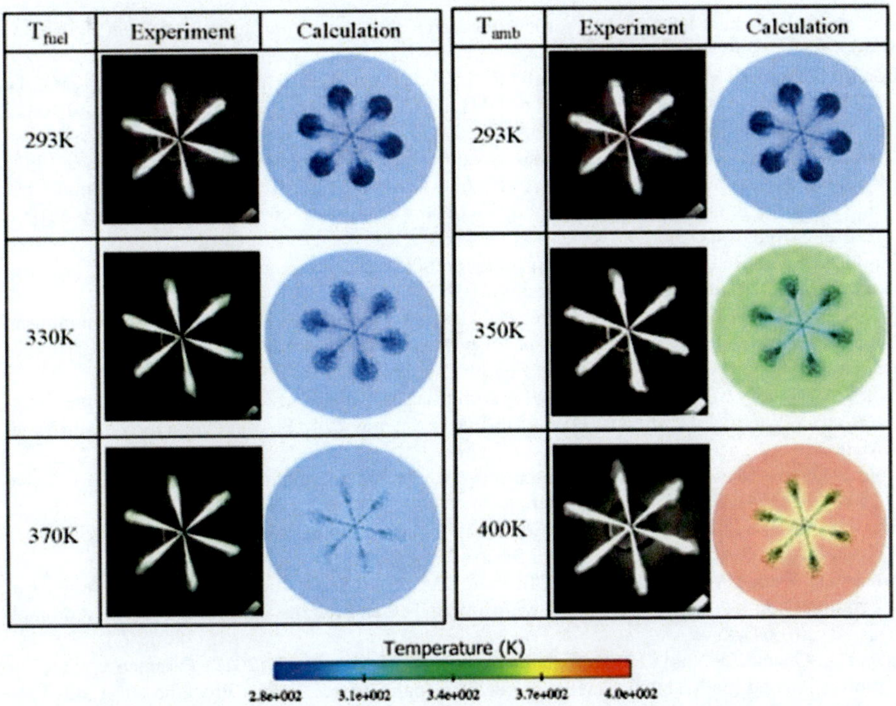

Fig. 4.19 Comparison between the calculated and experimental DME spray images with varying ambient gas temperature and fuel temperature (tasoi = 1.0 ms, P_{amb} = 1.5 MPa) (Park et al. 2010)

discussed the advantage and shortcomings of various catalysts such as γ-Al_2O_3, TiO_2/ZrO_2, TIO_2/Al_2O_3(Na+), TiO_2/Al_2O_3(NH4+), HZNS-5, which can be utilised in the production of DME from methanol. Discussions on macroscopic and microscopic spray characteristics of DME spray added further value to this chapter. Spray characteristics have a direct correlation with engine performance and emission characteristics. Hence, different parameters used for defining different spray characteristics are discussed comprehensively. These parameters include spray penetration length, spray cone angle, and droplet size distribution. Last, this chapter discussed the effect of fuel injection pressure, fuel temperature, ambient conditions, and injection timings on spray characteristics.

References

Arcoumanis C, Bae C, Crookes R, Kinoshita E (2008) The potential of dimethyl ether (DME) as an alternative fuel for compression-ignition engines: a review. Fuel 87(7):1014–1030

Ateah MA, Manufacturing of DME from Methanol

Azarabadi H, Lackner KS (2019) A sorbent-focused techno-economic analysis of direct air capture. Appl Energy 250:959–975

Bonura G, Cordaro M, Spadaro L, Cannilla C, Arena F, Frusteri F (2013) Hybrid Cu–ZnO–ZrO2/H-ZSM5 system for the direct synthesis of DME by CO_2 hydrogenation. Appl Catal B 140:16–24

Bos MJ, Kersten SRA, Brilman DWF (2020) Wind power to methanol: renewable methanol production using electricity, electrolysis of water and CO_2 air capture. Appl Energy 264:114672

BP Statistical Review of World Energy (2019) 68th edn. (https://www.bp.com/content/dam/bp/businesssites/en/global/corporate/pdfs/energy-economics/statistical-review/bp-stats-review-2019-full-report.pdf)

Cipolat D, Valentim D (2013) Comparison of theoretical and experimental diesel and DME injection spray characteristics. Fuel Process Technol 107:36–43

Farooqui A, Di Tomaso F, Bose A, Ferrero D, Llorca J, Santarelli M (2019) Techno-economic and exergy analysis of polygeneration plant for power and DME production with the integration of chemical looping CO2/H2O splitting. Energy Convers Manage 186:200–219

Fulk SM, Beaudry MR, Rochelle GT (2017) Amine aerosol characterisation by phase doppler interferometry. Energy Procedia 114: 939–951. (PDI) For Spray Drop Size and Velocity Measurement, Manual

Goto S, Oguma M, Suzuki S (2005) Research and development of a medium duty DME truck (No 2005–01–2194). SAE Technical Paper

Han SW, Shin YS, Kim HC, Lee GS (2020) Study on the Common Rail Type Injector Nozzle design based on the Nozzle flow model. Appl Sci 10(2):549

Huang ZH, Wang HW, Chen HY, Zhou LB, Jiang DM (1999) Study of combustion characteristics of a compression ignition engine fuelled with dimethyl ether. Proc Inst Mech Eng Part D: J Automob Eng 213(6):647–652

Inayat A, Ghenai C, Naqvi M, Ammar M, Ayoub M, Hussin MNB (2017) Parametric study for production of dimethyl ether (DME) as a fuel from palm wastes. Energy Procedia 105:1242–1249

International Energy Agency (1999)

Kapus P, Ofner H (1995) Development of fuel injection equipment and combustion system for DI diesels operated on dimethyl ether (No 950062). SAE technical paper

Kim HJ, Park SH, Suh HK, Lee CS (2009) Atomisation and evaporation characteristics of biodiesel and dimethyl ether compared to diesel fuel in a high-pressure injection system. Energy Fuels 23(3):1734–1742

Kim HJ, Park SH, Lee CS (2010) A study on the macroscopic spray behaviour and atomisation characteristics of biodiesel and dimethyl ether sprays under increased ambient pressure. Fuel Process Technol 91(3):354–363

Kim HJ, Park SH, Lee KS, Lee CS (2011a) A study of spray strategies on improvement of engine performance and emissions reduction characteristics in a DME fuelled diesel engine. Energy 36(3):1802–1813

Kim HJ, Lee KS, Lee CS (2011b) A study on the reduction of exhaust emissions through HCCI combustion by using a narrow spray angle and advanced injection timing in a DME engine. Fuel Process Technol 92(9):1756–1763

Masson-Delmotte V, Zhai P, Pörtner HO, Roberts D, Skea J, Shukla PR, Pirani A, Moufouma-Okia W, Péan C, Pidcock R, Connors S (2018) Global warming of 1.5 C. An IPCC Special Report on the impacts of global warming of, 1

Microscopic Characterization of Diesel and Biodiesel Spray Using Phase Doppler Interferometry A thesis submitted in partial fulfilment of requirements for degree of Master of Technology Krishna Kumar Yadav.

Microscopic and Macroscopic Spray Characterization of Biodiesel in a Constant Volume Spray Chamber A thesis submitted in partial fulfilment of requirements for degree of Master of Technology Harsh Goyal.

Naik SP, Ryu T, Bui V, Miller JD, Drinnan NB, Zmierczak W (2011) Synthesis of DME from CO_2/H_2 gas mixture. Chem Eng J 167(1):362–368

Park SH, Lee CS (2014) Applicability of dimethyl ether (DME) in a compression ignition engine as an alternative fuel. Energy Convers Manage 86:848–863

Park SH, Kim HJ, Lee CS (2010) Macroscopic spray characteristics and breakup performance of dimethyl ether (DME) fuel at high fuel temperatures and ambient conditions. Fuel 89(10):3001–3011

Parvez AM, Wu T, Li S, Miles N, Mujtaba IM (2018) Bio-DME production based on conventional and CO_2-enhanced gasification of biomass: a comparative study on exergy and environmental impacts. Biomass Bioenerg 110:105–113

Pfeifer C, Kuhn D, Class AG (2013) Influence of injection pressure on droplet atomisation and auto-ignition of dimethyl ether (DME) at elevated pressure. Fuel 104:116–127

Sezer I (2011) Thermodynamic, performance and emission investigation of a diesel engine running on dimethyl ether and diethyl ether. Int J Therm Sci 50(8):1594–1603

Singh BR, Singh O (2012) Global trends of fossil fuel reserves and climate change in the 21st century, vol 8. Chapter

Sorenson SC, Glensvig M, Abata DL (1998) Dimethyl ether in diesel fuel injection systems. SAE Trans 438–449

Sugiyama K, Kajiwara M, Fukumoto M, Mori M, Goto S, Watanabe T (2004) Lubricity of liquefied gas assessment of multi-pressure/temperature high-frequency reciprocating rig (MPT-HFRR)-DME fuel for diesel. SAE transactions, pp 866–876

Suh HK, Lee CS (2008) Experimental and analytical study on the spray characteristics of dimethyl ether (DME) and diesel fuels within a common-rail injection system in a diesel engine. Fuel 87(6):925–932

Suh HK, Park SH, Kim HJ, Lee CS (2009) Influence of ambient conditions on the droplet atomisation characteristics of dimethyl ether (DME). Fuel 88(6):1070–1077

Teng H, McCandless JC, Schneyer JB (2002) Viscosity and lubricity of (liquid) dimethyl Ether-An alternative fuel for compression-ignition engines. SAE Trans 325–334

Teng H, McCandless JC, Schneyer JB (2003) Compression ignition delay (physical + chemical) of dimethyl ether–an alternative fuel for compression-ignition engines. SAE Trans 377–389

Teng H, McCandless JC, Schneyer JB (2004) Thermodynamic properties of dimethyl ether—an alternative fuel for compression-ignition engines. SAE Trans 134–157

Tomatis M, Parvez AM, Afzal MT, Mareta S, Wu T, He J, He T (2019) Utilisation of CO2 in renewable DME fuel production: a life cycle analysis (LCA)-based case study in China. Fuel 254:115627

Tsuchiya T, Sato Y (2006) Development of DME engine for heavy-duty truck (No 2006–01–0052). SAE Technical Paper

Veselovskaya JV, Derevschikov VS, Kardash TY, Stonkus OA, Trubitsina TA, Okunev AG (2013) Direct CO_2 capture from ambient air using K2CO3/Al2O3 composite sorbent. Int J Greenhouse Gas Control 17:332–340

Xu M, Lunsford JH, Goodman DW, Bhattacharyya A (1997) Synthesis of dimethyl ether (DME) from methanol over solid-acid catalysts. Appl Catal a 149(2):289–301

Youn IM, Park SH, Roh HG, Lee CS (2011) Investigation on the fuel spray and emission reduction characteristics for dimethyl ether (DME) fuelled multi-cylinder diesel engine with common-rail injection system. Fuel Process Technol 92(7):1280–1287

Chapter 5
Spray Chamber Designs and Optical Techniques for Fundamental Spray Investigations

Sam Joe Chintagunti, Ankur Kalwar, Dhananjay Kumar, and Avinash Kumar Agarwal

Abstract In the present scenario, research in the area of Internal Combustion (IC) engines is mainly driven to address the alarming depletion of conventional fossil fuels and to control the tail-pipe emissions in order to comply with stringent emission norms. Combustion is one of the primary reasons for global warming; however, ~80% of total global energy production is based on combustion of conventional fuels. Hence, researchers have been trying to understand the in-cylinder combustion phenomenon to improve efficiency of energy conversion devices and new ways to utilize alternate fuels. Spray studies in engine-like environment play vital role in combustion and consequent heat loss to the cylinder walls. Fuel spray affects the air–fuel mixture formation, which is responsible for combustion and emission formation in the engine combustion chamber. To study mixing processes and spray distribution, in-cylinder conditions need to be simulated in constant volume combustion chamber (CVCC). Development of high-pressure high-temperature chambers and optical diagnostics involves lasers and high-speed cameras. These investigations enable us to understanding the insights into combustion that takes place in few milliseconds. This chapter starts with design of combustion chambers, followed by explanation of prominent optical techniques. This is followed by detailed discussions with the help of recent studies involving these chambers and techniques to understand spray atomization and combustion in different operating conditions. This chapter aims to give an understanding of different aspects of experimental spray studies and their impact in the field of IC engines.

Keywords Spray · Constant volume combustion chamber · Optical techniques · Spray morphology · In-nozzle flow · Spray–wall interaction · Spray flame

---

S. J. Chintagunti · A. Kalwar · D. Kumar · A. K. Agarwal (✉)
Engine Research Laboratory, Department of Mechanical Engineering, Indian Institute of Technology Kanpur, Kanpur-208016, India
e-mail: akag@iitk.ac.in

A. P. Singh and A. K. Agarwal (eds.), *Novel Internal Combustion Engine Technologies for Performance Improvement and Emission Reduction*, Energy, Environment, and Sustainability, https://doi.org/10.1007/978-981-16-1582-5_5

5.1 Introduction

Spray atomization plays a vital role in understanding the in-cylinder phenomenon in IC engines, which consequently affects the thermal efficiency. By employing an optimized fuel injection system, significant quantity of fuel can be conserved along with reduction of emissions in the transport sector. Fuel spray is mainly responsible for governing the fuel–air mixture quality, conversion of chemical energy of fuel to heat energy and formation of emission species (Sankar et al. 1999). Hence, spray must use optimum amount of air present in the cylinder, so that complete combustion can be achieved (Su et al. 1995). Utilization of new alternative fuels in IC engines is increasing and several new fuels are blended with conventional fuels. Hence, their spray and combustion investigations are of interest to many researchers. Various diagnostic tools are used to understand the underlying physics behind spray evolution and combustion processes. Some of these techniques are direct spray visualization, Phase Doppler Particle Anemometry (PDPA), Planar Laser-Induced Fluorescence (PLIF), Particle Image Velocimetry (PIV), etc. (Shao et al. 2003; Zama et al. 2017). These techniques are employed for both in-cylinder measurements, and constant volume vessel studies equipped with optical access. Through these optical techniques, spray and combustion processes can be improved by optimizing the injection parameters, so that the superior engine performance can be ensured with minimal emissions (Pierpont and Reitz 1995).

Development of injected fuel spray in the engine cylinder is a combination of many complex phenomena such as spray break-up (primary and secondary) and spray droplet collisions, coalescence and evaporation. At some distance from the nozzle exit, fine spray droplets coalesce to form larger droplets, which are not desirable (Duronio et al. 2020). Hence, understanding these phenomena and the conditions that drive them are very important to develop high-performance Fuel Injection Equipment (FIE). This can be understood by various phases in the life journey of an individual droplet right after its exit from the nozzle. Schematic of a single spray plume and associated processes taking place in the life of an individual droplet are shown in Fig. 5.1. High Fuel Injection Pressure (FIP) leads to the formation of a liquid fuel jet oozing out of the nozzle exit under the influence of cavitation in the nozzle, which forms spray ligaments upon injection. These ligaments further break into a large number of small droplets via bag breakup. A large fraction of liquid droplets mainly occupy the thick spray region with a conical shaped jet. As the distance from the nozzle tip increases, air fraction considerably increases, resulting in a larger spray. Relatively shorter distance between fuel droplets in the near nozzle region (dense spray) enhances droplet collision and coalescence. Beyond that, a dilute region comes with a very small fraction of liquid fuel mass present in the form of isolated droplets suspended in the air (Reitz 1996).

A primary breakup and a secondary breakup achieve disintegration of fuel jet, breaking it first into ligaments and then into small droplets. Primary breakup deals with breaking of liquid core into the first set of large droplets just after the injection, that is close to the nozzle exit. Furthermore, low pressure (lower than the vapour

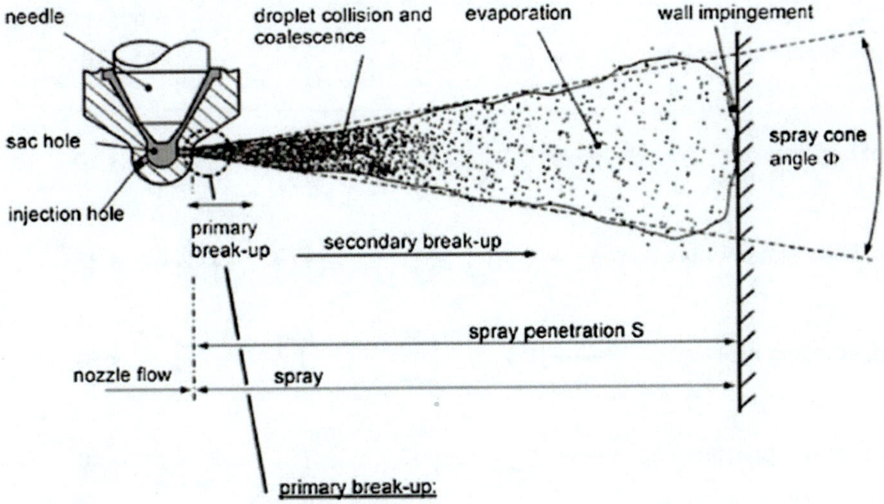

Fig. 5.1 Schematic of a liquid fuel spray and the processes involved (Baumgarten 2006)

pressure) at the exit of the nozzle because of reduction in the flow area gives rise to formation of bubbles due to cavitation in the nozzle. Large pressure difference in the FIP and ambient air results in large acceleration of flow jet, and the resulting cavitation in the nozzle generates perturbations at the liquid–air interface, causing breakup of the liquid jet into droplets. Final mechanism behind the breakup is aerodynamic drag acting on the high-velocity droplets, which break the droplet further into smaller droplets (Baumgarten 2006; Stiesch 2003).

Secondary breakup generally happens further downstream, when the droplets pass through the turbulent ambient air. The aerodynamic drag forces dominate the surface tension of these droplets and distort their shape, breaking them further into smaller droplets. This was determined by the Weber number (We), which related the inertia forces and the surface tension forces acting on the droplets. Figure 5.2 shows different regimes of secondary spray breakup based on the values of We (Lefebvre and McDonell 2017). Once the spray disintegrates into fine droplets, energy exchange takes place between the hotter ambient air and the droplets, and eventually evaporation of smaller droplets occurs first. This is a very critical stage, which defines the fuel–air mixture quality in an engine combustion chamber, which affects the combustion and formation of emissions (Reitz and Bracco 1982; Wierzba and Takayama 1988).

Subsequent sections of this chapter explain different chamber designs and associated techniques used for conducting the spray investigations and their visualization. Furthermore, the effects of various operating parameters on the spray characteristics and phenomenon have been discussed using some recent case studies.

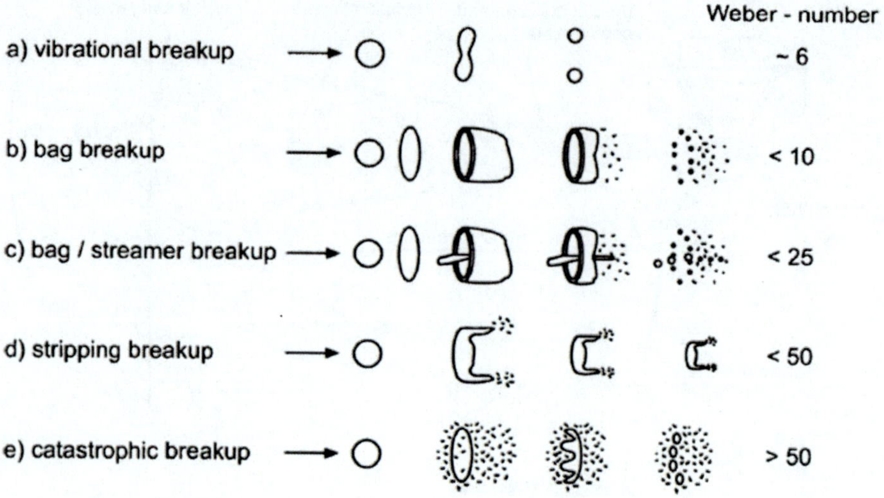

Fig. 5.2 Secondary breakup regime classifications based on Weber number (Wierzba 1990)

5.2 Constant Volume Spray/Combustion Chamber Designs

To conduct studies on possibility of application of new fuels and combustion strategies, a wide range of optically accessible chambers were developed and used. Over the conventional chambers such as Rapid Compression Machine (RCM), Constant Volume Spray Chamber (CVSC)/Constant Volume Combustion Chamber (CVCC) offer several advantages. These are compact, simple designs and do not offer any compromise in generating engine-like conditions, while making the entire system optically accessible. The experiments in CVCC also require lesser time (Oren et al. 1984). Cost-effectiveness and safety offered by a CVCC have made them a more popular used system for spray characterization over an optically accessible engine.

In the CVCC, the required high ambient pressure and temperature conditions could be obtained by two different methods, namely: (i) Precombustion and (ii) Electric heaters. The first method involves lean fuel–air mixture, which enters the chamber from a high-pressure cylinder. Required mixture composition is achieved by governing the partial pressures of the constituents. Swirl at the inlet and a mixing fan (if present in the design) help in producing a homogeneous combustible mixture. Once proper mixing is achieved, spark plug ignites the mixture (premixed combustion). Initially, a steep rise of pressure in the chamber is observed, as shown in Fig. 5.3. As the pressure drops in the combustion chamber and reaches the required ambient pressure, fuel spray to be investigated is injected (Baert et al. 2009). A quartz window in the chamber provides optical access for the optical diagnostic techniques. After the experiment is completed, an exhaust valve is opened for the gas in the chamber to escape, and then the chamber is cleaned. Now the chamber is ready for the next set

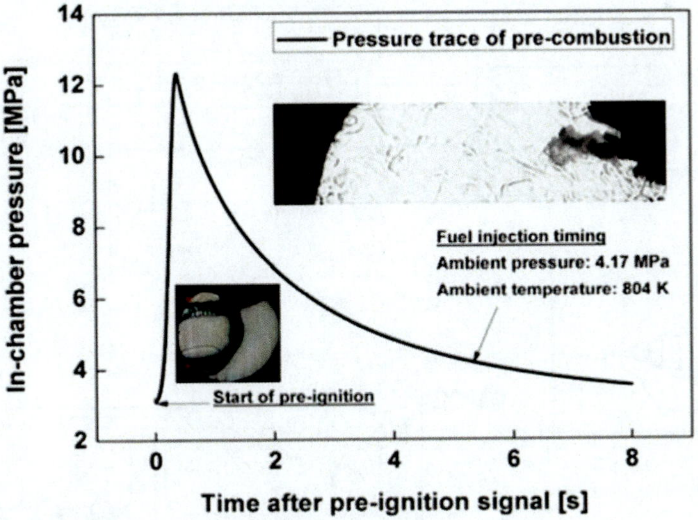

Fig. 5.3 Curve showing the working of premixed combustion-type CVCC for spray investigations (Hwang et al. 2017)

of experiment. Schematic of a CVCC experimental setup with premixed combustion is shown in Fig. 5.4.

Size of the chamber should be chosen carefully. In small chambers, wall impingement may affect the spray characteristics, and in a large chamber, the pressure rise would be small; hence, an intermediate size (usually larger than an engine cylinder) is selected (Wang et al. 2019; Cao et al. 2020). Literature suggests that the size of the CVCC depends on the kind of study that one aims to perform, e.g. for a study involving multiple injections, a small chamber of the size of a diesel engine cylinder could be used, so that the effect of first injection can be seen on the second injection (Hwang et al. 2017). The design of a typical CVCC is given in Fig. 5.5. The main components that have exposure to high pressure are the chamber and its optical windows. The most critical stress for a cylindrical chamber is hoop stress (Munsin et al. 2013). For a compression ratio between 16 and 28, the cylinder pressure before the fuel injection is normally <50 bar, and after the combustion, the cylinder pressure rises to >100 bar. Determination of chamber material and thickness are done considering a high factor of safety (as high as 15), since the CVCC must sustain high temperature in the chamber during experimental conditions. Materials such as SS-304 are used, which has a yield strength of 290 MPa. Quartz is the widely used material for the optical window. It has a yield strength of ~1130 MPa (Munsin et al. 2013). Also, the inlet for incoming gases is designed in such a way that it aids in fuel–air mixing. A tangential inlet is preferred for this reason (Oren et al. 1984). Strength of the CVCC is analyzed using Finite Element Methods (FEM), and mixing characteristics for the inlet flow are analyzed using Computational Fluid Dynamics (CFD). It is ensured that the operational stresses are well under the yield stress of

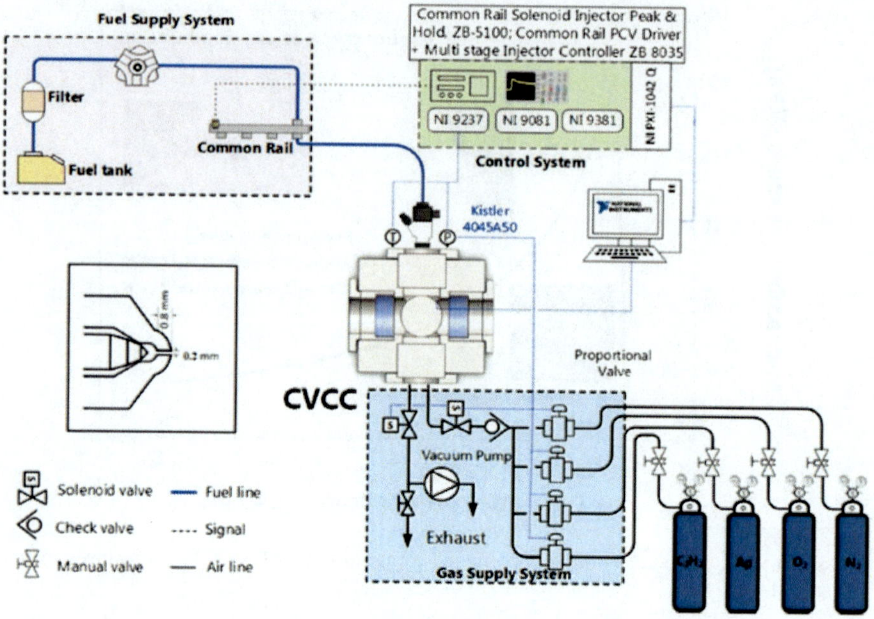

Fig. 5.4 Schematic of the premixed combustion-type CVCC setup for Spray Investigations (Vu and Ocktaeck 2019)

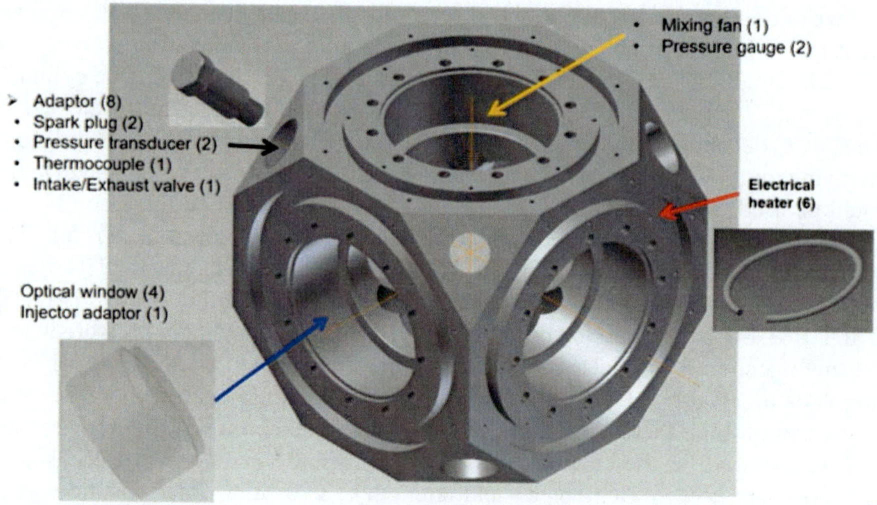

Fig. 5.5 Typical Design of a CVCC frame (Accessed from presentation on "ECN (Engine Combustion Network) Progress in KAIST (Korea Advanced Institute of Science & Technology) by Choongsik Bae)

the material. Even then, it is important to place the CVCC in an explosion-protected room for experiments such that one could operate it from a safe distance.

Another type of CVCC is an electrical heater-type model, in which, instead of premixed combustion of lean fuel–air mixture, an electrical heater is used, which heats up the chamber gases to the required temperature electrically (Wang et al. 2019; Cao et al. 2020). Significant numbers of studies have been reported using such chambers, which simulated engine-like conditions for spray investigations. Nitrogen and oxygen enter the chamber from high-pressure cylinders, and the heaters located at the base of the chamber heats the nitrogen–oxygen mixture. A feedback device controls the pressure and temperature of the chamber. Once the required chamber pressure and temperature are achieved, the fuel spray understudy is activated. Schematic for such a system is shown in Fig. 5.6.

In cases, where combustion of fuel spray is not under consideration, it is ensured that there is no oxygen in the chamber at the time of injection. Chambers specifically designed for such conditions do not need as much strength as the CVCC and these are termed as Constant Volume Spray Chambers (CVSC). Since combustion chambers cannot simulate the airflow and turbulence similar to an engine in-cylinder conditions, researchers use optical engines with certain modifications in the engine design to study the effect of airflow and turbulence on spray characteristics.

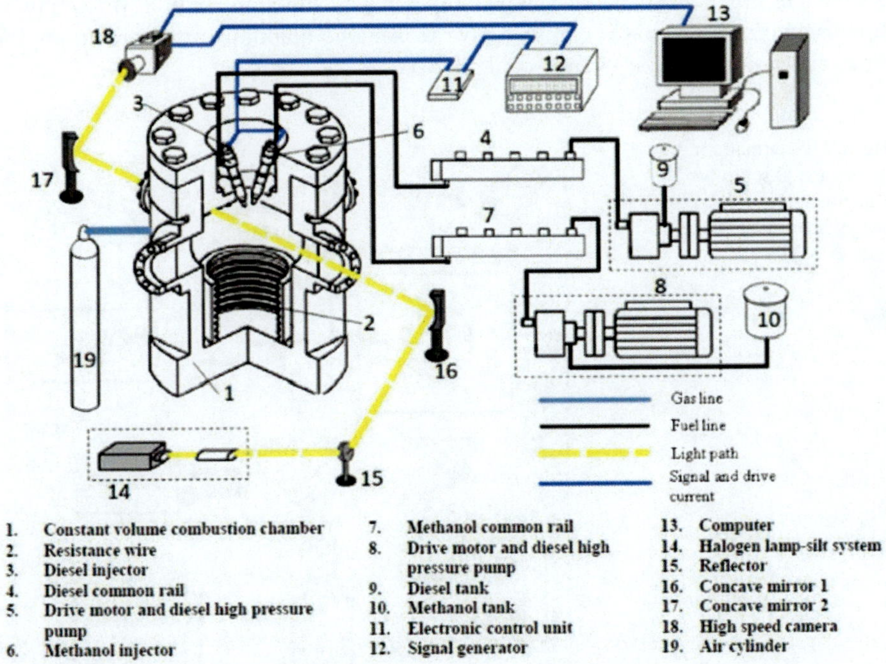

Fig. 5.6 Schematic of heater-type CVCC setup for Spray Investigations (Wang et al. 2019)

5.3 Optical Techniques for Spray Investigations

Broadly optical spray diagnostics techniques are of two types: (i) direct visualization, and (ii) laser-based techniques. Visualization techniques such as Shadowgraph, Schlieren and Mie-Scattering are widely used in macroscopic spray studies. Macroscopic spray parameters include spray penetration length, spray cone angle, spray width, etc. It is a property of light to refract and scatter, when passing through different density mediums and this forms the basis for these visualization techniques. All these optical techniques used for spray characterization are discussed in the following sub-sections.

Shadowgraph is a popular technique for spray visualization because of its simplicity and relatively lower cost. Different density regions in the spray area formed due to pressure, temperature and mixture gradients cause the formation of non-uniform refractive index regions. When light passes through these non-uniform refractive index regions, it undergoes refraction and phase shifts, which lead to variations in refracted light intensity. These effects create shadows on a screen, which are called a 'shadowgraph'. The light intensity originating from a light source (laser or non-laser) is detected using a camera (usually a high-speed camera) after passing through the test section. Shadowgraph method measures the second derivative of the density of the fluid. The schematic of the experimental setup is shown in Fig. 5.7. Shadowgraph technique is well suited to study flow from two-dimensional nozzles, but it can also give reasonable qualitative interpretations for three-dimensional nozzles (Yang et al. 2020).

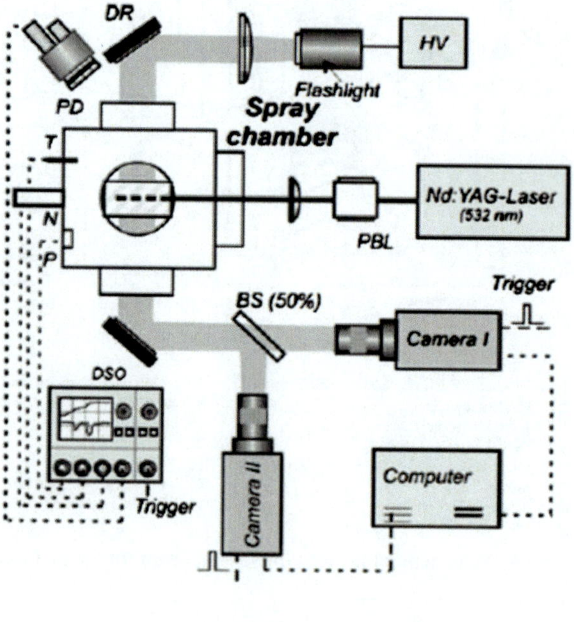

Fig. 5.7 Schematic of the experimental setup for Shadowgraph of dense sprays in a CVCC (Bougie et al. 2005)

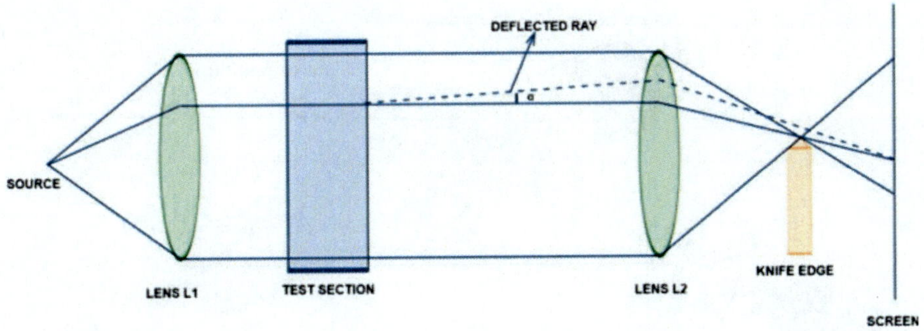

Fig. 5.8 Principle of Schlieren technique

Schlieren technique works on the same principle. When light beam passes through the test section, the denser part deflects the light beam shifting the image in the image plane. Unlike the Shadowgraph, Schlieren method measures the first derivative of the density of the fluid and uses a knife-edge for cut-off, as shown in Fig. 5.8. The position of the knife-edge affects the image on the screen, and it is usually positioned perpendicular to the nozzle, so that the density variation along the length of the spray is visualized (Goldstein 2017). Both these techniques are used to determine the integral quantity over the length of the light beam and are instrumental in imaging in a wide range of density regions.

Mie-Scattering is a popular method to visualize the liquid droplets of a fuel spray using the principle of light scattering. Mie-Scattering is observed when a photon incident on a particle/droplet is having a diameter similar to or greater than the wavelength of the incident light photon. The intensity of scattering is dependent on the size of the particle/droplet and the wavelength and frequency of the incident beam. The test area is illuminated using a High-Intensity Discharge (HID) lamp, and the images are taken using a high-speed camera. It is important to note that the camera is positioned perpendicular to the incoming light beam, unlike the position of the camera directly behind the light source in Shadowgraph and Schlieren techniques. The optical setup for a Mie-Scattering experiment is shown in Fig. 5.9. Mie-Scattering signal is proportional to the square of the particle/droplet diameter. Interference problem arises in dense regions and hence is used at a distance from the nozzle end. Mie-Scattering is usually used along with Shadowgraph or Schlieren because it cannot visualize spray vapours. It is a powerful tool to visualize the liquid droplet in a dense fuel spray.

Because of strong absorption and scattering of laser light beam in dense sprays near the nozzle (dense) region, the above-mentioned methods experienced difficulties in analyzing near-nozzle flows (Jeon and Moon 2018). Use of X-rays to study dense spray region near the nozzle and inside the nozzle was reported successfully because X-rays have very small wavelength, hence, they face little interference from liquid droplets in the dense spray regions (Duronio et al. 2020). When X-rays pass through the dense region, they undergo both phase shift as well as absorption. Incident rays

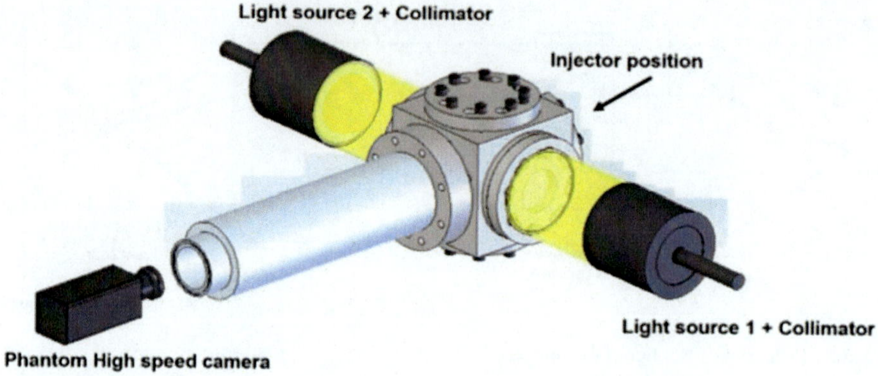

Fig. 5.9 Schematic for Mie-Scattering experimental setup in a CVCC (López et al. 2014)

and the phase-shifted rays intersect with each other and form fringe patterns at boundaries of different density (gas–liquid, liquid–liquid) regions. Intensity change is observed due to absorption by liquid fuel droplets. These are recorded by X-ray Phase-Contrast Imaging (XPCI), which give us information about these dense regions. Schematic of the XPCI experimental setup is shown in Fig. 5.10. X-ray-based techniques in transparent nozzles were used in few studies to understand the internal nozzle flow, which is crucial for in-nozzle studies (Salvador et al. 2018; Falgout and Linne 2015).

Phase Doppler Particle Anemometry (PDPA) is a laser optical diagnostic technique that measures both individual particle/droplet velocity and size simultaneously. The instrument is known as 'Phase Doppler Particle Analyzer' (PDPA) or 'Phase Doppler Interferometry' (PDI). It is a non-linear optical diagnostic method (Ofner 2001) that works on the principle of interference between two laser beams generated from a single laser source (produced using beam splitters, so that they have the same frequency and wavelength). One laser beam is phase shifted. When

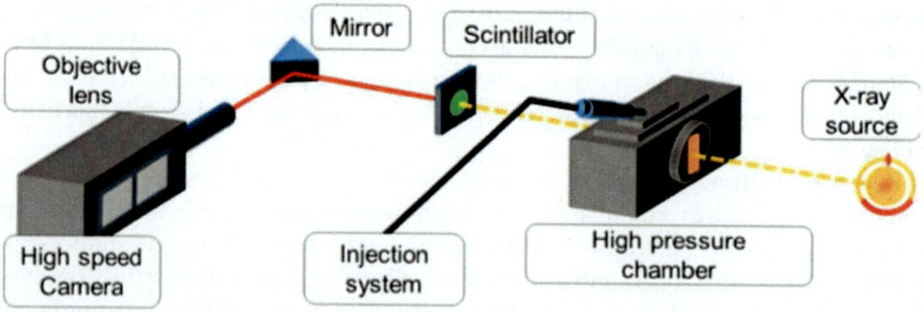

Fig. 5.10 Schematic of experimental setup for X-ray phase-contrast imaging of dense sprays (Jeon and Moon 2018)

these two laser beams intersect in the test section, a fringe pattern is produced at the intersection due to interference of light. When a particle/droplet passes through the interference pattern (called probe volume), a series of burst signal peaks are generated. The particle velocity in the probe volume is measured from its relationship with the measured frequency of these signals. For particle size measurements, the phase difference between the scattered light from the two laser beams is found to be proportional to the droplet size. With more simultaneous laser beam pairs, more velocity component measurements can be made. However, it is important to make sure that only one droplet is present in the probe volume at a time, and, hence, this method cannot be used in highly dense spray regions (near the nozzle exit) (Duronio et al. 2020). It is a point measurement technique, i.e. it cannot make instantaneous measurements over a spatial domain and requires a large number of measurements before averaging, at each measuring point. Figure 5.11 represents the PDI system configuration.

Planar Laser-Induced Florescence (PLIF) is relatively modern method of flow visualization, which provides data related to equivalence ratio distribution. It is useful in studying complex sprays qualitatively because it gives information at different points in a plane unlike an integrated value along the line of sight, which is the case with shadowgraph and Schlieren measurements. The principle behind this technique is the excitation and de-excitation of an atom or molecule, when a laser

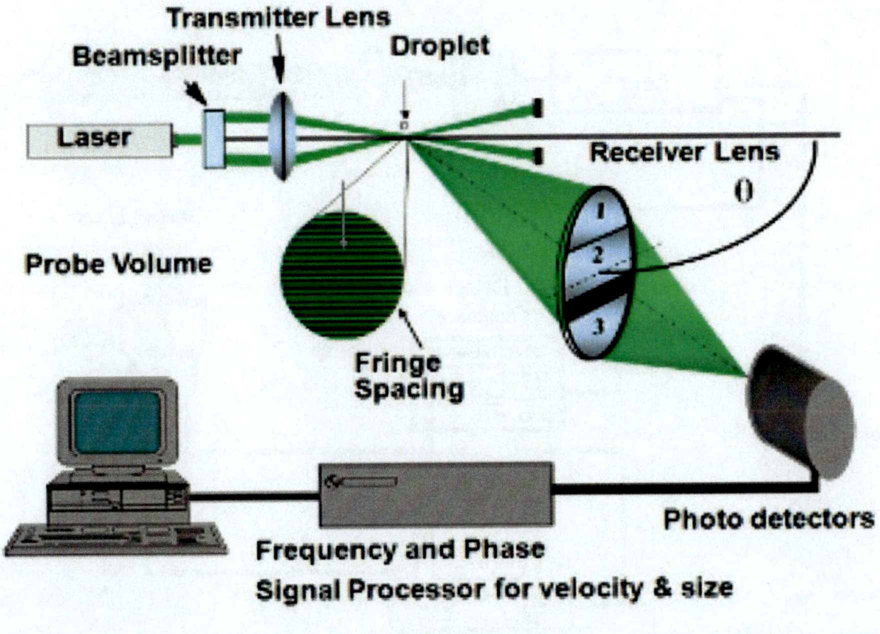

Fig. 5.11 Schematic of basic PDI measurement setup configuration (https://www.artium.com/pdi-theory-of-operation)

beam is an incident upon them, when it is accompanied by light emission. When a particle absorbs a photon from the incident laser light, it gets excited to an unsteady state and decays spontaneously, emitting a photon again. This is known as Laser-Induced Fluorescence (LIF) phenomenon. Schematic of the optical setup for PLIF is shown in Fig. 5.12. The wavelength of the laser light is to be selected such that it lies in the absorption spectrum of the molecules. It is therefore important to note that the fluorescence measured is not an absolute measure of the absorbed radiations. It comes with many unknown factors, which are difficult to determine. One such factor is quenching, which is an electronic energy transfer from the excited atom due to inelastic collisions. Quenching is predominant in presence of atmospheric oxygen, and, hence, nitrogen is used as the ambient gas in the experiments. Another important aspect is the selection of dopant. Most fuels used in IC engines do not produce fluorescence in the visible or ultraviolet (UV) region; hence, dopants are required to be added. The dopant selected must match the properties of the fuel and must produce satisfactory fluorescence in the available laser wavelength. Detailed information regarding PLIF and its adaptation for application in IC engine studies are given in the open literature (https://exact-sciences.m.tau.ac.il/sites/exactsci.tau.ac.il/files/media_server/Exact_Science/Chemistry/laboratories/3.%20laser%20induced%20fluorescence.pdf; Zhao and Ladommatos 1998).

Particle Image Velocimetry (PIV) is a technique to measure flow velocity using the images of the seeding particles. Role of PIV in spray studies is to analyze the

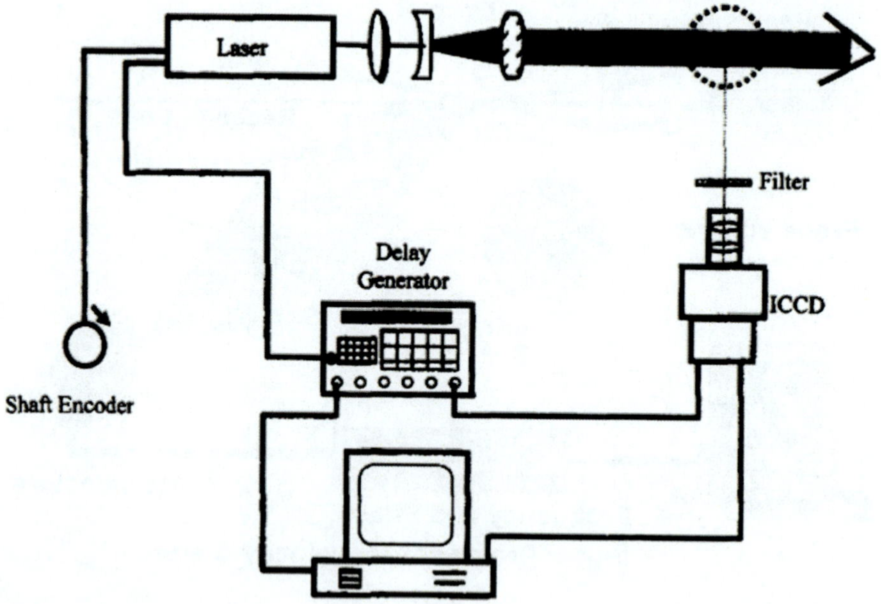

Fig. 5.12 Schematic of the experimental setup for PLIF (Zhao and Ladommatos 1998)

effect of spray–air interactions inside the chamber (Zama et al. 2017). PIV uses a double-pulsed laser to illuminate the plane in a test section in quick interval. A high-speed camera is used to capture the scattered light by the seed particles/droplets in the plane under study. The images obtained are processed to obtain instantaneous velocity vectors of the particles/droplets in the plane. A typical PIV experimental setup is shown in Fig. 5.13. In this technique, velocity is measured indirectly by obtaining the displacements in each time interval. Particles, popularly called 'seeding particles', are added to follow the flow faithfully. Tracer particles are assumed to move with the same velocity as that of the flow; hence, their correct selection is crucial. Tracer particles should be neutrally buoyant and small enough to follow the flow field. Also, they should be big enough to scatter sufficient light. A plane within the flow is illuminated twice in short interval using two different lasers or one dual-pulsed laser. Lasers such as Argon-ion laser and Nd:YAG laser are widely used because of their ability to emit monochromatic light with high energy density. A set of cylindrical lenses and mirrors are used to shape the light as a planar sheet. The particles/droplets scatter the light, which is captured on two separate frames by the high-speed close-coupled device (CCD) camera. A synchronizer is used for syncing the laser illumination and the camera. Given the small time interval between the two lasers and the magnification calibration of the camera, the velocity vector fields onto the plane of the light with the origin at the centre can be obtained. A significant error that occurs is due to gravitational force, when the density of the seeding particles differs from the fluid density. More information can be seen in reference (Tropea and Yarin 2007).

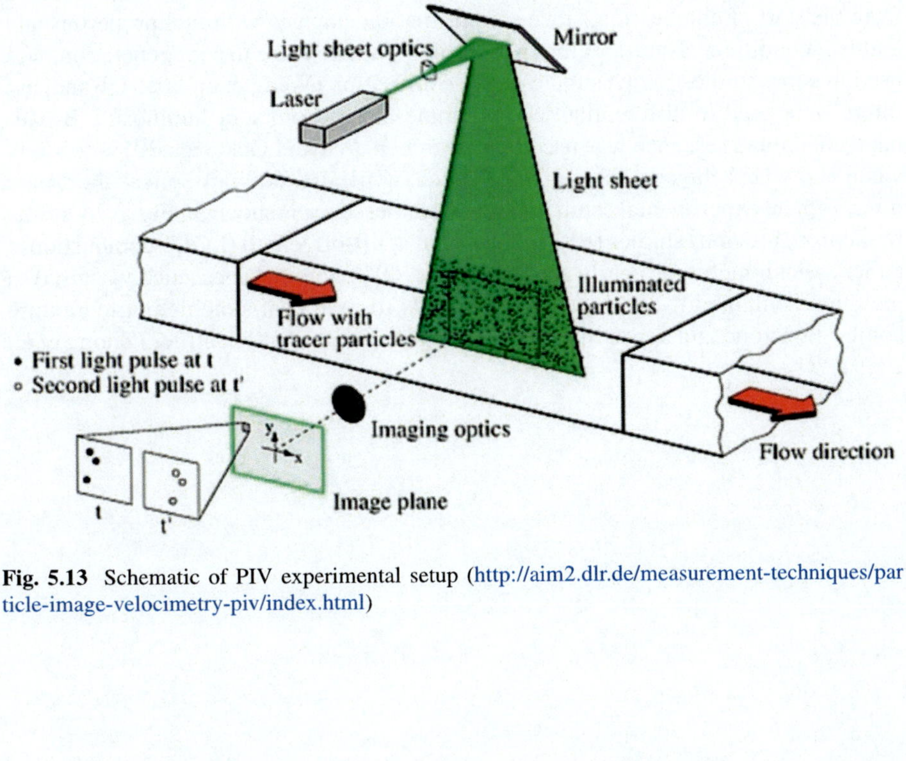

Fig. 5.13 Schematic of PIV experimental setup (http://aim2.dlr.de/measurement-techniques/particle-image-velocimetry-piv/index.html)

Table 5.1 Camera setting for chemiluminescence imaging (Dec and Espey 1998)

	Schlieren	Broadband Chemiluminescence
Camera	Phantom v7.1 CMOS	Phantom v7.1 CMOS
Lens	Nikon 85 mm (f/2.8)	Nikon 50 mm (f/1.2)
Filter	450 nm band pass	600 nm low pass
Framing period	45.25 μs	50 μs
Exposure	26 μs	46 μs
Resolution	512 * 128	512 * 128

Chemiluminescence is the phenomenon, where light is produced during a chemical reaction. When chemicals react, they form exciting intermediate called radicals. The radicals break down and release some of their energy in the form of light photons in order to reach the ground state. Since combustion is a chemical process, capturing the light emitted by chemiluminescence gives vital information about the combustion. Every molecule has a specific wavelength band for chemiluminescence, and these molecules change for different combustion regimes (Gaydon 2012). A conventional high-speed imaging system that is used for shadowgraph cannot capture chemiluminescence because it is a significantly weaker light signal (Zhao and Ladommatos 1998). The camera settings used by researchers (Dec and Espey 1998) for both chemiluminescence and schlieren are given in Table 5.1. Luminous soot formation during combustion dominates these signals (Dec and Espey 1995), though soot formation relates to the pressure rise to some extent (Heywood 1988). It is incorrect to observe from the start of combustion, since soot formation changes with fuel properties and ambient conditions. Natural luminosity, which is mainly due to soot generation, was used in some studies to investigate diesel combustion (Wang et al. 2019). Bandpass filters were used to differentiate chemiluminescence from soot luminosity. Broadband chemiluminescence was used by researchers (Vu and Ocktaeck 2019) to study when and where the combustion was happening using a 600 nm optical short pass filter. Typical experimental setup for chemiluminescence is shown in Fig. 5.14. It was found from previous studies (Peters 2001) that at lift-off length (LOL), combustion is either stoichiometric or nearly stoichiometric. OH-chemiluminescence was used to measure the flame lift-off length as OH radical form in near stoichiometric mixture combustion zone and hence this technique can be used as an identifier (Zhong et al. 2019).

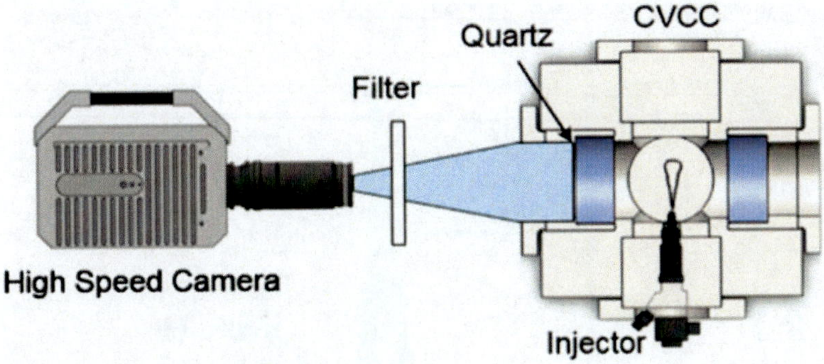

Fig. 5.14 Schematic setup (Vu and Ocktaeck 2019)

5.4 Spray and Spray Flame Investigations

Spray atomization is achieved by breaking up of liquid fuel jet into millions of small droplets, which enhance spray evaporation and forms fuel–air mixture in vapour form. This phenomenon plays a significant role in determining the quality of fuel–air mixture and eventually the combustion and emission performance of the engine. Figure 5.15 represents various characteristic features of free spray.

Many researchers have investigated spray morphology characteristics under different operating parameters in simulated engine like conditions in CVSCs and CVCCs. In-cylinder measurements are accompanied with several complexities such as turbulent flow, flame initiation, and changing thermophysical properties, hence, these are quite challenging to investigate. Microscopic and macroscopic spray characteristics completely define the spray morphology of the liquid fuel jet. Microscopic spray characteristics include the determination of droplet size and velocity distribution and their statistical average values. Macroscopic spray characteristics include parameters such as spray penetration length, spray cone angle and spray area. Determination of these parameters under different operating conditions certainly helps in deciding optimized parameters of the injection process.

5.4.1 *Microscopic Spray Characteristics*

Microscopic spray characteristics give quantitative information to evaluate the fuel injection process. From previous studies, it is evident that the jet velocity plays a vital role in determining the size of droplets in a spray. At higher jet velocities, surface waves proliferate, which helps the droplets to break-free due to dominance of surface tension forces, that gives rise to formation of smaller drops (smaller than the nozzle diameter). Apart from fuel injection parameters, other factors such as differences in the size distribution of droplets at the edge and the core of the spray,

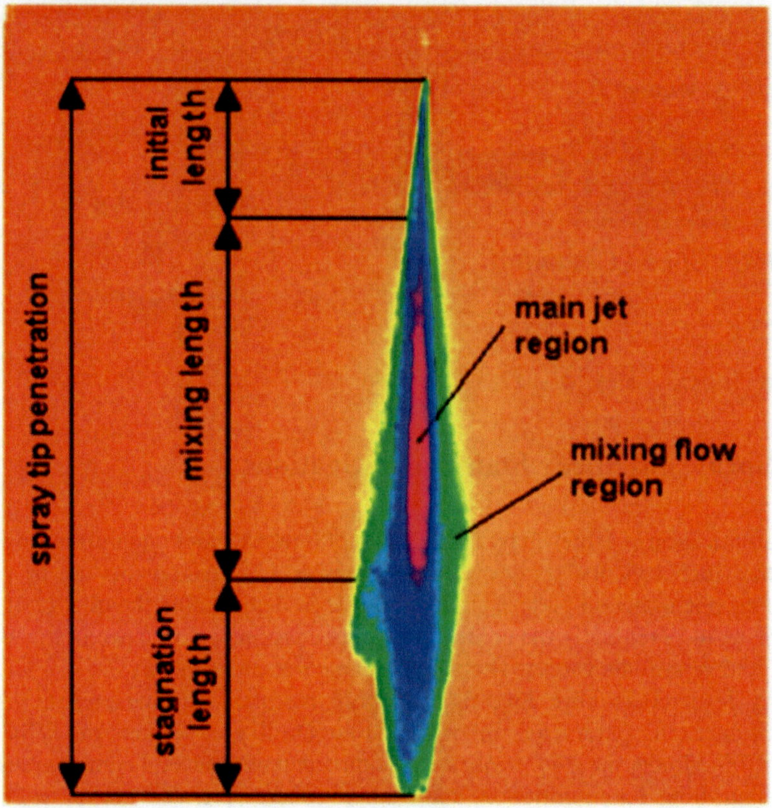

Fig. 5.15 Characteristics of free spray (Duronio et al. 2020)

droplets trajectories, and ambient conditions also affect the droplet size distribution at a particular location in the spray plume (Soid and Zainal 2011). Various studies have been performed to investigate microscopic spray characteristics for different operating parameters. During evaluation of biodiesel as an alternative to conventional diesel, spray studies have played a significant role. Researchers have examined spray characterization of different fuels to analyze the effect of fuel properties of spray formation. Hwang et al. (2017) compared three biodiesels obtained from waste cooking oil (WCO), Karanja and Jatropha oils, with baseline mineral diesel in non-evaporating conditions at different FIPs. Measurement of Sauter Mean Diameter (SMD) of spray droplets was done at 16 points downstream of the primary spray breakup length for different fuels, as shown in Fig. 5.16 for mineral diesel.

SMD in radial direction did not show any significant difference among the test fuels; however, in the axial direction, SMD decreased upon moving away from the nozzle. Karanja biodiesel showed the largest SMD, which was ~15% larger than that of the smallest SMD showed by baseline diesel, as shown in Fig. 5.17. Higher values of surface tension, viscosity and density shown by biodiesel delayed the spray

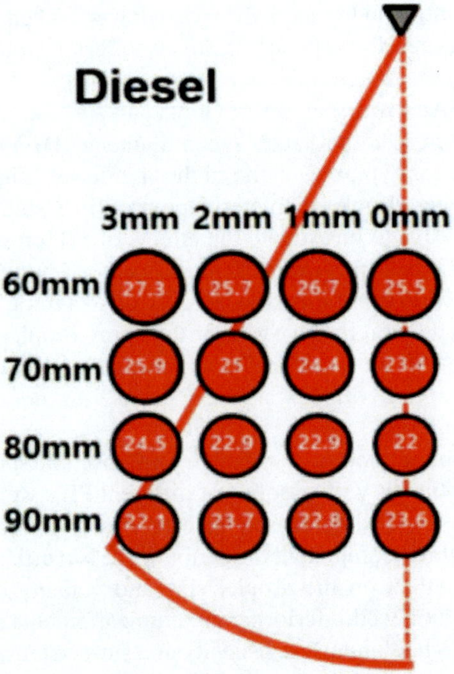

Fig. 5.16 Distribution of SMD for diesel for 80 MPaFIP (Hwang et al. 2017)

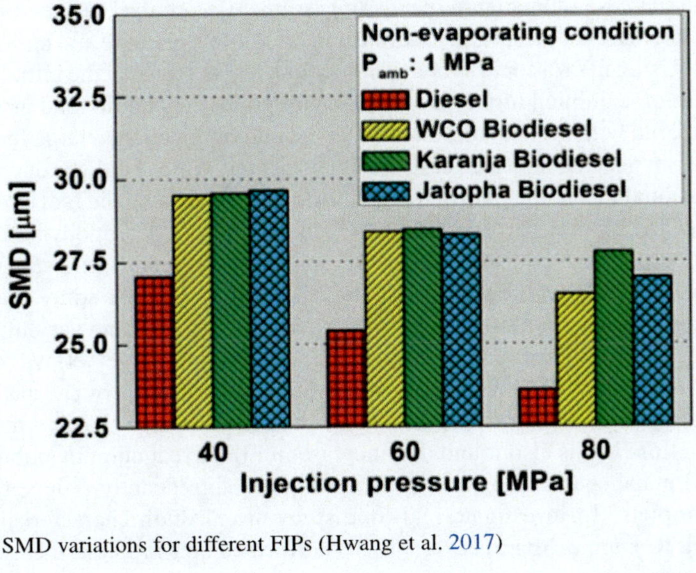

Fig. 5.17 SMD variations for different FIPs (Hwang et al. 2017)

break-up process, resulting in higher SMD. Also, lower Weber number of biodiesels (83% compared with diesel) explained relatively inferior atomization tendency of biodiesels.

The other studies (Agarwal et al. 2015; Gupta and Agarwal 2016) concluded that increasing the FIP decreased the Sauter mean diameter (D_{32}) and arithmetic mean diameter (D_{10}) of the spray plumes while higher biodiesel concentration in the test fuels increased the mean droplet diameter. However, in a study conducted by Lee et al. (Lee and Park 2014) to investigate the effects of FIP on spray breakup length in case of a Gasoline Direct Injection (GDI) fuel injector, it was found that increased FIP improved the spray droplet formation but only up to a limit. Increasing FIP from 5 to 20 MPa led to reduction in the SMD of the spray droplets almost linearly by approximately 10 μm. Above 20 MPa FIP, there was no significant reduction in the SMD of the spray droplets, which was due to very little noticeable change in jet velocity with increasing FIP.

In another study (Sharma and Agarwal 2016), microscopic spray characteristics of methanol and ethanol blends with gasoline at different FIPs were investigated using PDI technique. Increase in FIP decreased the droplet size distribution and increased the droplet velocity (three components) distributions. Nevertheless, fuel properties showed a significant effect on the droplet size and velocity distributions. It was observed that alcohols showed inferior spray atomization characteristics compared with gasoline due to higher kinematic viscosity and surface tension. Droplets having sizes ranging from 4 to 5 μm exhibited the highest probability for all test fuels. Similar observations were reported in a study conducted by Sonawane and Kalwar (2020), where spray droplet size distributions shifted toward higher diameters and velocities with an increase in methanol blends in gasoline. Lee et al. (2018) compared the spray characteristics of diesel and gasoline under high-temperature and high-pressure conditions. Using microscopic spray visualization, it was seen that the initial gasoline spray evolution exhibited higher dispersion with greater instability and asymmetric shape compared with mineral diesel. Also, at the end of injection, a large fuel vapour cloud was seen near the nozzle in case of gasoline spray, which could be due to evaporation of residual fuel in the sac volume of the nozzle. Increasing the fuel temperature also reduced the droplet size distribution (Chen et al. 2017). Thermodynamic conditions such as pressure and temperature of spray environment also influenced the atomization behaviour of the spray. Li et al. (2016) determined the spray characteristics of a single jet from multi-hole injector under different FIPs and varying ambient pressures in a CVCC. Deviation of the jet under observation (target jet) was observed throughout the experimental conditions due to imbalance created by the gas pressure between the inner and outer sides of the jet. This deviation also changed the droplet size distribution. It was also found that increase in FIP or reduction in ambient pressures could enhance deviations in jet trajectory and significantly reduce the SMD of spray droplets. To investigate real-time spray atomization characteristics in an actual firing IC engine, Sharma et al. (2020) performed non-intrusive measurements of spray droplets size and velocity distributions in an optical engine using PDI. They also compared the results with those obtained from a CVSC. Droplet mean diameters were always smaller in engine measurements as compared with ambient conditions in the CVSC. Droplet diameters were less scattered, and majority of droplets was confined to sizes < 10 μm in the engine cylinder.

Further, higher drag force experienced during the engine operation lowered the values of maximum velocity components, with $V_x = 29.8$ m/s and $V_y = 14.2$ m/s. On the other hand, maximum velocity components under ambient conditions in the CVSC were higher with $V_x = 78.41$ m/s and $V_y = 23.92$ m/s, indicating lower drag forces in the CVSC. These in situ measurements helped in analyzing the effect of highly dynamic and turbulent nature of reacting flows in the combustion chamber on the microscopic spray characteristics. Split injection strategy of diesel exhibited increased droplet collisions, which increased droplet size distribution compared with single injection case (Wang et al. 2016a, b). However, increased FIPs minimized differences in droplet sizes for these two cases. The spray characteristics also depend on the manufacturing technology used for nozzle. Behringer et al. (2014) analyzed spray formation from Spark-Eroded (SE) injector and its 'direct replacement' Laser Drilled (LD) injectors for Direct-Injection Spark-Ignition (DISI) engines for gasoline, ethanol and n-butanol usage. For gasoline, both injectors showed similar droplet size histograms; however, for ethanol and butanol, the LD injector produced larger spray droplets than SE injector, and this difference decreased as the ambient conditions reached closer to fuel's boiling point. The LD injector clearly showed the tip of the spray encompassed with a cluster of small droplets in all test fuels.

Several researchers employed microscopic imaging technique to analyze primary breakup of spray near the nozzle exit. Wang et al. (2016) investigated near-nozzle gasoline spray breakup using a high-speed microscope under flash boiling conditions. They observed that immediate collapsing of vapour bubbles due to flash boiling led to radial expansion of spray and considerable improvement in spray atomization. However, the explosion strength of bubbles was found to be mainly affected by internal flow in the nozzle. Also, increasing the FIP increased the spray cone angle during the start of injection, but due to drag force, cone angle decreased during the steady injection. Crua et al. (2015) studied initial diesel spray formation using microscopic imaging to understand physical processes during primary atomization. On identifying the dynamics behind mushroom-shaped structure of initial fuel injected from the nozzle, they found that residual fuel trapped inside the orifice during consecutive injections was mainly responsible for this. Vortex ring motion in the residual fuel produced slipstream effect, by which microscopic liquid fuel ligaments were propelled. They suggested that numerical models should consider the sac volume to be initially filled with liquid fuel only partially rather than fully filled by in-cylinder gases. A similar study was performed by another researcher to understand the mechanisms for initial spray morphology under varying FIPs (Wang et al. 2019).

5.4.2 Macroscopic Spray Characteristics

Macroscopic characteristics by fuel spray are represented by spray penetration length and cone angle, as shown in Fig. 5.18. After emerging from the nozzle, the spray

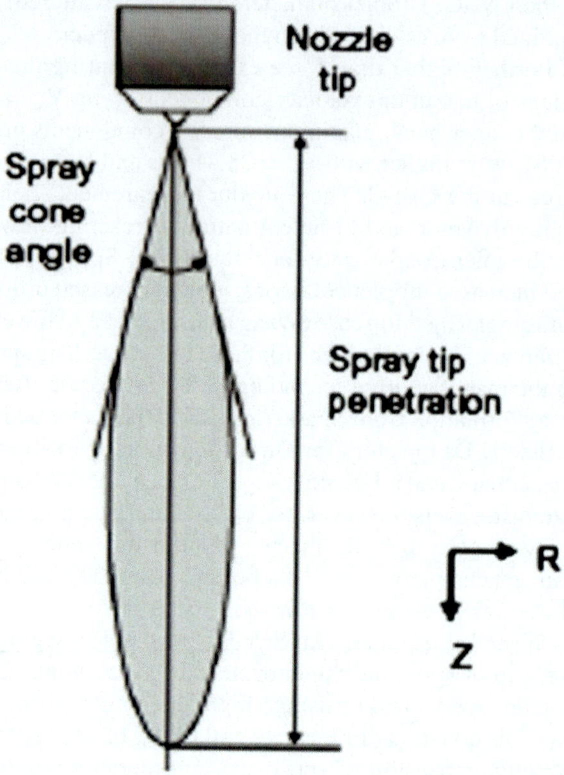

Fig. 5.18 Depiction of macroscopic spray characteristics (Suh et al. 2007)

jet undergoes turbulence due to its high velocity, eventually disintegrating into a large number of droplets. Eddies formed at shear layer during the turbulent mixing entraps the surrounding air, resulting in molecular mixing at the fuel–air interface (Kawano et al. 2002). The transport mechanism mainly exhibits turbulent mixing. Hence its study would help in determining the fuel jet development with liquid fuel concentration and vortex motion distributed along the jet (Kawanabe et al. 2010). As the spray jet continues to develop, air mass trapped in the spray increases, which increases the spray cone angle and spray area; however, it decreases the droplet velocity. This air entrainment process helps in evaporation of smaller fuel droplets.

To maximize the air utilization within combustion chamber, higher spray penetration, cone angle and spray area are always desirable because it enhances combustion. However, it further depends on the design constraints of the engine combustion chamber. The chamber design with high air swirl and hot piston walls is designed for spray impingement over the piston. On the contrary, quiescent combustion chamber with multi-nozzle injection is designed to avoid the spray impingement on the walls, as it decreases the fuel–air mixing rate and increases the unburnt hydrocarbon (HC) emissions.

Bao et al. (2014) studied spray evolution of ethanol, gasoline and iso-octane using Mie-scattering techniques, to analyze the impact of fuel properties on the spatial and temporal growth of spray. They characterized the spray in CVCC by simulating cold-start operating conditions. Moreover, they used optical DISI engine for simulating the warming up conditions. Ethanol resulted in the shortest spray penetration. Iso-octane exhibited the longest spray penetration at lower FIP of 4 MPa, which might be due to higher discharge losses and lower droplet velocity in case of ethanol due to its higher density and viscosity than iso-octane and gasoline. With an increased FIP, differences in spray penetration decreased, and the trend reversed with ethanol exhibited highest penetration length compared with gasoline at 15 MPa FIP. This is because of the change in the governing factors in the spray regime due to dominance of aerodynamic drag force. Higher FIP led to the formation of smaller droplets. Hence, smaller droplets of gasoline and iso-octane experienced higher drag, which limited their spray penetration. Since larger ethanol droplets have higher momentum, they overcame drag to exhibit longer spray penetration length. Similar trend was observed in both cold chamber and warming up conditions; however, the warmed-up conditions resulted in higher spray penetration due to lower air density. In another study (Hwang et al. 2017), spray characteristics of biodiesels were investigated in a CVCC under evaporating conditions using high-speed shadowgraph technique. Due to inferior volatility of biodiesels, it was observed that they had slower air–fuel mixing characteristics. Under these conditions, diesel evaporated much faster than biodiesel. This resulted in denser liquid region in the centre of the spray plume in case of biodiesel sprays. Inspite of this, mathematical calculations for the equivalence ratio showed lower values for biodiesel due to presence of oxygen in biodiesel. López et al. (2014) found that there was no significant difference in the liquid penetration length of gasoline and diesel in non-evaporating conditions; however in vaporizing conditions, diesel fuel penetration was1.8 to 2.4 times longer than gasoline.

Furthermore, the effect of FIP and ambient pressure and temperature conditions of the spray environment was investigated by researchers. Gupta and Agarwal (2016) studied the effect of chamber pressure and FIP on the macroscopic spray characteristics of Karanja biodiesel blends in a CVSC and compared it with baseline mineral diesel. The effect of blending on the spray characteristics was minimal compared with the effect of FIP and chamber pressure. Increase in FIP increased the spray tip penetration, spray area and spray cone angle. In contrast, increase in ambient pressure shortened the spray tip penetration, spray area and increased the spray cone angle. Higher ambient density caused higher resistance to flow, causing the spray to travel transversely. This increased the spray cone angle with an increased ambient pressure. Lee and Park (2014) studied gasoline spray breakup and atomization processes at high FIPs up to 30 MPa using laser visualization in atmospheric temperature and pressure. 'Branch-like structures' were observed with counter-rotating branches after the complete development of spray jets. It was also observed that as the FIP increased, the time to exhibit these structures decreased, whereas the width and number of branches increased. It was noted that these structures were a result of air-entrainment and facilitated the spray breakup process, as shown in Fig. 5.19. The effect of ambient density and FIP was also studied by Mohan et al. (2014). They

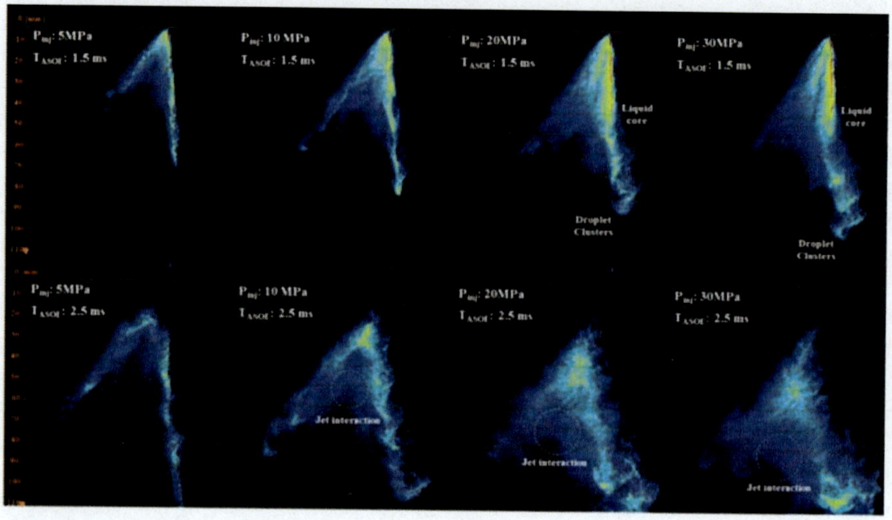

Fig. 5.19 Internal structures of showing spray breakup for different FIPs (Lee and Park 2014)

measured the spray momentum flux using a piezoelectric force sensor. High-speed camera images were used to study other macroscopic spray parameters like spray tip penetration and spray cone angle. It was concluded that the spray momentum flux increased and decreased with increasing FIP and ambient density, respectively. However, it was seen that the momentum coefficient (actual/theoretical momentum) was the same at different FIPs, which was possibly due to cavitation in the injection nozzle. Also, spray tip penetration increased and decreased with an increasing FIP and ambient density, respectively. Possible reasons were explained by the momentum flux trends. As in previous studies, quasi-steady spray angle increased with increasing ambient density due to increasing resistance offered to the spray plume in the axial direction. Spray volume was calculated by assuming the spray plume as a hemispherical cone. Higher FIPs and ambient pressures exhibited higher spray volume when calculated for same penetration length due to wider spray angle.

Wang et al. (2016) studied the effect of ambient air temperature on spray morphology in both low temperature (LT; −2 °C) and room temperature (RT; 25 °C) conditions. Due to higher viscosity of the fuel at lower temperatures, it was observed that the dispersion was inferior along with slower penetration and smaller spray area. In microscopic spray study, larger spray droplets with higher droplet velocity were observed due to higher surface tension of the fuel at the low temperatures. These differences in spray characteristics between LT and RT conditions decreased with increasing FIP.

Researchers also investigated the effect of design-related parameters on the macroscopic spray characteristics. Behringer et al. (2014) reported that out of LD and SE injector nozzles, LD nozzles had larger holes and, hence, showed wider sprays leading to superior fuel droplet dispersion. Delay in fuel sprays was observed in the

SE injectors and this difference was higher in alcohols than in gasoline sprays. This difference decreased with the increasing fuel temperature because relatively higher viscosity of alcohols decreased with increasing temperature. It was also observed that the difference in spray geometry for these fuels was lesser in LD injectors. Further spray penetration of LD injectors was lower compared with SE injectors because of their wider angle design. However, at low ambient pressures, higher spray penetration for LD injectors was reported because of concentrated axial momentum from more severe plume merging. Chang et al. (2020) investigated spray characteristics of two GDI injectors with different hole arrangements under flash boiling conditions using high-speed shadowgraph techniques. It was reported that the injector with smaller holes spaced closely (N) showed increased ability of sprays to collapse because of superior plume interactions and also finer spray atomization by smaller holes. These smaller droplets are more sensitive to pressure changes than the injector (W) with bigger holes spaced more loosely. The spray from injectors N and W showed lower spray tip penetration and spray width in the transition region (the region where increasing the superheat degree decreased the normalized rate of collapse), which makes them ideal for use in GDI engines because they exhibit lower wall wetting and better diffusion in radial direction. Jeon and Moon (2018) studied the effect of ambient density with varying needle lift (which alters the turbulence and spray sheet thickness near the nozzle) on the spray breakup and local spray droplet size distribution for an outwardly opening GDI Injector. These types of injectors produce a hollow cone spray, which has better atomization than the traditional hole type. Higher needle lift decreased the spray breakup length due to disturbance waves generated inside the nozzle. Furthermore, the effect of ambient air density on spray breakup length was found to be minimum with higher needle lift, as shown in Fig. 5.20. Furthermore, it was observed in a study (Geng et al. 2019) that with decreasing nozzle orifice diameter, spray penetration increased and spray cone angle decreased. Also, the droplet size distribution decreased, which showed that high FIPs and small orifice diameter led to superior spray atomization.

In another study (Lei et al. 2019), it was found that the spray tip penetration followed the FIP variations, when using dynamic injection behaviour (variable FIPs) in a CVCC using a common rail injector. Spray penetration length increased in the transition zone, when the FIP was increased and was constant in the stable zone, when the FIP was constant. Subsequently, an expression was developed to relate the spray penetration length with the FIP in both the zones. Wang et al. (2016) performed diesel spray studies under split injection strategy. The near-field spray morphology showed primary collision between the tip of second injection and the wake of the first injection, which caused poor dispersion axially. It was found that the second injection droplets had lower velocities than the first injection droplets and this difference reduced as the dwell decreased because with less dwell time, the time for opening of the injector for the second injection was more, leading to more fuel injection with higher droplet velocity. The second split injection was observed to have lower spray area and slower spray penetration during early stages due to primary collisions, but the spray penetration and spray area increased in later stages. Split injection showed higher SMD compared with single injection due to droplet collisions in split injections.

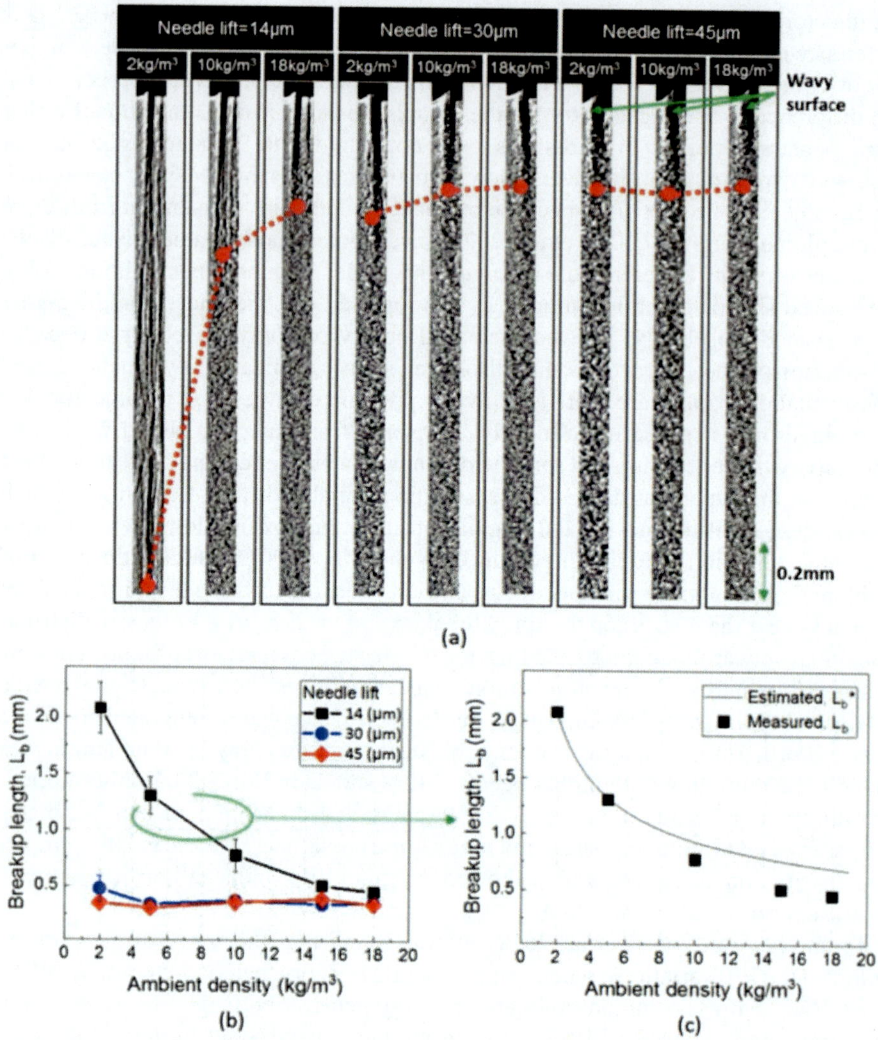

Fig. 5.20 Liquid sheet breakup length for different ambient densities and needle lifts (**a**) measured breakup length, (**b**) and (**c**) comparison between estimated and measured breakup lengths (Jeon and Moon 2018)

5.4.3 *In-Nozzle Flow Characteristics*

Internal flow in the nozzle is considered very important in defining the flow behaviour at the nozzle exit and subsequent spray formation, which in turn define quality of fuel–air mixture formation and combustion and emission performance of the engine. Nozzle's internal flow determines subsequent spray breakup processes and complex interactions between the fuel droplets and surrounding gases (Wang et al. 2016).

Hence, it is very important to investigate the fundamentals of the in-nozzle flow so as to determine related spray phenomenon and processes.

The flow in the nozzle is highly dynamic since the effective pressure increases and decreases during the injection process. Furthermore, it is greatly dominated by the two-phase flow along with different flow regimes, i.e. laminar to turbulent (Payri et al. 2012). Various studies have been performed to determine the effect of different parameters on the internal nozzle flow such as nozzle geometry, FIP, fuel type, needle movement, etc. (Zhao 2009; Wang et al. 2016). In a study by Wang et al. (2018), the effect of split injections on the nozzle's internal flow and its subsequent primary breakup under different FIP were studied. They found that the spray tip geometry was significantly influenced by the interactions of pressure shock waves with the liquid and gas phases of the spray. Lower FIP resulted in spray with mushroom-shaped tip. On the other hand, increasing the FIP resulted in adverse breakup of air bubbles and superior primary spray breakup, which resulted in a well-dispersed spray. They observed cavitation formation in two regions with different origins. The first region was at the needle seat, where throttling effect produced during the start and end of the injection caused cavitation. The second region was at the nozzle inlet where flow redirection was mainly responsible for cavitation during steady flow and at high effective pressures. Applying a split injection strategy reduced the effective FIP, injection duration and mass injected. In this case, only throttling-induced cavitation was observed, which also weakened upon reducing the first injection duration to 0.5 ms.

Furthermore, it deteriorated the spray dispersion due to lower effective FIP. Hence, they concluded that higher FIP was necessary for split injection strategy to achieve superiorspray quality. Lešnik et al. (2018) also reported similar observations. They studied nozzle's internal flow and the effect of cavitation on spray evolution with the help of computational tools. They reported that cavitation enhanced the process of disintegration of fuel jet at the nozzle exit. Furthermore, internal flows and cavitation were highly dependent on the injector geometry. However, the effect of fuel properties had minimal impact on the spray geometry at high ambient pressures. Cao et al. (2020) used a scaled-up nozzle to study the vortex cavitation characteristics of diesel and its effect on spray characteristics. A pressure acquisition system was used to investigate dynamic pressure fluctuations at the end of the nozzle. In addition, high-speed imaging using shadowgraph technique was used to visualize and study the vortex cavitation effect on the spray. A proportional increase in the spray angle was reported with growth in vortex cavitation. They reported that vortex cavitation originated from conical tip of needle was mainly responsible for inducing hollow spray cone, as shown in Fig. 5.21. Moreover, variations in the intensity of vortex cavitation caused pressure fluctuations due to differences in the effective cross-sectional area.

Various researchers conducted spray investigations under flash boiling conditions. Flash boiling can result in improved spray atomization without the need for high-pressure fuel injection as it increases the rate of evaporation of fuel droplets. Hence, it considerably enhances combustion and emission characteristics of an engine (Xu et al. 2015). Studies have reported wider spray geometry, smaller droplet size, early

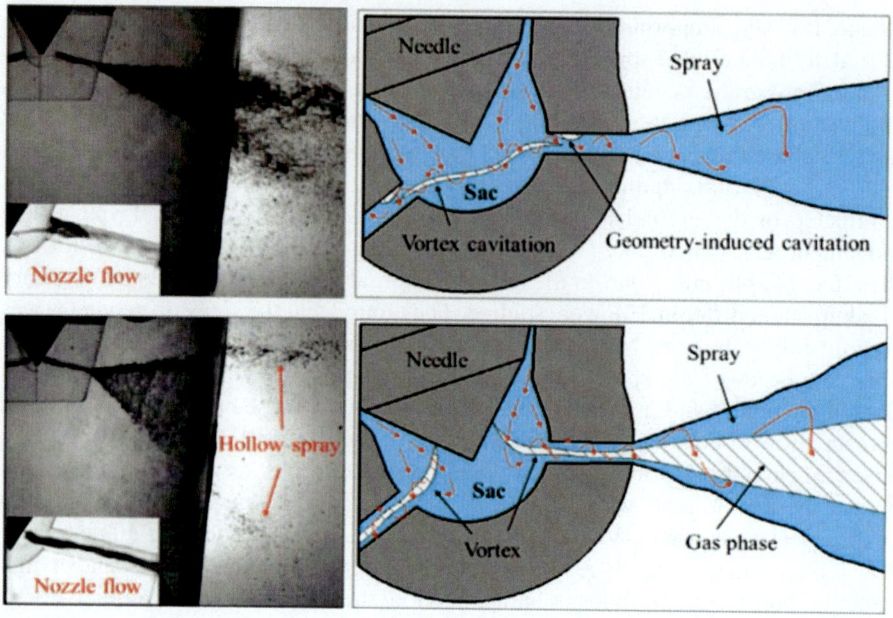

Fig. 5.21 Hollow spray generated by hole-to-hole vortex cavitation and needle-originated vortex cavitation (No.2 nozzle) (Cao et al. 2020)

primary breakup and increased rate of momentum transfer to the surrounding gases in case of spray with flash boiling conditions compared with high-pressure liquid fuel injection (Zhang et al. 2012, 2014).

Furthermore, to investigate the effect of nozzle design parameters on in-nozzle and near-nozzle flows under flash boiling conditions, Wu et al. (2017) used a transparent nozzle with slit thickness of 40 μm to study n-pentane at an FIP of 0.6 MPa, as shown in Fig. 5.22. Wide range of superheated conditions was achieved by varying the ambient pressure from 40 to 190 kPa and fuel temperature from 41 to 71 °C. It was observed that bubbles forming inside the nozzle at the exit in superheated conditions improved the atomization of fuel spray droplets in the near-nozzle region because the bubbles formed ruptured upon discharge. The bubble formation increased under intense superheated conditions. Moreover, extended nozzles gave more time for bubble formation and hence bigger bubbles were observed. Also, sharp inlet corners had a superior effect on the bubble formation compared with round inlet corners. Due to superior atomization under flash boiling conditions, it was seen that the width of the fuel jet decreased. More extended nozzles and nozzles with sharp inlet corners further reduced the width of the fuel jet near the nozzle.

In a study by Wang et al. (2018), the effects of bubbles generated at the nozzle exit due to cavitating flow and flash boiling conditions were studied. The formation of vapour bubbles enhanced the spray breakup phenomenon. They used transparent nozzle to visualize the evolution of bubbles and performed fuel injection in water.

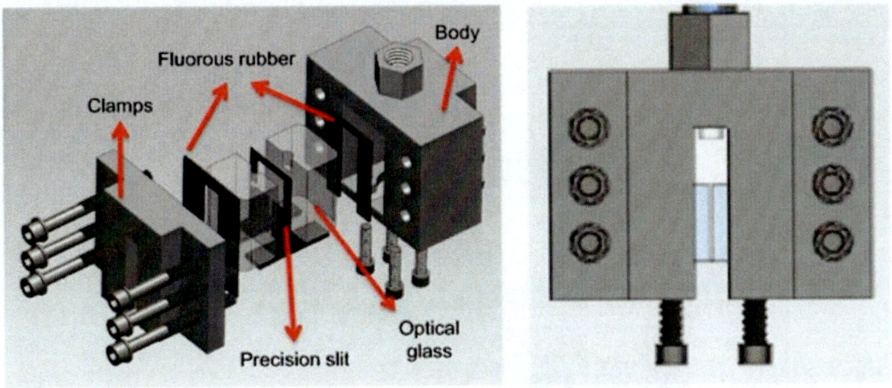

Fig. 5.22 Schematic of 2D transparent nozzle experiment (Wu et al. 2017)

After the primary breakup, the sizes of bubbles were found to be very small (~9 μm) at the nozzle outlet. In addition, higher fuel temperature reported superior atomization, though it reduced the breakup tendency during the start and end of injection due to low spray velocity. Significant radial propagation was indicated in the cold diesel spray (20 °C) under the influence of strong disturbances in the flow and in hot gasoline spray (140 °C) due to cavitating flow.

5.4.4 Spray–Wall Interactions

Spray–wall interactions refer to the behaviour of injected fuel spray, when it comes in contact with the walls of the combustion chamber. This phenomenon is very critical in determining the secondary evaporation of spray droplets or wall film, which further affected the soot and HC emissions from the engine (Moreira et al. 2010). Spray–wall interactions in case of high-pressure diesel injection can be found in several studies in the literature (Allocca et al. 2002; Li et al. 2015). However, employment of high-pressure injection in GDI and gasoline compression ignition (GCI) engines certainly requires in-depth investigations, which would be of high scientific relevance (Liang and Mudawar 2016).

Different regimes of spray–wall interactions have been discussed by many researchers. Stanton and Rutland (1996) and O'Rourke and Amsden (2000) described the classification of drop–wall interaction in four regimes, namely, stick, spread, rebound and splash, based on energy involved in the processes. Rioboo et al. (2001) further categorized the splash regime into four different mechanisms, namely, prompt splash, corona splash, receding breakup and partial rebound. They also explained fingering formation in the spread phase, which was further added by Moita and Moreira (2007) as finger breakup. Figure 5.23 shows a different related phenomenon on the scale of impact energy and time scale. Sticking of a spray droplet with a wall

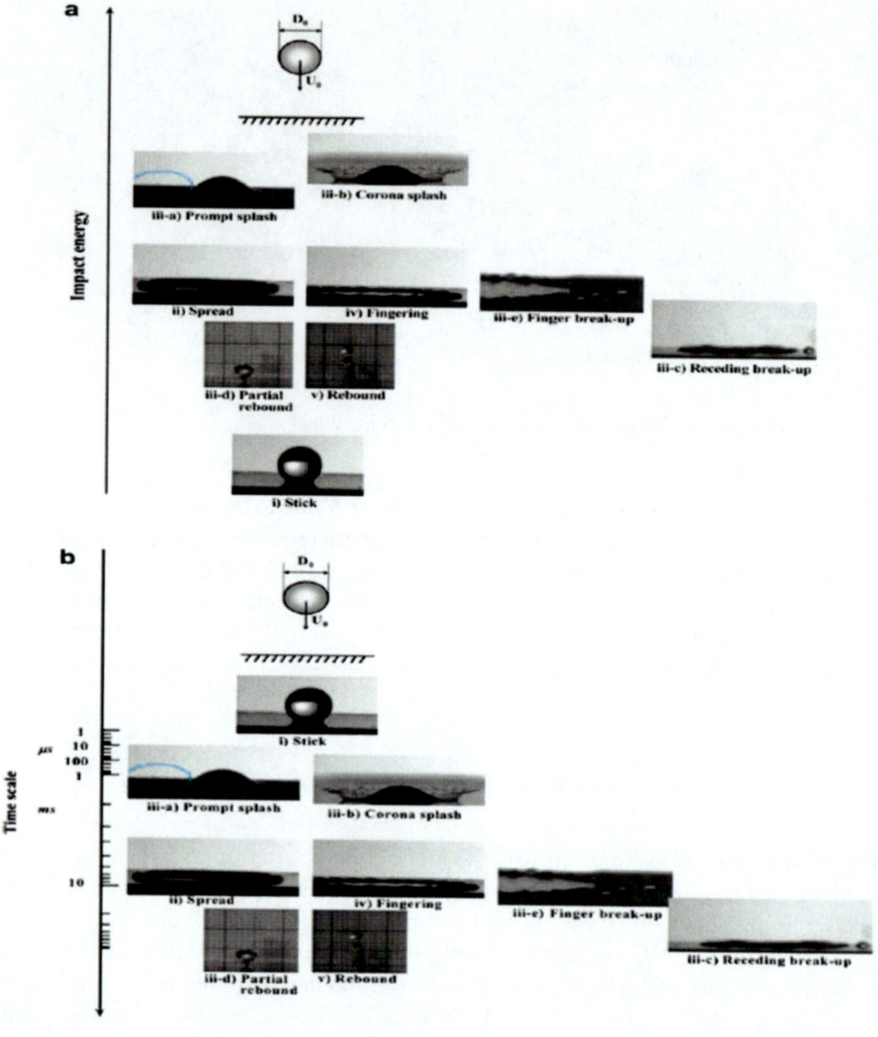

Fig. 5.23 Drop impact mechanisms classified on the basis of impact energy and time scales (Moreira et al. 2010)

is associated with shallow impact energy. On increasing the impact energy slightly, it spreads and recoils in the form of lamella. Moderate energy level makes lamella unstable, resulting in the formation of fingering. The growth in such structures is followed by a breakup at the end of spreading phase (Marmanis and Thoroddsen 1996). Increasing the impact energy to higher levels leads to disintegration of droplets. The transition from spread to disintegration is defined by prompt splash, where inertial forces dominate the capillary effects (Yarin and Weiss 1995). Moreover, the criteria for these impact regimes have been quantitatively defined by the relations comprising of non-dimensionless numbers, which are Reynolds, Ohnesorge, Laplace and Weber numbers. A broad spectrum of these impact situations can be defined by the algebraic equations (Bai and Gosman 1995; Han et al. 2000).

Furthermore, consideration of wall temperature is also important in determining the impact regimes, as heated wall enhances secondary evaporation of spray droplets. As shown in Fig. 5.24, the transition between different impact regimes can be shown in terms of *We* of impacting droplets and wall temperature (Bai and Gosman 1995).

Secondary droplets formed after the disintegration of impacting droplets exhibit entirely new characteristics. Hence, investigations of spray droplets before and after the interactions with both heated and unheated walls are crucial. Many researchers have widely investigated this. Luo et al. (2018) studied the microscopic behaviour of spray droplets before and after impingement over flat wall using a Particle Image Analysis (PIA). Larger droplets were seen at the centre of the spray compared with the edge, showing that the spray has lower atomization at the core. After impingement, it was observed that the droplet sizes were smaller, and the droplet density was higher near the wall, especially at low ambient pressure conditions. But at high ambient pressures, coalescence effect was observed away from the walls, giving rise to larger droplets. This coalescence effect seemed to reduce as the droplet velocity increased (higher FIP, lower ambient density).

In another study (Li et al. 2019), authors performed the experimental investigations to understand the effect of impingement geometry, cross-flow and wall temperature on a port fuel injection (PFI) spray system using a high-speed camera under ambient temperature and pressure conditions. Cross-flow was obtained using a centrifugal blower and a transparent acrylic wind tunnel. On increasing the impingement distance, it was seen that the spray tip penetration increased, and the spray height

Fig. 5.24 Different impact regimes varying with wall temperature (Moreira et al. 2010)

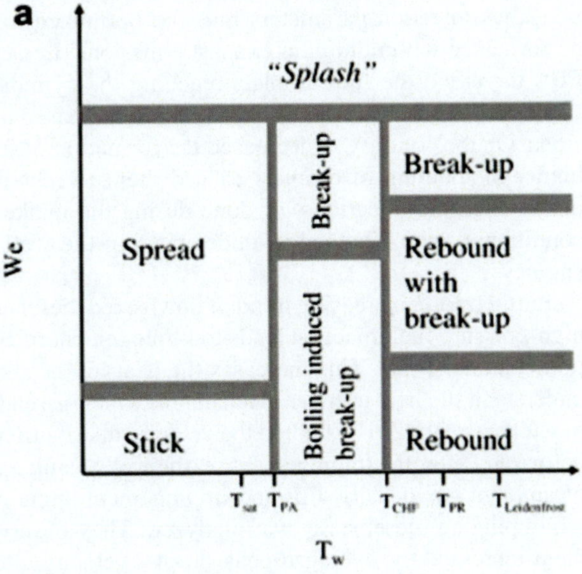

decreased, i.e. the spray tip became flatter and thinner, once it impinged on the wall. As the cross-flow introduced and its velocity increased, apart from spray impinging on the walls decreased and spray tip penetration decreased significantly because of the resistance offered to the spray by the cross-flow. The effect of wall temperature was not significant on the spray tip penetration. However, the spray height increased slightly with an increasing wall temperature below 420 K and increased sharply with the increasing wall temperature above 420 K. This was because at this temperature, the Leidenfrost limit of the gasoline fuel film, which caused the formation of vapour film above the wall, increased the vertical momentum of the droplets after impingement.

As spray impingement problem is very severe in GDI engines, hence, split injection strategy is implemented to reduce the injection duration, which can limit the wall–film formation. To investigate it, Wang et al. (2018) studied the effect of flash boiling on the wall impingement characteristics of a split injection strategy aimed for GDI engines using a long-distance microscope and high-speed imaging. With no flash boiling, it was seen that the presence of the fuel film caused by the first injection on the wall surface caused quicker fuel film formation by the second injection. Increase in dwell time had no significant effect on the fuel film formation in this case. With marginal flash boiling, the rate of film formation was still high, but there was a significant reduction in film build-up rate with the increasing dwell time. The fuel film reduction rate significantly increased with further intensifying flash boiling conditions. At intense flash boiling conditions, split injection did not affect the film formation because of rapid evaporation. A similar study was conducted using a single injection by Wang et al. (2019). Strong flash boiling conditions (ambient pressure = 0.2 bar, fuel temperature = 100 °C) were observed to reduce the film thickness by ~75% due to superior spray evaporation and dispersion. As already discussed, wall film formation due to spray impingement significantly contributes to PM formation, thus injection parameters must be optimized to obtain excellent combustion performance with minimum exhaust emissions. Lee et al. (2020) found that higher FIPs increased the fuel droplets reaching the cylinder walls, which increased the fuel film formation on the walls leading to increased particulate number (PN) emissions. On the contrary, it decreased the formation of fuel film on the piston because higher momentum of droplets caused them to rebound. These observations were made when the injection was done during the intake stroke (low ambient density conditions), which helped in understanding the wall impingement effect in DISI engines.

Furthermore, increasing trend of downsized diesel engines with the application of high-pressure fuel injection leads to impingement of spray flames over the combustion chamber walls. This increases the heat losses to the chamber walls. Hence, to understand the heat transfer mechanism, which is related with the fuel flow properties, Zama et al. (2017) studied the velocity distributions of diesel spray droplets just before and after the impingement on the walls using a time-resolved PIV. The effect of ambient gas density, FIP and impingement angle on the flow characteristics of the impinging diesel spray was analyzed. They observed that the oblique impingement increased the averaged peak droplet velocity after impingement as compared with normal impingement because of the radial velocity component of the impinging spray for oblique impinging spray. Furthermore, averaged peak spray velocity after the impingement was proportional to the spray tip velocity (before impingement) for tip velocities up to 40 m/s, but above this, the average peak velocity was proportional to the root of the tip velocity. This 40 m/s mark on the spray tip velocity was equivalent to 1000 mark on *We* as the trend in the averaged peak velocity behaved the same at both these points. Therefore, it was concluded that the velocity postimpingement was dependent on droplet adhesion on the walls. This inturn depends on the spray tip velocity.

In another study, Chen et al. (2019) investigated the effect of wall temperature and ambient conditions on mixture strength and the characteristics of wall-impinging diesel spray flame in a CVCC. Laser Induces Exciplex Fluorescence (LIEF), and flame natural luminosity imaging was used to visualize the liquid and vapour phases of the spray. It was observed that the vibrant vapour part of the mixture moved closer to the wall, as its temperature increased because the hot walls caused the hot gases to move towards the wall, as shown in Fig. 5.25.

Increase in wall temperature decreased the ignition delay and increased the flame luminosity because high wall temperature supported the formation of vapour and combustible mixture, and this effect was more prominent at low ambient pressures. The effect of wall temperature on flame area and flame height was negligible at high ambient pressure (4 MPa) and had a visible impact at low ambient pressure (2 MPa). Low wall temperature and ambient density conditions reduced the effectiveness of combustion process due to liquid spray impingement on the wall. Soot formation was more at high wall temperatures in both low and high ambient pressure conditions. At low ambient pressure conditions, high wall temperatures generated richer mixture at the wall, which was the source of higher soot formation. At high ambient pressure conditions, ignition started closer to the wall, and, hence, there was less air entrapment, which caused more soot production.

5.4.5 *Spray Flame Characteristics*

Employing optical measurement techniques to directly visualize macroscopic spray morphologies such as spray penetration, geometry, wall interaction and eventually, the combustion helps in improving our understanding of the in-cylinder events, sequencing from fuel injection to the start of combustion. Flame characterization obtained from visualization methods further provides useful information on the formation of soot, emission species, and radicals. These results significantly contributed to optimizing the injection parameters to achieve superior engine performance. CVCC is used for conducting spray and combustion investigations in simulated engine-like conditions. Two methods mostly achieve the heated environment in the CVCC. The first is applying electrical heaters on the walls of the chamber, and the second is by combusting a mixture of gases i.e. premixed combustion under lean

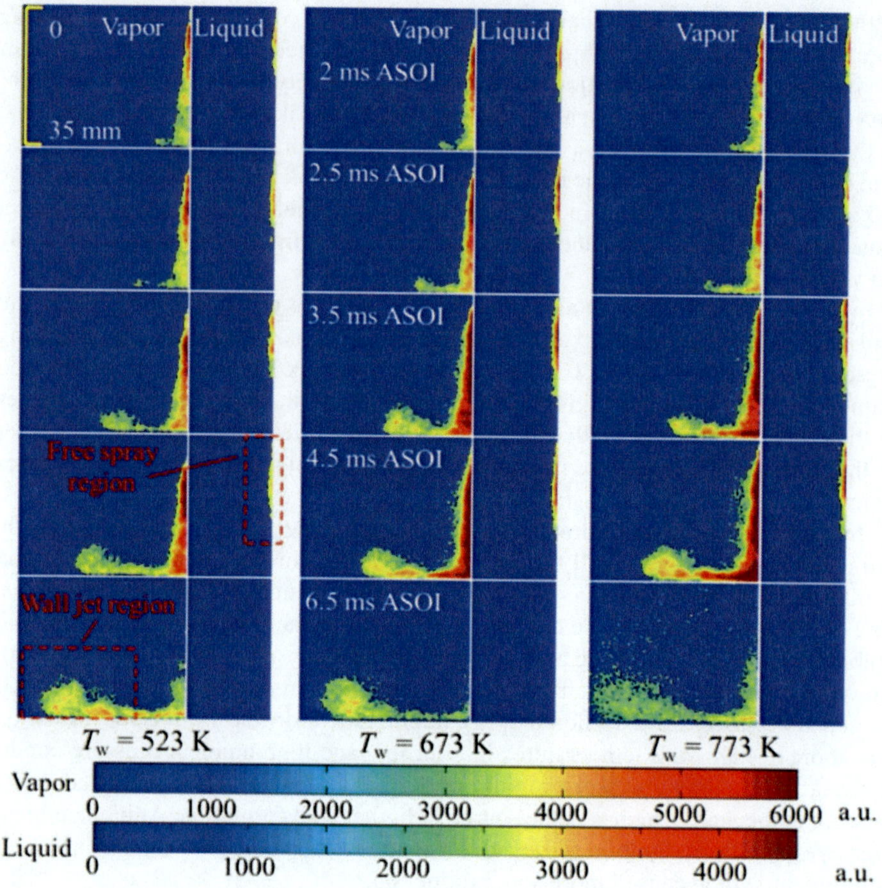

Fig. 5.25 Fluorescence signal distribution for vapour phase and liquid phase ($T_a = 773$ K, $P_a =$ 4 MPa, $P_i = 60$ MPa) (Chen et al. 2019)

conditions. The details of these methods have already been covered comprehensively in the preceding sections of this chapter.

Researchers have investigated the effect of different fuels and operating parameters on the spray combustion studies. For applications in Low-Temperature Combustion (LTC) in diesel engines, Vu and Ocktaeck (2019) studied the ignition and combustion characteristics of GB20 (20% biodiesel and 80% gasoline) blends in a CVCC by varying the ambient conditions and FIP. The ignition and combustion parameters, ignition delay, and flame lift-off length (FLOL) were measured by broadband chemiluminescence imaging using a high-speed camera. Figure 5.26 shows the sequence of spray combustion flame evolution. It was observed that the ignition was delayed and flame lift-off length increased with increasing FIP. At higher FIP, a higher number of spray droplets were formed. Hence, the local fuel–air mixture temperature decreased upon vaporization of these droplets, causing an increase in ignition delay.

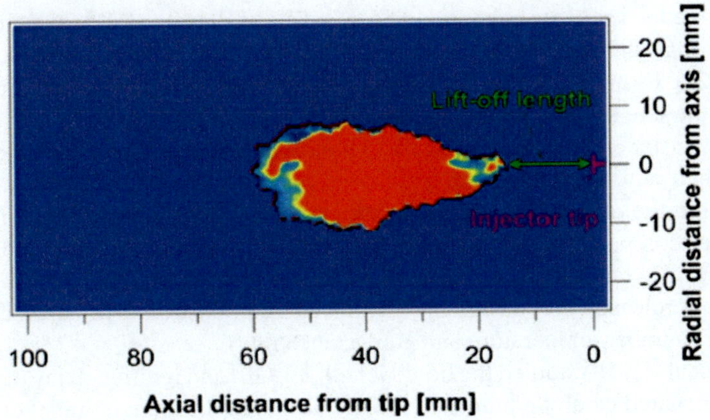

Fig. 5.26 Broadband chemiluminescence imaging showing Lift-off length (Vu and Ocktaeck 2019)

Higher momentum of these droplets at these high FIPs caused increase in the FLOL. Increasing the injection duration increased the amount of combustible mixture in the CVCC, creating a reduction in local temperature and hence increasing the ignition delay. This increase in ignition delay gave more time for the spray to travel downwards, causing an increase in FLOL. Increase in ambient temperature and ambient density decreased the combustion duration and consequently, the FLOL, which is a consistent observation in line with other studies. Upon increasing biodiesel content in the biodiesel–diesel blend, the ignition delay and FLOL increased. The effect of oxygen percentage in the fuel mixture was investigated, and it was found that at lower oxygen percentage, local equivalence ratio achieved the flammability limit slower, causing an increase in ignition delay and FLOL.

In another work, combustion characteristics of diesel injection with pilot methanol (DIPM) were studied and compared with conventional diesel injection (CDI) by Wang et al. (2019) using natural luminescence technique in a CVCC. It was observed that the ignition delay first decreased and then increased, when diesel FIP exceeded 120 MPa. The increase in the later half was due to consumption of OH radicals by methanol, and this was more prominent at higher diesel FIPs. At lower FIP, due to improper mixing of diesel and methanol, the inhibition effect of methanol was not prominent. In the case of CDI, the ignition delay continuously decreased with an increasing FIP because of superior sprayatomization and better air entrapment. The same inhibition effect of methanol was found in the study of FLOL. At high FIPs of diesel (140 MPa), there was a reduction in the FLOL because the inhibition effect of methanol was so high that ignition could only happen in the diesel-rich region. Hence, the flame appeared close to diesel nozzle though the ignition delay increased. The FLOL of both injection strategies first increased and then decreased with time in an injection. Spatial Integration (SINL) obtained by summing up of the pixel value of every single natural luminescence image was used to obtain amount and temperature of soot. The SINL of both the injection strategies decreased with increasing diesel FIP, indicating better spray atomization. At lower FIP of diesel, SINL values of DIPM were very high and appeared faster. The SINL of DIPM dropped faster than CDI after reaching its maximum value. They concluded that late combustion of methanol enhanced the rate of soot oxidation.

Further, to understand the combustion kinetics of blends of n-heptane and butanol isomers, their auto-ignition characteristics were studied in a CVCC by Liang et al. (2019). From the pressure and the heat release curve, their ignition and combustion delays were compared. It was found that auto-ignition tendency decreased because of addition of butanol to n-heptane, which was evident from the delay in pressure rise and heat release curves. Further, ignition and combustion delay decreased with increasing chamber temperature and equivalence ratio.

Zhong et al. (2019) studied the liquid length, FLOL, and ignition delay of gasoline and hydrogenated catalytic biodiesel (HCB) blends in different ambient conditions for their applicability in GCI engines. They simulated in-cylinder conditions in a CVCC and performed experiments using Mie-scattering and natural luminosity. It was observed that the liquid spray penetration decreased with increasing oxygen percentage and ambient temperature. The liquid length decreased with blending. This proved that these blends expand the usage of GCI to small engines. As HCB proportion increased in the blend, ignition delay decreased because of higher cetane number of biodiesels. Also, with increasing temperature and oxygen percentage, the ignition delay further reduced because the fuel could evaporate and form combustible mixture quickly. Blending decreased FLOL because HCB blends reduced the autoignition temperature and the ignition delay. Also, G50H50 with 15% and 21% oxygen showed similar properties as that of diesel and it could be an excellent replacement to diesel in compression ignition (CI) engines. Moreover, at higher exhaust gas recirculation (EGR) (oxygen percentage < 10%), blends exhibited higher cycle-to-cycle variations, and hence the combustion was unstable under high EGR conditions.

Though multiple injection strategies in a diesel engine are studied by many researchers to access its feasibility in achieving Homogeneous Charge Compression Ignition (HCCI) combustion, controlling combustion phasing and making successful combustion still require further investigations. Hence, in an attempt to understand the phenomena involving interactions of subsequent injections and their resulting combustion and soot characteristics, Cao et al. (2020) used broadband luminosity to study the injection delay and measured soot properties. They used diffused back illumination extinction imaging at different ambient conditions for double injection strategies in a CVCC. In double injection strategy, second injection had lower ignition delay and FLOL with large combustion area due to high-temperature conditions created by the first spray. This also gave rise to more homogeneous mixture formation, which was useful in reduction in soot formation. Flame propagation rate of the second injection was faster than the first and also because of the oxidation of fuel droplets in the spraytips, spray width of the second injection tended to be narrower than the first. The soot formation from the second injection increased at low ambient temperature conditions. However, increased amount of pilot fuel injection quantity showed negligible effect on the total soot mass produced, though the initial sooting location came closer to the nozzle. In another study, researchers (Hwang et al. 2017) found that multiple injections strategy improved engine combustion in cold-start conditions. However, it increased the fluid flow towards the crevices, which could possibly increase unburned hydrocarbons, if it misfires. Therefore, it could be concluded that spray optimization is necessary for bringing out the best compromise.

From these studies, it can be inferred that the spray and its ignition analysis in CVCC provides an insight into the fundamental processes related to spray evolution as well as combustion, which could be utilized in optimizing the fuel injection parameters and strategies.

5.5 Conclusions

Development of advanced IC engine technologies along with adaptation of newer alternate fuels demands an improved understanding of fundamental processes/phenomenon in the engine cylinder right from ‘start of injection’ to the ‘end of combustion’. Key factor affecting this overall process is ‘quality of spray formation’. Fuel spray governs various processes such as fuel–air mixture quality, time scale of chemical reactions during delay period and combustion, heat release rate, formation of pollutants, etc. This chapter provides an overview of spray breakup processes, various chamber designs, and various optical measurement techniques used for spray investigations. Detailed discussions on understanding the effects of various operating parameters on spray and combustion characteristics have also been incorporated with the help of several recent case studies.

CVCCs are considered to be most important in spray studies. These chambers provide flexibility in maintaining different thermo-physical conditions required for spray investigations. Different spray-related phenomena such as spray morphology, wall interactions, flame analysis and emission species formation can be investigated in these chambers. Two types of CVCC designs were discussed, which included premixed combustion type and heater type. Features of various spray measurement techniques were described, along with their working principles and applications. Optical visualization techniques such as Shadowgraph, Schlieren and Mie-scattering are mainly used for determining macroscopic spray characteristics. PDPA/PDI techniques determine microscopic characteristics of the spray, such as droplet velocity, and droplet size distributions. PIV helps understand the effect of spray on the air motion in the cylinder, while PLIF provides the information about fuel–air mixture distribution. Near nozzle dense flow region is mainly analyzed by XPCI. Natural luminosity and chemiluminescence from spray flames are used mainly for studying the flame development and formation of pollutants and radicals.

The next section focuses on understanding the influence of different operating parameters such as FIP, injection strategy, injector design, fuel type, ambient condition, etc. on the microscopic, macroscopic, in-nozzle flows, spray-wall interactions and spray combustion characteristics. Atomization of spray significantly improves with increasing FIP, use of lower viscosity fuels and higher ambient temperature conditions. However, these parameters are further optimized so as to obtain adequate spray penetration and wider dispersion so that maximum coverage of the combustion chamber can be achieved without causing spray impingement on the walls. In addition, nozzle's internal flows for different nozzle geometries, cavitating flows, multiple injection strategy, and flash boiling conditions were discussed. Cavitation bubbles were improved the spray breakup in the near nozzle spray regime. Different regimes of spray–wall interactions and optimization of operating parameters according to applications were discussed, where enhancement of secondary evaporation from the wall-film formation was required and minimization of the spray impingement was needed to limit the soot and unburnt HC emissions. Finally, factors affecting spray combustion characteristics such as ignition delay, FLOL, flame-wall interactions, soot generation, etc. were discussed in detail. In summary, a thorough review of recent studies on various aspects related to spray behaviour for its better understanding has been attempted in this chapter.

5.6 Way-Forward

Spray studies have played an important role in the development of IC engines in the past and are playing an essential role in refining this technology for achieving further higher thermal efficiency and reduced emissions at the same time. Although the development of high-end laser-based optical diagnostics tools has helped in obtaining far better understanding of highly dynamic and complex phenomenon of spray formation, complete understanding of fuel sprays in an engine environment still remains elusive. There is a lot left to investigate and understand in this broad field of spray research related to IC engines. Fuel–air mixture formation in an IC engine is a complex phenomenon involving numerous synergistic factors and spray studies throws some light in understanding some of these phenomena. Few areas still remain unclear e.g. surface disturbances due to cavitation, and turbulence in liquid jet leading to disintegration of spray droplets are some of the challenging aspects to be understood physically. There is a highly dense spray region near the nozzle exit, which remains hard to investigate experimentally and still remains largely opaque. Internal spray mechanisms responsible for mixture formation and combustion are also not fully understood. Numerical models give improved insights into different processes occurring in an engine cylinder in milliseconds time-scale, and more comprehensive spray studies are required to develop better and more accurate models. Hence, further understandings of the spray fundamentals are required for future developments in fuel injection systems and engine combustion technology for use in more advanced IC engines of the future.

References

Accessed from presentation on "ECN (Engine Combustion Network) Progress in KAIST (Korea Advanced Institute of Science & Technology) by Choongsik Bae

Agarwal AK, Dhar A, Gupta JG, Kim WI, Choi K, Lee CS, Park S (2015) Effect of fuel injection pressure and injection timing of Karanja biodiesel blends on fuel spray, engine performance, emissions and combustion characteristics. Energy

Allocca L, De Vita A, Di Angelo L (2002) Wall-impingement analysis of a spray from a common rail injection system for diesel engines. In THIESEL 2002 Conference on Thermo-and Fluid Dynamic Processes in Diesel Engines (pp 67–76)

Baert RS, Frijters PJ, Somers B, Luijten CC, de Boer W (2009) Design and operation of a high pressure, high temperature cell for HD diesel spray diagnostics: guidelines and results (No. 2009-01-0649). SAE Technical Paper

Bai C, Gosman AD (1995) Development of methodology for spray impingement simulation. SAE transactions, pp 550–568

Bao Y, Chan QN, Kook S, Hawkes E (2014) Spray penetrations of ethanol, gasoline and iso-octane in an optically accessible spark-ignition direct-injection engine. SAE Int J Fuels Lubr 7(3):1010–1026

Baumgarten C (2006) Mixture formation in internal combustion engines. Springer Science & Business Media

Behringer M, Aleiferis P, OudeNijeweme D, Freeland P (2014) Spray formation from spark-eroded and laser-drilled injectors for DISI engines with gasoline and alcohol fuels. SAE Int J Fuels Lubr 7(3):803–822

Bougie B, Tulej M, Dreier T, Dam NJ, Ter Meulen JJ, Gerber T (2005) Optical diagnostics of diesel spray injections and combustion in a high-pressure high-temperature cell. Appl Phys B 80(8):1039–1045

Cao J, Leng X, He Z, Wang Q, Shang W, Li B (2020) Experimental study of the diesel spray combustion and soot characteristics for different double-injection strategies in a constant volume combustion chamber. J Energy Inst

Cao T, He Z, Zhou H, Guan W, Zhang L, Wang Q (2020b) Experimental study on the effect of vortex cavitation in scaled-up diesel injector nozzles and spray characteristics. Exp Thermal Fluid Sci 113:

Chang M, Park JH, Kim HI, Park S (2020) Flash boiling macroscopic spray characteristics of multi-hole direct injection injectors with different hole arrangement. Appl Therm Eng 170:

Chen Z, Yao A, Yao C, Yin Z, Xu H, Geng P, Dou Z, Hu J, Wu T, Ma M (2017) Effect of fuel temperature on the methanol spray and nozzle internal flow. Appl Therm Eng 114:673–684

Chen B, Feng L, Wang Y, Ma T, Liu H, Geng C, Yao M (2019) Spray and flame characteristics of wall-impinging diesel fuel spray at different wall temperatures and ambient pressures in a constant volume combustion vessel. Fuel 235:416–4

Crua C, Heikal MR, Gold MR (2015) Microscopic imaging of the initial stage of diesel spray formation. Fuel 157:140–150

Dec JE, Espey C (1995) Ignition and early soot formation in a DI diesel engine using multiple 2-D imaging diagnostics. SAE transactions, pp 853–875

Dec JE, Espey C (1998) Chemiluminescence imaging of autoignition in a DI diesel engine. SAE transactions, pp 2230–2254

Duronio F, De Vita A, Allocca L, Anatone M (2020) Gasoline direct injection engines–a review of latest technologies and trends. Part 1: Spray break-up process. Fuel, 265, p 116948

Falgout Z, Linne M (2015) Cavitation inside high-pressure optically transparent fuel injector nozzles. In Journal of Physics: Conference Series (Vol. 656, No. 1, p. 012082). IOP Publishing

Gaydon A (2012) The spectroscopy of flames. springer science & business media

Geng L, Wang Y, Wang Y, Li H (2019) Effect of the injection pressure and orifice diameter on the spray characteristics of biodiesel. J Traffic Transp Eng (English Edition)

Goldstein R (2017) Fluid mechanics measurements. Routledge

Gupta JG, Agarwal AK (2016). Macroscopic and microscopic spray characteristics of diesel and Karanja biodiesel blends (No. 2016-01-0869). SAE Technical Paper

Han Z, Xu Z, Trigui N (2000) Spray/wall interaction models for multidimensional engine simulation. Int J Engine Res 1(1):127–146

Heywood JB (1988) Combustion engine fundamentals. 1ª Edição. Estados Unidos

https://www.artium.com/pdi-theory-of-operation

https://exact-sciences.m.tau.ac.il/sites/exactsci.tau.ac.il/files/media_server/Exact_Science/Chemistry/laboratories/3.%20laser%20induced%20fluorescence.pdf

http://aim2.dlr.de/measurement-techniques/particle-image-velocimetry-piv/index.html

Hwang J, Bae C, Patel C, Agarwal RA, Gupta T, Agarwal AK (2017a) Investigations on air-fuel mixing and flame characteristics of biodiesel fuels for diesel engine application. Appl Energy 15(206):1203–13

Hwang J, Park Y, Kim K, Lee J, Bae C (2017b) Improvement of diesel combustion with multiple injections at cold condition in a constant volume combustion chamber. Fuel 197:528–540

Jeon J, Moon S (2018) Ambient density effects on initial flow break-up and droplet size distribution of hollow-cone sprays from outwardly-opening GDI injector. Fuel 211:572–581

Kawanabe H, Kondo C, Kohori S, Shioji M (2010) Simultaneous measurements of velocity and scalar fields in a turbulent jet using PIV and LIF. J Environ Eng 5(2):231–239

Kawano D, Senda J, Kawakami K, Shimada A, Fujimoto H. Fuel design concept for low emission in engine systems 3rd report: analysis of spray characteristics for mixed fuels. SAE Technical paper; 2002 Mar 4

Lee S, Park S (2014) Experimental study on spray break-up and atomization processes from GDI injector using high injection pressure up to 30 MPa. Int J Heat Fluid Flow 45:14–22

Lee MY, Lee GS, Kim CJ, Seo JH, Kim KH (2018) Macroscopic and microscopic spray characteristics of diesel and gasoline in a constant volume chamber. Energies 11(8):2056

Lee Z, Kim D, Park S (2020) Effects of spray behavior and wall impingement on particulate matter emissions in a direct injection spark ignition engine equipped with a high pressure injection system. Energy Convers Manag 213:

Lefebvre AH, McDonell VG (2017) Atomization and sprays. CRC press

Lei Y, Liu J, Qiu T, Mi J, Liu X, Zhao N, Peng G (2019) Effect of injection dynamic behavior on fuel spray penetration of common-rail injector. Energy 188:

Lešnik L, Kegl B, Bombek G, Hočevar M, Biluš I (2018) The influence of in-nozzle cavitation on flow characteristics and spray break-up. Fuel 222:550–560

Li K, Nishida K, Ogata Y, Shi B (2015) Effect of flat-wall impingement on diesel spray combustion. Proc Inst Mech Eng Part D J Automob Eng 229(5):535–549

Li Y, Guo H, Ma X, Wang J, Xu H (2016) Droplet dynamics of DI spray from sub-atmospheric to elevated ambient pressure. Fuel 179:25–35

Li Y, Huang Y, Yang S, Luo K, Chen R, Tang C (2019) A comprehensive experimental investigation on the PFI spray impingement: effect of impingement geometry, cross-flow and wall temperature. Appl Therm Eng 159:

Liang G, Mudawar I (2016) Review of mass and momentum interactions during drop impact on a liquid film. Int J Heat Mass Transf 101:577–599

Liang X, Zhong A, Sun Z, Han D (2019) Autoignition of n-heptane and butanol isomers blends in a constant volume combustion chamber. Fuel 254:

López JJ, García-Oliver JM, García A, Domenech V (2014) Gasoline effects on spray characteristics, mixing and auto-ignition processes in a CI engine under Partially Premixed Combustion conditions. Appl Therm Eng 70(1):996–10

Luo H, Nishida K, Uchitomi S, Ogata Y, Zhang W, Fujikawa T (2018) Microscopic behavior of spray droplets under flat-wall impinging condition. Fuel 219:467–476

Marmanis H, Thoroddsen ST (1996) Scaling of the fingering pattern of an impacting drop. Phys Fluids 8(6):1344–1346

Mohan B, Yang W, Tay KL, Yu W (2014) Macroscopic spray characterization under high ambient density conditions. Exp Therm Fluid Sci 59:109–117

Moita AS, Moreira ALN (2007) Drop impacts onto cold and heated rigid surfaces: morphological comparisons, disintegration limits and secondary atomization. Int J Heat Fluid Flow 28(4):735–752

Moreira ALN, Moita AS, Panao MR (2010) Advances and challenges in explaining fuel spray impingement: How much of single droplet impact research is useful? Prog Energy Combust Sci 36(5):554–580

Munsin R, Shing BCL, Phunpheeranurak K, Phongphankasem T, Laoonual Y, Jugjai S, Chanchaona S (2013) Design of constant volume combustion chamber (CVCC) with Pre-Combustion Technique for Simulation of CI Engine Conditions. In 4th TSME In

O'Rourke PJ, Amsden AA (2000) A spray/wall interaction submodel for the KIVA-3 wall film model. SAE transactions, pp 281–298

Ofner B (2001) Phase doppler anemometry (PDA). In Optical Measurements (pp 139–152). Springer, Berlin, Heidelberg

Oren DC, Wahiduzzaman S, Ferguson CR (1984) A diesel combustion bomb: proof of concept. SAE transactions, pp 945–960

Payri R, García A, Domenech V, Durrett R, Plazas AH (2012) An experimental study of gasoline effects on injection rate, momentum flux and spray characteristics using a common rail diesel injection system. Fuel 97:390–399

Peters N (2001) Turbulent combustion

Pierpont DA, Reitz RD (1995) Effects of injection pressure and nozzle geometry on DI diesel emissions and performance. SAE transactions, pp 1041–1050

Reitz RD (1996) Computer modeling of sprays. Spray Technology Short Course, Pittsburgh, PA

Reitz RD, Bracco FV (1982) Mechanism of atomization of a liquid jet. Phys Fluid 25(10):1730–1742

Rioboo R, Tropea C, Marengo M (2001) Outcomes from a drop impact on solid surfaces. At Sprays 11(2)

Salvador FJ, Gimeno J, De la Morena J, Carreres M (2018) Comparison of different techniques for characterizing the diesel injector internal dimensions. Exp Tech 42(5):467–472

Sankar SV, Maher KE, Robart DM, Bachalo WD (1999) Rapid characterization of fuel atomizers using an optical patternator

Shao J, Yan Y, Greeves G, Smith S (2003) Quantitative characterization of diesel sprays using digital imaging techniques. Meas Sci Technol 14(7):1110

Sharma N, Agarwal AK (2016) Microscopic and macroscopic spray characteristics of GDI injector using gasohol fuels at various injection pressures (No. 2016-01-0868). SAE Technical Paper

Sharma N, Bachalo WD, Agarwal AK (2020) Spray droplet size distribution and droplet velocity measurements in a firing optical engine. Phys Fluids 32(2):

Soid SN, Zainal ZA (2011) Spray and combustion characterization for internal combustion engines using optical measuring techniques–a review. Energy 36(2):724–741

Sonawane U, Kalwar A (2020) Microscopic and macroscopic spray characteristics of gasohols using a port fuel injection system (No. 2020-01-0324). SAE Technical Paper

Stanton DW, Rutland CJ (1996) Modeling fuel film formation and wall interaction in diesel engines. SAE transactions, pp 808–824

Stiesch G (2003). Modeling engine spray and combustion processes. Springer Science & Business Media

Su TF, Chang CT, Reitz RD, Farrell PV, Pierpont AD, Tow TC (1995) Effects of injection pressure and nozzle geometry on spray SMD and DI emissions. SAE transactions, pp 975–984

Suh HK, Park SW, Lee CS (2007) Effect of piezo-driven injection system on the macroscopic and microscopic atomization characteristics of diesel fuel spray. Fuel 86(17–18):2833–2845

Tropea C, Yarin AL (2007) Springer handbook of experimental fluid mechanics. Springer Science & Business Media

Vu DN, Ocktaeck LIM (2019) Ignition and combustion characteristics of gasoline-biodiesel blend in a constant volume chamber: effects of the operation parameters. Fuel 255:

Wang Z, Xu H, Jiang C, Wyszynski ML (2016a) Experimental study on microscopic and macroscopic characteristics of diesel spray with split injection. Fuel 174:140–152

Wang Z, Jiang C, Xu H, Wyszynski ML (2016b) Macroscopic and microscopic characterization of diesel spray under room temperature and low temperature with split injection. Fuel Process Technol 142:71–85

Wang Z, Wang B, Jiang C, Xu H, Badawy T (2016c) Microscopic characterization of iso-octane spray in the near field under flash boiling condition. Appl Energy 180:598–606

Wang Z, Ding H, Ma X, Xu H, Wyszynski ML (2016d) Ultra-high speed imaging study of the diesel spray close to the injector tip at the initial opening stage with single injection. Appl Energy 165:335–344

Wang Z, Li Y, Wang C, Xu H, Wyszynski ML (2016e) Near-nozzle microscopic characterization of diesel spray under cold start conditions with split injection strategy. Fuel 181:366–375

Wang Z, Dai X, Li F, Li Y, Lee CF, Wu H, Li Z (2018a) Nozzle internal flow and spray primary break-up with the application of closely coupled split injection strategy. Fuel 228:187–196

Wang Z, Dai X, Liu F, Li Z, Wu H (2018b) Break-up of fuel sprays under cavitating and flash boiling conditions. Appl Therm Eng 143:22–33

Wang Z, Li Y, Guo H, Wang C, Xu H (2018c) Microscopic and macroscopic characterization of spray impingement under flash boiling conditions with the application of split injection strategy. Fuel 212:315–325

Wang Y, Wang H, Meng X, Tian J, Wang Y, Long W, Li S (2019a) Combustion characteristics of high pressure direct-injected methanol ignited by diesel in a constant volume combustion chamber. Fuel 254:

Wang Z, Dai X, Liu F, Li Y, Wu H, Wang C, Li Y (2019b) Microscopic study on the mechanisms for formation of the initial spray morphology. Fuel 235:715–722

Wang Z, Dai X, Liu F, Li Z, Li F (2019c) Microscopic characterization of spray impingement under flash boiling conditions. J Energy Inst 92(3):640–652

Wierzba A (1990) Deformation and break-up of liquid drops in a gas stream at nearly critical Weber numbers. Exp Fluids 9(1–2):59–64

Wierzba A, Takayama K (1988) Experimental investigation of the aerodynamic break-up of liquid drops. AIAA J 26(11):1329–1335

Wu S, Xu M, Hung DL, Pan H (2017) Effects of nozzle configuration on internal flow and primary jet break-up of flash boiling fuel sprays. Int J Heat Mass Transf 110:730–738

Xu M, Hung DLS, Yang J, Wu S (2015) Flash-boiling spray behavior and combustion in a direct injection gasoline engine. In Proceedings of the Australian Combustion Symposium, Melbourne, Australian

Yang S, Ma Z, Li X, Hung DL, Xu M (2020) A review on the experimental non-intrusive investigation of fuel injector phase changing flow. Fuel 259:

Yarin AL, Weiss DA (1995) Impact of drops on solid surfaces: self-similar capillary waves, and splashing as a new type of kinematic discontinuity. J Fluid Mech 283:141–173

Zama Y, Odawara Y, Furuhata T (2017) Experimental investigation on velocity inside a diesel spray after impingement on a wall. Fuel 203:757–763

Zhang G, Xu M, Zhang Y, Zhang M, Cleary DJ (2012) Macroscopic characterization of flash-boiling multi-hole sprays using planar laser induced exciplex fluorescence technique. Part I. On-axis spray structure. At Sprays 22(10)

Zhang M, Xu M, Hung DL (2014) Simultaneous two-phase flow measurement of spray mixing process by means of high-speed two-color PIV. Meas Sci Technol 25(9):

Zhao H ed (2009) Advanced direct injection combustion engine technologies and development: diesel engines (vol 2). Elsevier

Zhao H, Ladommatos N (1998) Optical diagnostics for in-cylinder mixture formation measurements in IC engines. Prog Energy Combust Sci 24(4):297–336

Zhong W, Li B, He Z, Xuan T, Lu P, Wang Q (2019) Experimental study on spray and combustion of gasoline/hydrogenated catalytic biodiesel blends in a constant volume combustion chamber aimed for GCI engines. Fuel 253:129–138

Part III
Technologies for Performance Improvement and Emissions Reduction

Chapter 6
Efficiency Improvement of Internal Combustion Engines Over Time

Sarthak Baweja and Rajan Kumar

Abstract The demand for more efficient engines is increasing gradually and will continue to rise in the coming decades. In the present era, with the continuous rise in a global energy crisis and increasing ecological awareness, waste heat recovery became a common concern in various sectors of energy. The goal of this chapter is to address the role of waste heat recovery methods such as turbocharging, turbo-compounding, organic Rankine cycle, and thermoelectric generators, to enhance the thermal efficiency of the internal combustion (IC) engine over the past few decades. The maximum efficiency achieved in turbocharging is 44.1%, which is more than the conventional engines by 9.1–14.1%. The efficiency of turbo-compounding can be brought closer to 50%. The concept of compounded Rankine cycle increases the combined engine and waste heat recovery efficiency almost by 10%. The highest increment in the efficiency of the IC engine can be achieved with the thermoelectric generators which have tremendously increased the efficiency by almost 15–20%. This chapter provides some of the important basic information on IC engines and demonstrates some of the latest advanced technologies, such as engine downsizing, advance engine controls, variable valve timing, variable geometry engine design, advanced fuel injection, advanced compression ignition engines, advanced spark-ignition engines, alternative fuels that keep IC engines competitive due to their ability to improve the fuel economy and better performance of IC engines with near-zero emissions. Furthermore, the chapter presents opportunities, challenges, and technical barriers related to the future areas of IC engines.

Keywords Internal combustion engine · Thermal efficiency · Waste heat recovery techniques · Advanced technologies

---

S. Baweja · R. Kumar (✉)
Department of Mechanical Engineering, Dr B R Ambedkar National Institute of Technology, Jalandhar 144011, Punjab, India
e-mail: rajank@nitj.ac.in

A. P. Singh and A. K. Agarwal (eds.), *Novel Internal Combustion Engine Technologies for Performance Improvement and Emission Reduction*, Energy, Environment, and Sustainability, https://doi.org/10.1007/978-981-16-1582-5_6

6.1 Introduction

6.1.1 Current Scenario of Energy Utilization by IC Engines

The current utilization of fossil fuel in running the IC engines is 25% of the world's power, i.e., about 3000 out of 13,000 million tons of oil equivalent per year release about 10% of the world's greenhouse gases (GHG) emissions (Reitz et al. 2020). There are lots of efforts been made to reduce the dependency on fossil fuels due to environmental issues and also to reduce the oil import which is the cause of trade deficit in India. There are lots of efforts to make electrical vehicles to further reduce the consumption of fossil fuels as well as to reduce GHG emissions for environmental aspects, but researchers need to brainstorm that there is still potential in advanced IC engines. There is still no perfect explanation of flame speed in SI engines with respect to engine rpm (Reitz 2013). The practical limitations of engines still need some refinement and have lots of research scope (Splitter et al. 2013). To begin with, on the advancements in the IC engines, let us first discuss traditional IC engines and different types of IC engines and their working and efficiencies. Furthermore, energy and exergy distributions of various transfer and destruction seen in IC engines are presented for more insight into them.

6.1.1.1 Conventional IC Engines

Conventional IC engines have been the primary propulsion system in vehicles since the turn of the 20th century and as the world's primary powertrain technology, they continue to be used today. Basically, these are classified into two categories. One is SI engines, which are spark ignition engines and the other is CI engines which are compression ignition engines. Both these types of engines have their advantages and disadvantages. SI engine uses petrol, ethanol blends, LPG, or natural gas as a fuel which is ignited by the spark plug premixed earlier, i.e., the mixture of fuel and air at the correct time. On the other hand, the CI engine uses mineral diesel and biodiesel as fuel, and the compression ratio of SI engines is low as compared to CI engines.

The octane number (ON) represents the resistance of gasoline and other fuels used in SI engines to premature detonation when exposed to heat and pressure in the combustion chamber. Premature detonation is indicated by knocking or ringing noises that occur when the engine is operating. If such noise is caused by an engine running on a particular gasoline, it can be reduced or eliminated by using a gasoline with higher octane number. The ON of the fuel sample shall be determined by gasoline combustion in the SI engine under controlled conditions, e.g., compression ratio, spark timing, the engine speed, and the load, until a standard level of knock happens. The ON is the number that gives the amount, by volume, of iso-octane in a mixture of iso-octane and normal heptane that will have the same anti-knocking capacity as the fuel under consideration (Viswanathan 2017). A high tendency to auto-ignite, or a low octane rating, is undesirable for a SI engine but is desirable for a CI engine. A diesel fuel with a high cetane number (CN) has a high tendency to auto-ignite. The fuel which has high CN has low ON and vice-versa, hence diesel fuel is not used in SI engines and vice-versa. The CN is an empirical parameter related to the ignition delay (ID) time of diesel fuel, determined through the standard tests based on the ASTM D613 standard (Annual Book of ASTM Standards 1980). The ID is the time interval between the start of injection and the start of combustion. The ID period begins with fuel injection and includes physical and chemical delay periods until auto-ignition occurs (Taşkiran and Ergeneman 2011). Fuels with a low CN have a large ID time and most of the injected fuel is collected in the cylinder before auto-ignition. This leads to fast combustion. In severe situations, this creates an audible knocking sound referred to as "diesel knock". Fuels with a high CN have a very short ID time, i.e., ignition happens within a very short period after injection, and very little fuel is injected before auto-ignition. The heat release rate is controlled by the rate of fuel injection and fuel–air mixing—smoother engine operation. The higher the CN, the better the ignition quality, i.e., shorter ID. CI engines are much more efficient than gasoline engines as they convert 45% of the fuel's chemical energy into mechanical work, compared to only 30% by SI engines (EERE U 2003). Also, we know that lots of work have already been done on biodiesel that is found to be very costly as compared to our conventional fuels besides having lesser emissions, but it is not uniformly distributed, so there are lots of infrastructure improvement required and also it is facing lots of problem in cold climate condition where biodiesel gets solidified and forms wax which is difficult to pump. So, there is a need to think differently, may be in terms of modifying the engines with advanced equipments which require the current information of the existing engines and it is more important to bring change in the existing one. Therefore, this study has been done on the advanced IC engine technologies, which are one of the best solutions to bring the change in the conventional one without modifying the fuel. One of the ways is to do by smart engine management systems which have tremendously enhanced the combustion quality and also reduced the toxic emissions with the help of exhaust treatment, majorly the GHG and a little bit increment in the efficiency too. Nowadays, turbochargers are also being used to enhance the fuel and air in the cylinders which have a direct impact on the performance and efficiency of the engine.

The fuel used by an IC engine has a great impact on the environment globally. It is found that when combustion occurs in a diesel engine at ideal conditions, three times more air is required than for petrol, or 43.5 kg (3 times 14.5 kg) air for each kg of diesel and the combustion is auto-ignited as a result of compression. After the combustion, the exhaust gas contains 3.18 kg of carbon dioxide (CO_2) which is one of the major GHG (Bernodusson 2018). In 2018, fossil fuel carbon emissions hit 10 gigatons per year (or just under 37 gigatons of CO_2) for the first time, more than double the level of the 1970s, and the average per capita emissions were 4.8 tons of fossil-fuel CO_2 per person (Levin and Lebling 2019).

The International Energy Agency has developed a technology roadmap to achieve a 30–50% reduction in fuel consumption per kilometer by 2030 for new road vehicles around the world and by 2050 for all vehicles on the road (International Energy Agency 2012). We all know that it requires lots of investment and time for a new technology to penetrate the market, as 2050 is almost 30 years from now but still the process of the intervention of new technology and its complete implementation on all vehicles on road is a challenging task. Hence, in order to find technological advances that maximize the efficiency of the engine, minimize emissions of the pollutants like CO_2, sulfur oxides (SO_x), etc., and optimize its capability to a wider variety of fuels in power generation and transportation systems is a challenging task for the automotive sector. So, the first step to do this is, to begin with, awareness of the properties of the fuel, compression ratio, parameters governing the air–fuel mixture, etc. (Heywood 1988).

6.1.1.2 Classification and Working of IC Engines

The classification of an IC engine is very vast. In order to make it simplified it is presented in the form of a paragraph. Based on the application these may be an automobile, aircraft, locomotive, marine engines, etc. Sometimes the design of the engines is also changing; hence on the basis of design, it may be reciprocating in which it is further classified as single and multi-cylinder in-line, V, radial, etc. In some applications, rotatory engines are used that may have the single or multi-rotor. The cycle on which the engine is operating is also one of the essential parameters to know. So, on the basis of the operation of the cycle, it may be Atkinson, Diesel, Dual, Miller, or Otto. Due to advancement in the technology, it may be naturally aspirated or turbocharged and may be operating on two-stroke or multi, i.e., four-stroke. Based on valve/port design and location, it may be a poppet valve or rotary valve and T, F, L head on the basis of location. It may be conventional, alternative, or blended on the basis of fuel. It may also be categorized as bi-fuel or dual-fuel. On the basis of mixture preparation, it is carburetion or fuel injection. The name SI engines and CI engines are itself on the basis of ignition, i.e., spark or compression. Based on stratification of charge, it may be homogeneous charge or a stratified charge with carburetion or fuel injection. It may be air-cooled or water-cooled. It may have a swirl chamber or pre-chamber based on chamber design for CI engines, or it may have a compound vortex-controlled chamber or other designs for SI engines (Heywood 1988). The working principle of SI and CI engines is almost the same. First, the charge is ignited either by the spark in case of SI engines or by compression in case of CI engines as mentioned above in the classification, the result of which is a tremendous amount of heat in the form of low-grade energy that causes the piston to reciprocate, and with the help of crankshaft this is converted into a high-grade form of energy, i.e., work in the form of rotary motion which make the vehicle to propel via a transmission system. Obviously, due to contact between various components, some of the energy gets lost in the friction which also affects efficiency.

6.1.1.3 Efficiency of IC Engines

To measure the performance of the given engine that has undergone a power cycle, may be Otto cycle or Diesel cycle or any other type of cycle, which can be measured in terms of the extent to which the energy added, generally in the form of heat gets converted into the network output of the cycle, is expressed in terms of a dimensionless ratio known as thermal efficiency. Generally, it is expressed by W_{net}/Q_{in}. Thermal efficiency is always less than 100%; this concept is applied to all the cycles whatever may be the operations included in it. Carnot cycle is an ideal concept which has maximum efficiency, since in Carnot cycle analysis heat is considered as the function of temperature. This is the standard cycle generally used to compare the efficiency of the other cycles (Moran et al. 2010). A typical CI engine operates around 30–35% of thermal efficiency. During the effective conversion of low-grade energy, i.e., heat to high-grade energy—this work is still a research topic—it has been found that 65–70% of the heat is wasted, i.e., not converted into work. It is also found that in more than 50% efficient engines, i.e., low-speed diesel engines used in ships, the maximum recorded efficiency is 51.7% for the largest diesel engines (Thermal Efficiency for Diesel Cycle 2021).

The various factors which influence and deviate the real engine from the ideal one that is operating on the Otto cycle are internal friction and incomplete combustion. All these are forms of irreversibility (Cengel and Boles 2007). The first basic parameter which affects the efficiency of the SI engine is the compression ratio. It is found that the curve of efficiency flats at a compression ratio of 8. Moreover, higher compression ratios auto-ignite the fuel, which causes engine knocking. Due to the fixed value of specific heat ratio of air, i.e. 1.4, this parameter is fixed; otherwise, this also affects the engine efficiency. For example, monoatomic gases have higher efficiencies because of the higher specific heat ratio, i.e., 1.67. The thermal efficiency of actual SI engines ranges from 25 to 30% (Cengel and Boles 2007). Efficiency can be analyzed easily by a proper understanding of the energy and exergy distribution, which is briefly discussed in the next section.

6.1.1.4 Energy and Exergy Distribution of Various Transfer and Destruction

The first basic idea of the available energy to do the useful work also known as "exergy" was introduced by Keenan (1951). The concept of exergy is the useful amount of work that is produced by the system until it reaches equilibrium condition with the environment. The main reason behind the loss of useful work is entropy and it is defined as the disordered form of energy that cannot be converted into useful work.

Entropy generation is the main cause for the exergy destruction as seen in IC engines due to which lot of energy which has the potential to be converted into useful work gets wasted, and the main reasons behind this are friction, combustion, expansion through the unrestricted passage, and transfer of energy due to the finite-difference in the temperatures. The heat transfer which is viewed as loss still has the potential to be converted into work. The crystal-clear analysis by which the energy and exergy get affected is done by Caton (2000) and Rakopoulos and Giakoumis (Rakopoulos and Giakoumis 2006). It was found that 30.7% of the fuel energy is converted into useful work without exhaust gas recirculation (EGR) and it is 31.25% with 20% EGR. Almost 29% of energy is wasted in the coolant and 40% of energy lost in the exhaust which is due to underutilization of thermal energy which has the potential to be converted into work in the engines operating without EGR. The effects of friction and brake fuel conversion efficiency are not inculcated in the analysis because friction offers no useful work as found by (Shyani and Caton 2009). The loss of exergy which is due to the high temperature of the exhaust gases can be recovered with the help of advanced technologies, i.e., wasted heat recovery system (Shyani and Caton 2009). Finally, it is concluded that there are three main causes of the destruction of the energy and these are the exchange of thermal energy among the various particles in the system, due to the diffusion of the fuel/oxidizer particles, and the last one is due to the mixing of the different product species (Dunbar and Lior 1994). Hence in order to enhance efficiency, two different ways, i.e., waste heat recovery techniques and advanced technologies, are discussed in detail below in the subsequent sections.

6.2 Efficiency Improvement with Waste Heat Recovery Techniques

The IC engine uses about one-third of the thermal energy produced during the combustion of the fuel. The leftover two-thirds of thermal energy is discarded from the engine through the exhaust, after-coolers, water-jackets, friction, and inter-coolers (Aladayleh and Alahmer 2015). Due to the lack of technological advancements, the thermal energy at low temperature appears to be difficult to recover. This causes a large amount of low-temperature thermal energy to be discharged because of its poor management. In the present era, with the continuous rise in a global energy crisis and increasing ecological awareness, waste heat recovery became a common concern in various sectors of energy. The recovery techniques for waste heat include turbocharging, turbo-compounding, organic Rankine cycle, and thermoelectric generators. Such techniques have shown significant increases in engine thermal efficiency, depending on system design, energy recovery quality, component performance, and implementation.

6.2.1 Turbocharging

Turbocharging is a process that is accomplished with the help of a device known as turbocharger. It is a device that increases the air pressure through compression by forcing the extra air in the combustion chamber which increases the volumetric efficiency and enhances the performance and efficiency of the engine (Karim 2000). The engine generates a large amount of exhaust gases carrying a massive amount of heat energy. This energy is usually wasted because it is emitted directly to the atmosphere. However, turbochargers are getting their power out of exhaust gases. This energy is converted into shaft power by a turbine coupled to a compressor with a shaft. The compressor pushes more air into the engine by compressing it. Usually, radial compressor and turbine are used, although some of the axial and mixed-flow models have been studied. Figure 6.1 indicates the direction of the fresh airflow through the compressor and the exhaust gas flow through the turbine. The compressor then compresses air and supplies it to the engine at high pressure. The turbine produces work from the engine exhaust gases which drive the compressor (Subramani et al. 2020).

In turbocharging experiment study, it is found that the boost pressure is varied from 1.65 to 3.3 bar and back pressure (BP) has also increased from 1.23 to 2.2 bar with the increase in the volumetric efficiency from 100 to 183% as well as an increase in the temperature from 125 to 156 °C. The boost pressures are the absolute pressure.

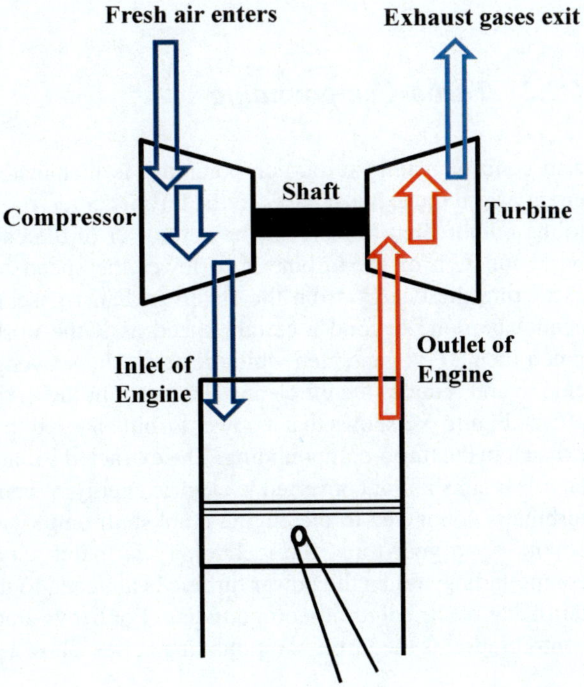

Fig. 6.1 Working of a turbocharger

This study showed that there is a reduction in the heat losses from 27.8 to 18.5% as the boost pressure varies from 1.65 to 3.3 bar and indicated mean effective pressure (IMEP) varied from 8.24 to 18.26 bar and increase in the work output from 29.3 to 32.8%, respectively. On the other hand, there is somewhat increment in the exhaust energy distribution, i.e., from 42.9 to 48.8%. On the production engine, as the IMEP varied from 14.45 to 28.41 bar, it was found that the work output varied from 39.9 to 42.2% and the heat energy distribution from 24.7 to 21.1% while keeping constant the range of boosting pressure, i.e., 1.65 to 3.3 bar. There is an increment in the exhaust gas distribution from 35.4 to 36.1%, respectively. The energy distribution in the above study is measured in % lower heating value (LHV). The efficiency varies from 39 to 42.8% when IMEP varies from 8.24 to 18.26 bar, whereas it varies from 40.5 to 44.1% for IMEP variation 14.45 to 28.41 bar on the production engine. This clearly shows that the maximum achieved efficiency is 44.1% which is more than the conventional engines by 9.1–14.1% which varies from 30 to 35% (Johnson 2015).

It is found that the combustion efficiency is almost greater than 95% but still less than 99%. This results in the formation of soot particles that increase as the IMEP is increased and reaches a maximum value, and then decreases. The soot value ranges from 0.05 to 0.08 g/hp-hr which is higher than the recommended value, i.e., 0.01 g/hp-hr (Taymaz 2006). CO_2 and carbon mono-oxide (CO) emissions are almost constant, whereas oxygen (O_2) emissions are increased with the increase in load. Unburnt hydrocarbon (UBHC) emissions decrease up to 12 bar IMEP and then increased drastically and nitrogen oxides (NO_x) reduced almost up to 16 bar and then increased.

6.2.2 *Turbo-Compounding*

The concept behind turbo-compounding is that instead of sending the recovered energy from the exhaust gases to the turbocharger, the energy is transferred directly to the output shaft to increase the net power of the engine but as we know that the basic function of the turbine is to lower the speed of the fast-moving stream by extracting the energy from the steam (exhaust gases) which has a limitation that cannot be done beyond a certain speed as if the upstream pressure increases too much then BP is generated which restricts the scavenging from the cylinder of the engine and reduces the efficiency (Facts about the Wright Turbo Compound (PDF) 1956). Figure 6.2 shows that a power turbine is used to extract extra energy from the exhaust in the turbo-compounding. The extracted exhaust energy can be added to the engine crankshaft or converted to electric energy. When the output shaft of the power turbine is connected to the engine crankshaft employing a mechanical connection, commonly a gear train, the technology is mainly stated to as mechanical turbo-compounding. When the power turbine is attached to the generator, the technique is termed as electrical turbo-compounding. For heavy-duty engines, the most essential configuration is the turbo-compounding series, shown schematically in Fig. 6.2.

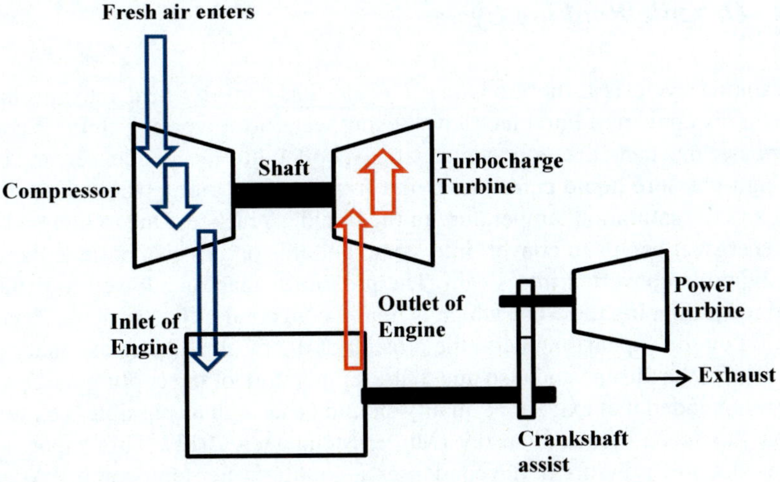

Fig. 6.2 Schematic representation of mechanical turbo-compounding

Turbo-compounding is more efficient than turbocharging because of the second exhaust turbine that allows for extra work (that would otherwise be lost) to be extracted from the exhaust. The turbo-compounding experiments were performed at constant boost pressure, i.e., 1.65 bar and four different BPs are used on the production engine, i.e., 1.57 bar, 1.88 bar, 2.09 bar, and 2.37 bar with constant volumetric efficiency and constant intake temperature, i.e., 100% and 125 °C respectively. It is seen that at the constant average IMEP of 9.5 bar and 100% volumetric efficiency and by varying the BP there is a reduction in the work output from 27 to 25% as the BP is increased from 1.57 to 2.37 bar and there is increment in the exhaust energy interaction and heat from 31.3 to 33.3% and 41.6 to 41.7%, respectively, which is minimal. As the IMEP is increased to 14.8 bar in the production engine, work output is increased from 41 to 41.8% whereas heat interaction increased from 26.9 to 28.6% and reduction in exhaust interaction from 32.1 to 29.6%. It was further found that for a single boosted turbo-compounded with volumetric efficiency 203%, intake temperature 125 °C, and BP of 2.32 bar the work output is increased from 21.9 to 42.6% with a reduction in heat and exhaust energy interaction variation from 36.2 to 23.9% and 44 to 33.5%, respectively, as the IMEP varied from 14.82 to 33.59 bar. Here all the pressures are absolute and energy interaction in % is LHV. It is seen that work output is almost double from the previous comparison which is due to increased air mass flow rate and turbine work. By increasing the BP, the efficiency can be brought closer to 50% LHV which is higher than as obtained in turbocharging, proving that turbo-compounding is more efficient than turbocharging. Combustion efficiency is increased to 98–99% where it is about 95–96% during turbocharging. NO_x and UBHC emissions were reduced whereas CO_2 and O_2 emissions are almost constant and CO emissions are increased initially and then decreased (Johnson 2015).

6.2.3 Organic Rankine Cycle

The Rankine cycle is a thermodynamic cycle that converts heat into mechanical work (that is converted into electricity), using water as a working fluid. It consists of four basic mechanical components, namely, boiler, turbine, condenser, and pump. First, high-pressure liquid enters the boiler from the feed pump where the energy is added, and the saturation temperature of the liquid is reached. Due to more addition of the energy, it begins to convert into steam and this process is continued until the whole liquid is converted into steam. The produced steam is allowed to expand in the turbine producing the work which is finally converted into electricity. There is a limitation on the expansion due to the erosion of the turbine blades as binary phase fluids exist in the turbine and also due to the temperature of the cooling medium. So, it is recommended that exit vapor quality should be as high as possible, i.e., dryness fraction should be greater than 0.9 (Muller-Steinhagen 2011). This vapor is then condensed at low pressure in the condenser at cooling water temperature. After that the pressure is raised in the feed pump where the work done by the pump is neglected because of the low specific volume of water. In this way, the cycle is repeated. Now, this concept can be applied to IC engines. Organic Rankine cycle (ORC) technology uses an organic working fluid instead of water. The ORC thermodynamic cycle begins at the pump where the operating fluid is pumped to the evaporator as shown in Fig. 6.3. The working fluid here receives heat from the waste heat source. The

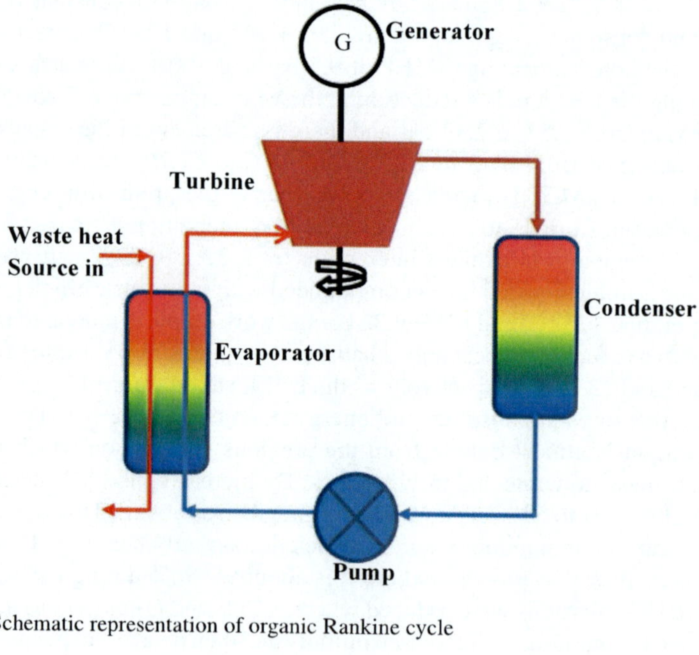

Fig. 6.3 Schematic representation of organic Rankine cycle

gas at the evaporator outlet is sent to the turbine/expander, whereby its expansion of mechanical work is produced and then electricity by means of a generator.

Generally, this technology is used in low-temperature applications, due to the low boiling point of the working fluid. It is also desirable at low temperatures because the working fluid generally does not condense during turbine expansion and it has also eliminated the requirement of the superheating of the vapor which is found from the shape of the liquid–vapor dome in T-S diagram. But we know that every fluid has its bright side and a dull side. We have discussed the bright side of the organic fluids, i.e., n-pentane which is generally used as organic fluid but the dull side is that organics can be easily disintegrated as compared to water and also a lot of care is required while operating with organics as leakage of it causes lots of pollution which also shows that steam Rankine cycle is more efficient (Quoilin and Lemort 2009). This has been overcome by a low-pressure steam bottoming cycle.

Combined cycle technology, based on gas turbine topping and steam turbine topping, is expected to remain attractive for several years to come. The exhaust gas of the SI/CI engine has a temperature of between 300 and 500 °C. As a consequence, the exhaust heat can be recovered in a waste heat boiler. However, the engine emits fewer exhaust gases compared to a gas turbine as it does not require sufficient airflow for cooling. The excess air for a diesel engine is usually 30–40%, compared with 200–350% for a gas turbine. As a result, the 10 MW engine will produce approximately 1–1.5 MW of electrical power by the use of the bottoming steam Rankine cycle. Furthermore, the performance of the Rankine cycle decreases rapidly as the exhaust temperature decreases (Korobitsyn 1998). It was found that by a little compromise in the efficiency, i.e., about one and a half percent have simplified the system and has reduced the challenges associated with the bottoming cycle (it is the one in which electricity is generated from the waste heat) and also the difficulties related with the handling of supercritical steams. As the name suggests, a low-pressure steam bottoming cycle but again the question arises in the mind that by how much pressure is reduced? The modeling reveals that pressure is reduced from 300 to 100 bar which is exactly one-third. Rankine cycle devices are reported to increase the combined engine and waste heat recovery efficiency almost by 10% (Srinivasan et al. 2008). The concept of the combined cycle is very common in the steam power plants but there are some restrictions with the IC engines as it makes the system bulkier but still, Teng et al. (2006) proposed such a concept for heavy-duty engines, and a similar idea has been considered by Cummins (Wall 2012). Besides regenerating energy from exhaust streams by exchanging heat with the secondary fluid, there are other alternatives such as thermoelectric regeneration.

6.2.4 *Thermoelectric Generators*

The thermoelectric effect was first observed by Seebeck in 1821, and he found by various experiments that when two dissimilar materials are brought in contact

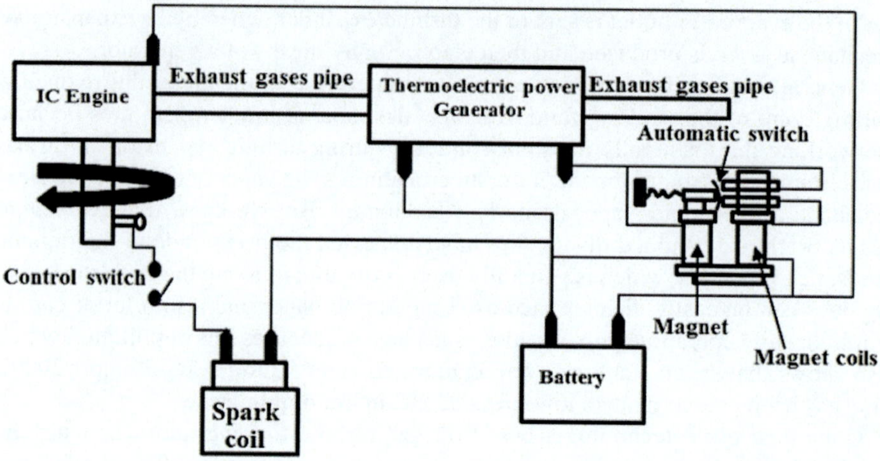

Fig. 6.4 Schematic diagram of converting waste heat into electrical power using a thermoelectric power generator

at the junction then the voltage is produced due to the difference in the temperatures (Goldsmid 1960). Practically, this effect produces electricity with the help of semiconductor materials which has suitable properties to transmit electricity with little resistance heating, i.e., Joule heating and thermal conductivity. The Seebeck effect has come into the picture after Peltier's effect which is just opposite to it and first applied to refrigerators. In this effect, voltage is supplied to create temperature difference (Goldsmid 1960). The idea of utilizing energy usually wasted or lost in the operation of IC engines operated by heat was explored and patented in 1914 (Creveling 1914). The exhaust gases from the engine supply the heat source to the thermopile or thermoelectric generator which converts heat taken from the gases into electrical energy, while the heat sink (cold side) is proposed to be supplied by the circulation of cooling water. Figure 6.4 shows a schematic diagram showing this patent of converting waste heat into electrical power applied to an IC engine using a thermoelectric generator (Creveling 1914; Ismail and Ahmed 2009).

Thermoelectric devices have found a suitable place to utilize the available thermal energy in an IC engine's exhaust for its conversion into electricity (Hussain et al. 2009). There are two types of semi-conductors, i.e., p-type and n-type, which are electrically connected in series and thermally connected in parallel. The hot side of the thermoelectric generator receives heat from the source and transfers it to the colder side with the excitation of electrons and holes present on the p-type and n-type. The colder side is made cooler with the help of air cooler so that the temperature gradient is maintained and a large amount of current starts flowing. Generally, semiconductors are made of high heat-absorbing materials like lead telluride which can absorb 200–600 °C and silicon–germanium which has a capacity of up to 900 °C. But there is one problem associated with it, i.e., thermoelectric generator can have efficiency range between 6 and 12%, so the choice of material is an important criterion. It is found that the efficiency of the IC engine is increased by 15–20% with the use of thermoelectric generators (Nandhakumar and Dinesh 2017).

Thermoelectric regeneration is the process of converting the temperature difference and heat flux into electricity and the device which does this process is known as a thermoelectric generator. It has been found that even with a figure of merit (ZT) of 3 and operating between a 1000 K exhaust temperature and the environment, a thermoelectric device would only be 30% efficient. Here, ZT is the parameter that depends on the Seebeck coefficient, electrical resistivity, and thermal conductivity as shown in (Kim et al. 2015). This is very-very less than the 37% efficiency of the bottoming cycle in the same configuration. Moreover, there are no thermoelectric devices with a ZT > 3 recently which means thermoelectric is not a suitable technique to achieve maximum efficiency (Tritt and Subramanian 2006). There is another technique which is known as thermochemical recuperation. This technique uses exhaust energy to reform the incoming fuel stream by transferring heat from the exhaust gases which can help to drive the endothermic reactions that ultimately convert the fuel/steam mixtures into a more energetic resource. This approach has been shown to increase the efficiency of engines by 5–10% (Chakravarthy et al. 2010).

6.3 Efficiency Improvement with Other Advanced Technologies

In addition to cost and emissions regulations, fuel economy has been an important factor for the development of IC engines. It is important to identify the appropriate package of technology for each market according to the different requirements of the region. Therefore, innovations are also required to boost the combustion efficiency of traditional engines. Different advanced technologies to boost the combustion efficiency for traditional engines are used and discussed here.

6.3.1 Engine Downsizing

The main idea behind the engine downsizing is, as the name suggests, to reduce the size of the engine without the loss of power, which means we have to increase the power density to do so (Clenci et al. 2007). There are lots of parameters that affect the power produced by the engine. These parameters include fuel conversion and volumetric efficiencies, fuel–air (F/A) ratio, inlet air density, fuel heating value, etc. (Heywood 1988). If the size of the engine is reduced then displaced volume is also reduced but this is not only the parameter which decides the overall performance of the engine, thereby other parameters are also discussed above. It is found that vehicle provides better fuel economy with lighter engines. Light engines have better efficiency near the mid-speed and at 75% peak loading condition when a light engine is operated at the same power as the heavy one with turbocharging. The motto behind this technology is to find the parameters other than displaced volume which affects the engine output and tries to improve it.

6.3.2 Advanced Engine Controls

Due to the shift in the engine controls from mechanical to electrical has shown lots of improvement in the IC engines. Earlier in mechanical controls, there is a facility to control power only, i.e., throttle or rack position or controlling variable load with constant speed like in governors. By the coming of advancement in the engine controls which are electric-based and can even sense least sense parameters like cam position, throttle position, manifold temperatures, and pressures and airflow and can also control most other aspects of the engine like spark timing, injection timing, injection pressure, EGR level as well as boost pressure. Cylinder pressure which is the main parameter to analyze the energy release rate of the engine is also controlled with electrical signals which can control the parameters affecting the combustion according to its burn profile (Leithgoeb et al. 2003; Corti et al. 2007). Also due to the development in the modeling-based engine, it requires less time to make an engine that can control the advanced control parameters (Blumberg et al. 1979; Filipi and Assanis 1991; Kamimoto and Kobayashi 1991; Reitz and Rutland 1995; Assanis and Heywood 1986).

6.3.3 Variable Valve Timing

It means the ability to change the events of the valve as the name signifies. The events may be opening and closing of the intake and exhaust valves etc. at any given point during the operation of an engine (Gray 1988). The valve events are fixed in conventional engines as the position is fixed on the camshaft with the help of mechanical lobes and the effectiveness with which inlet and exhaust operations are performed of the engine depends on various parameters like engine load, speed, spark timing, injection timing and also on the use of EGR which clearly shows that not only valve events which affect the performance and efficiency of the engine, there are other parameters too (Payri et al. 1988; Ma 1988). By the coming of variable valve timing (VVT) systems, it is easy to change the valve timing as per the requirement at any point in the operation of the engine which provides flexibility over the mechanical VVT systems. This technology also allows the implementation of Millers' approach which means variable compression ratio at part loads and conventional operation at peak loads (Sugiyama et al. 2007a). It also allows the control of the amount of residual fraction in the engine cylinder which further improves the emission characteristics (Meacham 1970). By controlling the amount of residual fraction has enabled the advanced mode of combustion and has also replaced the throttle for load control which allows controlling of the trapped mass without inculcating the pump work (Tuttle 1980).

6.3.4 Variable Geometry Engine Design

Variable geometry simply means to change the geometry as per the requirement. Conventional engines have fixed geometries which means fixed compression ratio and fixed displaced volume but due to advancement in the technology, it can be found that by varying the geometry of the engine with the new design has a lot of impact on the performance and efficiency of the engine. For example, a variable turbocharger has overcome the limitations of the fixed geometry turbocharger in such a way that it has eliminated turbo lag which is the dynamic issue created when the engine is accelerated. This is only possible with the help of widening the range of benefits which is narrow in case of fixed geometry turbines and hence fixed geometry turbines (FGT) only provide benefits to the engine at peak loads of performance. Earlier to boost the engine flow rates, a large size of the turbine has to be designed which requires substantial inertia to rotationally accelerate for maximum benefits, but this has also been eliminated with the help of variable geometry turbines (VGT). As we have seen that the "variable" word seems to be somewhat attractive until now. Just like variable geometry, variable turbine, the variable compression ratio is also one of the important parameters to enhance engine efficiency. J. Atkinson was the first person to propose the concept of a variable compression ratio. He was a very famous personality and we all have heard about the famous cycle which is frequently used in thermodynamics is Atkinson's cycle. Atkinson's original concept consists of linkages that are used to vary the compression ratio and expansion ratio in such a way that the compression ratio should be less than the expansion ratio which yields more expansion work and higher efficiency (Caton 2008). This similar idea was also implemented by Miller but the way of implementation was different (Caton 2008). It uses a late intake valve opening to shorten the compression and late exhaust valve opening to extend the expansion. Both these concepts are used in modern technology, i.e., variable compression ratio (Clenci et al. 2007; Wirbeleit et al. 1990; Sugiyama et al. 2007b) and alternative valve timing techniques (Boggs et al. 1995). The alternative valve timing is made possible with the help of variable valve timing (Clenci et al. 2007). It was found that with the help of variable displacement there is no pumping work required as in SI engines which has increased the efficiency besides having the friction and heat losses. The engine can deliver high power using full displacement. At part load instead of using the throttle, the cylinders are deactivated (Leone and Pozar 2001). These deactivated cylinders do not provide power but are still operating having friction loss and heat loss which is less as compared to the gain we have achieved by saving the pumping work. Besides having different geometrical designs, various advanced fuel injection techniques have also contributed to the enhancement of the efficiency of IC engines.

6.3.5 Advanced Fuel Injection

Fuel injection also plays an integral part in combustion and emission analysis of the engine. Due to favorable cost to benefit ratio, port fuel injection (PFI) is more popular in Asia for SI engines till 2015, but after that, due to more stringent norms on emissions, other techniques are also developed. There is always a trade-off between cost and benefit to the environment. But still, Bosch's have achieved improved preparation of mixture with fine fuel particles and reduced droplets size which has reduced HC emissions (Busch et al. 2012). A new technology, i.e., twin injection is developed. It is found that most of the emissions occur during the initial phase of the engine operation of cold start and catalyst heating. As found in the emission tests that most of the HC and NO_x emissions occurred initially in the first 20 s. The twin injector system is a renowned method for improving combustion stability by reducing the diameter of fuel droplets. The twin injector system uses two narrower injectors per cylinder and installs near to the intake valve, the injector may have smaller nozzle holes and wider spray cone angle, which helps to improve the fuel spray homogeneity in the intake valve area. The use of a twin injector shows great potential for optimizing the spray pattern and enhances the behavior of fuel vaporization. Besides the quality of the fuel spray, spray targeting also has a significant impact on the emissions during cold start. The improved homogeneity of the charge around the spark plug reduces HC emissions in the combustion chamber.

The weighted mean droplet diameter of the fuel used plays a great role in HC emissions. Smaller droplet size makes better mixture and enhances stable combustion which has also contributed to reducing the HC emission by almost 30% as the droplet size is reduced from 100 to 35 μm. Twin injector has targeted twin valves and a demand-driven fuel pump that has increased the fuel pressure in the beginning phase. Twin injection can alone cut the HC emissions by 30% (Busch et al. 2012). It is also found that if this technology is combined with other technique, i.e., demand control fuel supply system (DCFSS) which increases the fuel injection pressure from 3 to 9 bar has alone reduced the HC emissions by 30% and the combination of both, i.e., twin injection + DCFSS system has reduced drastically the HC emissions by 50% (Busch et al. 2012).

The fuel is directly injected into the combustion chamber above the piston in direct injection (DI) diesel engines. DI is the advance injection technique that plays a key role in the reduction of CO_2. As compared to the PFI technique DI technique has great potential of combining the turbocharger with engine downsizing. It has also lots of potential of combining VVT with the co-ordination of injection timing which has improved the torque performance at low engine speeds. Emissions are affected by various parameters like combustion chamber geometry, fuel spray pattern, and also on the location of the injector and spark plug. DI plays a significant role which is essential to turbocharged downsizing and other advanced combustion concepts like spray-guided direct injection (SGDI) and homogeneous charge compression ignition (HCCI). To accomplish the norms of emissions and to maximize the combustion, a new technology named controlled valve operation (CVO) is developed. It employs feedback control of the injection valve. It estimates the injected fuel quantity based on the injector control signal forming a close loop control. It also integrates a smart control strategy with specially developed engine control unit (ECU), i.e., hardware. This allows the very precise delivery of the fuel quantity by the lifting of the valve. It can also compensate for the change of environmental conditions, change of system pressure, fuel temperature, and influences of multiple injections. It uses HDEV injectors which are high-pressure injectors, and nowadays, HDEV6 which has injection up to 350 bar has been developed by Bosch's (Robert Bosch GmbH 2021). The current amount of fuel injection is greater than 3.5 mg/injection which can be reduced to 2 mg/injection with the help of more advanced injector, i.e., piezo-injectors which are even more accurate, i.e., around 12–15% as compared to HDEV injectors.

As we all know, heavy-duty engines worldwide are CI engines and play a key commercial role to target this DENSO International Europe, a German company, is targeting up to 3000 bar injection pressure for commercial vehicles. It has developed the fourth-generation solenoid injection system. The achievement of this huge target is due to optimized pump design and also due to leakage proof solenoid injector. This can be made more accurate with the help of a hydraulic-based optimized design of the injector with flexible multi-injection calibration which can be further accurate with the help of an in-injector pressure sensor (Busch et al. 2012; DENSO 2013). The brief discussion on the advancement in the CI and SI engines is discussed in the next sections.

6.3.6 Advanced Compression Ignition Engines

In CI engines lots of soot emissions were reduced with the development of technology, i.e., with the help of advanced fuel injection systems along with the common rail systems and electronic injectors in 1980. This technology has more precise control as compared to conventional engines in the combustion process. The other such technology which has increased the mileage of the engine and parallel reduced the emissions is turbocharging. It allows recovering energy that is wasted in the exhaust gases to run the turbine which ultimately pressurizes the cylinder air to boost the power of the engine. It is well known that NO_x increases as the temperature increases (Heywood 1988). To reduce these emissions another advanced technology, i.e., EGR is developed. The exhaust gases that are released after the combustion is allowed to again recirculate in the cylinder and mixed with the fresh charge which makes the mixture dilute in terms of A/F ratio so that temperature gets reduced and lesser emission of NO_x takes place. To promote more uniform combustion and to reduce NO_x along with particulate matter (PM), one more technology is introduced called homogeneous charge compression ignition (HCCI). In this technique mixture of fuel and air is prepared and then it is injected into the engine which reduces the emissions by reducing temperature as well as also helped in uniform combustion of the fuel. HCCI technique has reduced PM emissions up to some extent but according to the EPA standards, the commercial vehicles which emit lots of PM are not reduced only by the HCCI technique, so one more technology is introduced which is DPF which means diesel particulate filters that physically filters the exhaust gases. It has been found from the last decade that chemical compounds are used to reduce the NO_x emissions which transform NO_x into nitrogen gas and the name of this catalyst technology is a selective catalytic reduction (SCR) and lean NO_x catalysts (Rissman and Kennan 2013). Depending upon the ID if the fuel is injected during the compression stroke of the piston then the technique is termed as premixed charge compression ignition (PCCI) or as partially premixed charge compression ignition (PPCI) (Lu et al. 2011). Phasing of combustion is a very complex phenomenon and it depends on lots of parameters in which one is affecting the other. There are lots of ways for controlling the combustion phasing, i.e., for creating a premixed charge primarily with gasoline injection and then combining with secondary diesel injection for controlling the combustion timing is reactivity-controlled compression ignition (RCCI) (Kushch et al. 2002). Spark-assisted compression ignition (SACI) concept uses a spark ignited flame propagation for triggering the auto-ignition.

6.3.7 Advanced Spark-Ignition Engines

In SI engines, on the other hand, the lean-burn combustion technique is used which is different as used in conventional SI engines. In this technique low F/A ratio is used which has both advantages and disadvantages. The advantage is that it improves the fuel efficiency because of reduced throttling at part load and lower heat losses from the walls of the cylinder. The disadvantage is that due to misfiring and cyclic dispersion, this technique fails in the after-treatment of the exhaust gases, especially NO_x because it is difficult for three-way catalysts to reduce it in the excess of O_2. Due to excess air in the above technique, there is a need to stratify the charge and this is done with the help of new technique, i.e., stratified gasoline combustion. In this technique, fuel is injected at a very high pressure which makes the charge nearly stoichiometric in the region near to the spark plug and the rest of the portion of the cylinder is filled with the fresh air. The technique named state-of-the-art technology which is used to reduce the spray penetration and increase the homogeneity of the charge is surrounded by the spark plug with the help of multi-pulse and spray-guided strategies. VVT technologies are widespread in SI engines according to (EPA 2011). However, full-flexible continuous valve timing systems are still in an early development phase. Although cam-less systems are not still in production, engines using electro-hydraulic systems have been already marketed (Palma et al. 2011) and provide high flexibility of the valve timing with the cycle-to-cycle response time. Whenever the term homogeneity is used, then the technique is named as controlled auto-ignition in the case of SI engines and it is said as HCCI in case of CI engines as discussed earlier (Husted et al. 2009).

6.3.8 HCCI, PCCI, and RCCI Engines

HCCI means homogeneous charge compression ignition, which is different from conventional CI engines in which the charge is heterogeneous. Now the question arises in mind that how this homogeneity can be achieved? The answer is by the premixing of fuel and air in advance before inducting it into the engine cylinder. A quicker burn rate performed by compression of homogeneous charge by ignition allows heat release to occur under conditions of near-constant volume. Lean mixtures further promote higher efficiencies in HCCI because it uses compression rather than spark like in SI engines to ignite the fuel. This has also eliminated the use of the throttle. It is because the load is controlled by the amount of fuel participating which is in advance added to the mixture which also helped to reduce PM emissions as compared to conventional CI engines. There are also some problems associated with HCCI, i.e., HCCI does not have the direct trigger for ignition as in conventional SI engine, i.e., spark and conventional CI engine, i.e., direct fuel injection. HCCI concept depends on whether the mixture is controlled precisely from an initial point to the ignition point. Several factors such as heat transfer, turbulence, and the history of preceding combustion events which are even present in a firmly controlled research environment make the practical application of HCCI very challenging. The very first person to report the practical implementation of HCCI was Onishi et al. (1979) with theoretical developments experimentally provided by Najt and Foster (1983). Now the question arises in the mind that which technique is the best suitable technique to control the combustion in HCCI engines? It is by altering the kinetics of the combustion which can be done by precisely controlling the amounts of residual fraction and as a result also lowered the combustion temperatures. Most of the HCCI combustion is characterized by low-temperature mechanisms commonly referred to as low-temperature combustion (LTC). In order to implement LTC and derive its benefits, high levels of EGR and strategic injection timings are used to extend ID and create a nearly all-premixed combustion event. The long ID, coupled with low-temperature mechanisms, establishes the phenomenological observation of two-stage ignition characterized by the presence of cool-flame reactions (Fish et al. 1969). Interestingly, because of the attainment of LTC, soot formation is substantially reduced, and the engine is made to operate with very low emissions of NO_x and PM (Takeda et al. 1996).

HCCI combustion is an advanced LTC technique using the auto-ignition of a fully premixed charge. The LTC strategy is a flame-less, staged combustion of fuel in a combustion chamber at a lower temperature, as related to a conventional combustion chamber (Bharj et al. 2019).

PCCI combustion is another form of LTC that provides greater control over combustion compared to HCCI combustion. The basic operating method of PCCI combustion is to prolong the period for mixing the fuel–air mixture by separating the end of the injection and the start of combustion. PCCI combustion generates a partially premixed charge within the cylinder before the ignition takes place. The homogeneity of the fuel–air mixture in the case of PCCI combustion is relatively lower than that of HCCI combustion but is higher than that of CI combustion. PCCI combustion is also a single-stage combustion process similar to HCCI combustion, in which most fuel burns during the premixed combustion phase. This results in locally high but overall low combustion temperatures within the combustion chamber due to the lack of combustion diffusion phase. This is the main cause of considerably lesser NO_x emissions from PCCI combustion without considerably affecting the engine power output and efficiency. PCCI combustion also suffers from the issue of high HC and CO emissions. However, HC and CO emissions are relatively low compared to HCCI combustion. PCCI combustion also leads to high HC and CO emissions, but HC and CO emissions are comparatively low from HCCI combustion. Hence, PCCI combustion can be seen as an intermediate combustion technique between HCCI and traditional CI combustion, which provides lower NO_x and soot emissions and increases the regulation of combustion (Singh and Agarwal 2019a, b; Sun et al. 2016; Jain et al. 2017). RCCI is a type of HCCI that provides greater control over the combustion process and has dramatically reduced fuel consumption and emissions. RCCI uses in-cylinder fuel blending with at least two different reactivity, namely, a high-reactivity fuel (HRF) and a low-reactivity fuel (LRF) in a CI engine and multiple injections to regulate in-cylinder fuel reactivity in order to optimize combustion phase, duration, and magnitude. In RCCI combustion, HRF was considered to be high cetane fuel, such as mineral diesel or biodiesel. LRF is considered to be low cetane fuel, such as alcohol, gasoline, or compressed natural gas (CNG). The low-reactivity fuel is completely premixed with the intake charge and the high-reactivity fuel is injected directly into the cylinder before ignition. Combustion is started at high-reactivity fuel–air mixture areas, and subsequent reactions proceeded from high- to low-reactivity fuel–air mixture areas, enabling a controlled sequential ignition event. In order to generate clean and efficient combustion, RCCI combustion depends primarily on the fuel reactivity gradient; therefore, a broad range of fuels can be used. RCCI combustion produces outstanding results in terms of performance and emissions throughout the whole working range of the engine. In order to improve thermal efficiency and reduce NO_x and soot emissions, RCCI combustion also uses multiple injection strategies in a single cycle, along with EGR (Singh and Agarwal 2019b; Singh et al. 2019, 2020a, , b). Alternative fuels have also played a vital role in the improvement of engine efficiency.

6.3.9 Alternative Fuels

Majorly alternative fuels can be classified into four broad categories: firstly, diesel which is based on biomass, biofuels which are non-cellulosic and advanced, biofuels which are cellulosic and advanced, and conventional biofuels. There is a wide range of liquid and gaseous chemicals available that can be used in IC engines. The term alternative signifies that the fuel can be used in the substitute of conventional fuel, i.e., gasoline or diesel. Some of the fuels which can be used as an alternative in SI engines are ethanol-blended gasoline, ethanol-blended CNG, liquefied natural gas (LNG), and liquefied petroleum gas (LPG). For CI engines biodiesel blended in diesel, emulsion blended diesel, etc. This concept is not new, it is as old as the diesel engines as considered by Rudolf Diesel by using a gaseous, liquid, and solid fuels in "Theory and Construction of a Rotational Heat Motor" in 1894 (Diesel 1894). It plays a significant role by regulating emissions through cleaner combustion. The production of alternative fuels and their challenges is the other vast topic for discussion which is not included in this chapter. But the readers who are enthusiastic and eager to know can refer to (MacLean and Lave 2003). The term alternative should not be confused with renewable. Both the terms are neither mutually exclusive nor inclusive. The majority of renewable fuels are considered alternative fuels but not all alternative fuels are renewable. For example, methanol is renewable fuel only if it is produced from biomass, but it is not commonly used, so it may be considered as an alternative. Similarly, ethanol is renewable fuel if it is produced from corn or sugarcane, but it is commonly used, so it may or may not be an alternative. But in our context, it is still alternative because it is used as a substitute, i.e., in place of diesel or gasoline. Blends with conventional fuels like ethanol which is used in low-level gasoline/ethanol blends (e.g., E10) or biodiesel in low-level blends with diesel such as B5, B10, B20 or CNG or LPG are used as an alternative in IC engines. Methanol has lots of advantages like combustion stability, IMEP, and reduces misfires and has also reduced 70% CO as well as 40% HC emissions. These things are possible due to high oxygen content, vapor pressure, and low boiling point (Song et al. 2008). In the study, it is found that a 25% NO_x emissions reduction can be expected with methanol compared to the conventional fuel which is mainly caused by oxygenating which is prepared with methanol acts as an octane booster and also has reduced process temperatures (Caton 2009) but due to some limitations like contamination of groundwater it is not used. Some of the most important parameters that are used to compare the amount of energy contained inside the alternative fuel are, namely, gasoline gallon equivalent (GGE) and diesel gallon equivalent (DGE). GGE refers to the amount of alternative fuel needed to meet the energy content of 1 gallon of gasoline and DGE refers to the amount of alternative fuel needed to meet the energy content of 1 gallon of diesel fuel. GGE and DGE are somewhat controversial due to variations in the energy content of alternative fuels as well as gasoline or diesel (Boyd 2009). But still, it is used here to simplify the discussion regarding the energy content of alternative fuels.

6.4 Opportunities, Challenges, and Technology Barriers

On the one hand, rules are strictest in terms of pollutants and CO_2, though globalization, on the other hand, decreased time to market, and the fuel quality variability has switched the global situation around. Though these are threats, a crucial position in the coming decades, even though technologically significant, is required to evolve for improving the efficiency of the engine.

The recent scenario is that lots of economic efforts have been made for a wide variety of technologies and their combination. Still many of the advanced concepts need accurate control especially at system integration level as we know that engine control is the enabling tool at the subsystem level. There is a need for a holistic approach and high-level control for multiple operations and also for close interaction between different subsystems. The biggest challenge is to manage the subsystems requirement, operations of the engines, and also their limits. There is a need of providing procedure for calibration of the system and control system adoption to different environmental variations. There are also lots of problems associated with the calibration of the ECU and the variation in the present condition of the engine and the way it is operating is also a big challenge. There are discrepancies in manufacturing an individual unit to other and also lots of combinations available and aging of IC engine, which makes the situation more complex, so there is a need for strong solutions to these problems. Adaptive capabilities are extended from the system level to the subsystem level (Guardiola et al. 2013). Hybrid power plants can decouple the torque in IC engines on request (Rizzoni et al. 1999), which provides an additional degree of freedom for the smooth operation of the engine (Lindenkamp et al. 2009). Some other major issues are concerned with the inability of LTC concepts in the complete engine range. This is due to the huge differences in the required charge composition and combustion instabilities, misfiring, or significant torque disturbances (Ferrari et al. 2011).

Still, there is a dilemma between the two things, one is that to use the fuel which exists in the market in the modified engines with the advanced technology and the other is to use alternative fuels like dimethyl ether (DME) or biodiesel, etc. with the existing engines. Whatever may be the thing it has to be realized that if there are two given paths and we don't have that much clarity till now on which one is the best then it makes difficult to implement any one of them, and also we know that it took a lot of time for a new technology to penetrate into the market as a three-way catalytic converter took about 20 years to penetrate into the automobile industry. There is no limitation in the scope of research, but the main problem is the confusion between the two paths. As it is known, there is still no correct explanation about the variation of flame speed in the SI engine with the rpm (Reitz 2013). Also, in engine efficiency, many things are undefined and lots of corrections can be done (Splitter et al. 2013). Boyd's paper clearly shows the future areas of IC engines, many of which are still unexplored (Boyd 1950).

6.5 Conclusions

The maximum achieved efficiency in turbocharging is 44.1%, which is more than the conventional engines by 9.1–14.1% and varies from 30 to 35%. By increasing the BP, the efficiency of turbo-compounding can be brought closer to 50% LHV which is higher than as obtained in turbocharging, proving that turbo-compounding is more efficient than turbocharging. It has also achieved 99% combustion efficiency and reduced emissions as compared to turbocharging. It is also concluded that using the concept of compounded Rankine cycle, increases the combined engine and waste heat recovery efficiency almost by 10%. The highest increment in the efficiency of the IC engine can be achieved with the thermoelectric generators which have tremendously increased the efficiency by almost 15–20%. Besides improving efficiency from wasted heat we have also studied the advanced technologies which have also improved efficiency. First, engine downsizing which has increased power density with light engines achieved the best performance at higher loads. Due to the shift in the controls from mechanical to electrical has also enhanced the efficiency by controlling the combustion according to burn profile. VVT systems have also provided flexibility and boosted the performance by controlling residual mass fraction. With the help of variable geometry engine design, it is possible to achieve higher efficiency by saving the pump work besides having friction and heat losses. Twin injection + DCFSS system has reduced drastically the HC emissions by 50%. The current amount of fuel injection is greater than 3.5 mg/injection which can be reduced to 2 mg with the help of more advanced injector, i.e., piezo-injectors which are even more accurate, i.e., around 12–15% as compared to HDEV injectors. EGR technique has reduced NO_x, whereas HCCI reduces both NO_x along with PM. SCR and Lean NO_x are also some of the techniques to reduce NO_x. RCCI and SACI techniques are used to control combustion phasing to improve the performance of the engines. Methanol improves combustion stability, IMEP (torque), and reduces misfires. Also, a 70% reduction in CO emissions as well as a 40% reduction in HC emissions was observed. Biodiesels and DME also helped to reduce emissions to a great extent at the expense of cost.

References

Aladayleh W, Alahmer A (2015) Recovery of exhaust waste heat for ICE using the beta type stirling engine. J Energy 2015:1–8. https://doi.org/10.1155/2015/495418

Annual Book of ASTM Standards (1980) D613: standard test method for cetane number of diesel fuel oil. American Society for Testing and Materials, Philadelphia, PA, USA

Assanis D, Heywood J (1986) Development and use of a computer simulation of the turbocompounded diesel system for developments in engine performance and component heat transfer studies. SAE Trans 95 (SAE Paper No. 860329), 2.451–2.476. https://doi.org/10.4271/860329

Bernodusson J (2018) Combustion of fossil fuels. https://www.samgongustofa.is/media/siglingar/skyrslur/Combustion-of-fossil-fuels-2018-en-1.pdf. Accessed 14 Apr 2020

Bharj RS, Kumar R, Singh GN (2019) On-board post-combustion emission control strategies for diesel engine in India to meet Bharat Stage VI norms. In: Advanced engine diagnostics. Springer, Singapore, pp 105–125

Blumberg P, Lavoie G, Tabaczynski R (1979) Phenomenological models for reciprocating internal combustion engines. Prog Energy Combust Sci 5:123–167. https://doi.org/10.1016/0360-1285(79)90015-7

Boggs D, Hilbert H, Schechter M (1995) The Otto-Atkinson cycle engine: fuel economy and emissions results and hardware design. SAE Trans J Engines 104 (SAE Paper No.950089):220–232. https://www.jstor.org/stable/44633213

Boyd T (1950) Path finding in Fuels and Engines, vol.4. Society of Automotive Engineers, Warrendale, PA, pp 182–195. SAE paper 500175. https://doi.org/10.4271/500175

Boyd RA (2009) Proposed method of sale and quality specification for hydrogen vehicle fuel. Summary of current information. Standards Fuel Specifications Subcommittee (FSS). US National Work Group for the Development of Commercial Hydrogen Measurement

Busch R, Hammer J, Herynek R, Wehmeier K (2012) Advanced fuel metering gasoline–requirements and Bosch system solutions. In: Fuel Systems for IC Engines. Woodhead Publishing, pp 77–85

Caton J (2000) A review of investigations using the second law of thermodynamics to study internal-combustion engines. SAE Trans J Engines 109 (SAE Paper No. 2000–01–1081):1252–1266. https://doi.org/10.4271/2000-01-1081

Caton J (2008) Results from an engine cycle simulation of compression ratio and expansion ratio effects on engine performance. J Eng Gas Turbines Power 130(5):052809–1–052809–7. https://doi.org/10.1115/1.2939013

Caton JA (2009) A thermodynamic evaluation of the use of alcohol fuels in a spark-ignition engine. SAE Paper No. 2009-01-2621. https://doi.org/10.4271/2009-01-2621

Cengel YA, Boles MA (2007) Thermodynamics: an engineering approach 6th editon (SI Units). The McGraw-Hill Companies Inc., New York

Chakravarthy VK, Daw CS, Pihl JA, Conklin JC (2010) Study of the theoretical potential of thermochemical exhaust heat recuperation for internal combustion engines. Energy Fuels 24:1529–1537. https://doi.org/10.1021/ef901113b

Clenci A, Descombes G, Podevin P, Hara V (2007) Some aspects concerning the combination of downsizing with turbocharging, variable compression ratio, and variable intake valve lift. Proc Inst Mech Eng D J Automob. Eng 221:1287–1294. https://doi.org/10.1243/09544070JAUTO449

Corti E, Moro D, Solieri L (2007) Real-time evaluation of IMEP and ROHR-related parameters. SAE Paper No. 2007-24-0068. https://doi.org/10.4271/2007-24-0068

Creveling JL (1914) Means for utilizing waste energy. United States Patent US 1(118):269

DENSO develops a new diesel common rail system with the world's highest injection pressure (2013). https://www.denso.com/global/en/news/news-releases/2013/130626-01?

Diesel R (1894) Theory and construction of a rotational heatmotor. Spon & Chamberlain, London

Dunbar WR, Lior N (1994) Sources of combustion irreversibility. Combust Sci Technol 103:41–61. https://doi.org/10.1080/00102209408907687

EERE U (2003) Just the basics: diesel engine. US Department of Energy, Washington, USA. https://www1.eere.energy.gov/vehiclesandfuels/pdfs/basics/jtb_diesel_engine.pdf. Accessed 14 Apr 2020

EPA (2011) Light-duty automotive technology, carbon-dioxide emissions, and fuel economy trends: 1975 through 2011, EPA420-R-12-001a. Transportation and Climate Division. Office of Transportation and Air Quality, US Environmental Protection Agency

Facts about the Wright Turbo Compound (PDF). Wood Ridge New Jersey: Curtiss-Wright Corporation: Wright Aeronautical Division, October 1956. Archived from the original (PDF) on 16 February 2010. Accessed 21 June 2020

Ferrari A, Chiodi M, Bargende M, Roberti P, Millo F, Wichel-Rans D (2011) Virtual set-up of a racing engine for the optimization of lap performance through a comprehensive engine-vehicle-driver model. SAE paper 2011-24-0141. https://doi.org/10.4271/2011-24-0141

Filipi Z, Assanis D (1991) Quasi-dimensional computer simulation of the turbocharged spark ignition engine and its use for 2 and 4-valve engine matching studies. SAE Trans J Engines100 (SAE Paper No. 910075):52–68. doi: https://doi.org/10.4271/910075

Fish A, Read I, Affleck W, Haskell W (1969) The controlling role of cool flames in two-stage ignition. Combust Flame 13:39–49. https://doi.org/10.1016/0010-2180(69)90026-1

Goldsmid H (1960) Principles of thermoelectric devices. Br J Appl Phys 11:209–217. https://doi.org/10.1088/0508-3443/11/6/301

Gray C (1988) A review of variable engine valve timing. SAE Trans J Engines 97(SAE Paper No. 880386):6.631–6.641. https://doi.org/10.4271/880386

Guardiola C, Pla B, Blanco-Rodriguez D, Eriksson L (2013) A computationally efficient kalman filter based estimator for updating look-up tables applied to NOx estimation in Diesel engines. Control Eng Pract 21(11):1455–1468. https://doi.org/10.1016/j.conengprac.2013.06.015

Heywood JB (1988) Internal combustion engine fundamentals. McGraw-Hill, New York

Hussain Q, Brigham D, Maranville C (2009) Thermoelectric exhaust heat recovery for hybrid vehicles. SAE Int J Engines 2(1):1132–1142. https://doi.org/10.4271/2009-01-1327

Husted HL, Piock W, Ramsay G (2009) Fuel efficiency improvements from lean, stratified combustion with a solenoid injector. SAE Paper 2009-01-1485. https://doi.org/10.4271/2009-01-1485

International Energy Agency (2012) Technology roadmap: fuel economy of road vehicles. Technical report. https://webstore.iea.org/technology-roadmap-fuel-economy-of-road-vehicles

Ismail BI, Ahmed WH (2009) Thermoelectric power generation using waste-heat energy as an alternative green technology. Recent Pat Electr and Electron Eng (Formerly Recent Pat Electr Eng) 2(1):27–39

Jain A, Singh AP, Agarwal AK (2017) Effect of split fuel injection and EGR on NOx and PM emission reduction in a low temperature combustion (LTC) mode diesel engine. Energy 122:249–264. https://doi.org/10.1016/j.energy.2017.01.050

Johnson BH (2015) Exploring the pathway to high-efficiency, high-power ic engines through exergy analysis and stoichiometric direct injection (Doctoral dissertation, Stanford University). https://purl.stanford.edu/cp633zj5935

Kamimoto T, Kobayashi H (1991) Combustion processes in diesel engines. Prog Energy Combust Sci 17:163–189. https://doi.org/10.1016/0360-1285(91)90019-J

Karim N (2000) How turbochargers work. https://auto.howstuffworks.com/turbo.htm. Accessed 1 June 2020

Keenan J (1951) Availability and irreversibility in thermodynamics. Br J Appl Phys 2:183–192. https://doi.org/10.1088/0508-3443/2/7/302

Kim HS, Liu W, Chen G, Chu CW, Ren Z (2015) Relationship between thermoelectric figure of merit and energy conversion efficiency. Proc Natl Acad Sci 112:8205–8210. https://doi.org/10.1073/pnas.1510231112

Korobitsyn MA (1998) New and advanced energy conversion technologies. Analysis of cogeneration, combined and integrated cycles. Febroduck BV, Enschede

Kushch AS, Bass JC, Ghamaty S, Elsner NB (2002) Thermoelectric development at HI-Z technology. In: 2002 Diesel Engine-Efficiency and Emissions Research (DEER) conference. San Diego, California

Leithgoeb R, Henzinger F, Fuerhapter A, Gschweitl K, Zrim A (2003) Optimization of new advanced combustion systems using real-time combustion control. SAE Paper No. 2003-01-1053. https://doi.org/10.4271/2003-01-1053

Leone T, Pozar M (2001) Fuel economy benefit of cylinder deactivation–Sensitivity to vehicle application and operating constraints. SAE Trans J Fuels Lubricants 110 (SAE Paper No. 2001-01-3591):2039–2044. https://doi.org/10.4271/2001-01-3591

Levin K, Lebling K (2019) CO_2 Emissions climb to an all-time high (again) in 2019: 6 takeaways from the latest climate data, World Resources Institute. https://www.wri.org/blog/2019/12/co2-emissions-climb-all-time-high-again-2019-6-takeaways-latest-climate-data. Accessed 14 Apr 2020

Lindenkamp N., Sto¨ ber-Schmidt C., Eilts P. (2009) Strategies for reducing NOx and particulate matter emissions in diesel hybrid electric vehicles, SAE Paper 2009-01-1305. https://doi.org/10.4271/2009-01-1305

Lu X, Han D, Huang Z (2011) Fuel design and management for the control of advanced compression-ignition combustion modes. Prog Energy Combust Sci 37:741–783. https://doi.org/10.1016/j.pecs.2011.03.003

Ma T (1988) Effect of variable engine valve timing on fuel economy. SAE Trans J Engines 97 (SAE Paper No.880390):6.665–6.672. https://www.jstor.org/stable/44547403

MacLean HL, Lave LB (2003) Evaluating automobile fuel/propulsion system technologies. Prog Energy Combust Sci 29(1):1–69. https://doi.org/10.1016/S0360-1285(02)00032-1
Meacham G (1970) Variable cam timing as an emission control tool. SAE Trans 79 (SAE Paper No. 700673):2127–2144. https://www.jstor.org/stable/44723726
Moran MJ, Shapiro HN, Boettner DD, Bailey MB (2010) Fundamentals of engineering thermodynamics. John Wiley & Sons
Muller-Steinhagen HMG (2011) Rankine cycle. Thermopedia. https://doi.org/10.1615/AtoZ.r.rankine_cycle
Najt P, Foster D (1983) Compression-ignited homogeneous charge combustion. SAE Trans 92 (SAE Paper No. 830264):1.964–1.979. https://doi.org/10.4271/830264
Nandhakumar S, Dinesh K (2017) Increasing the efficiency of IC engine and thermal comfort using thermoelectric generator. Int J Adv Res Sci Eng 6:439–444
Nuclear Power for Everybody. Thermal Efficiency for Diesel Cycle (2021) https://www.nuclear-power.net/nuclear-engineering/thermodynamics/thermodynamic-cycles/diesel-cycle-diesel-engine/thermal-efficiency-for-diesel-cycle/
Onishi S, Jo S, Shoda K, Jo P, Kato S (1979) Active thermos atmosphere combustion (ATAC)–A new combustion process for internal combustion engines. SAE Trans 88 (SAE Paper No. 790501):1851–1860. https://www.jstor.org/stable/44658187
Palma A., Del Core D, Esposito C (2011) The HCCI concept and control, performed with multi-air technology on gasoline engines, SAE paper 2011-24-0026. https://doi.org/10.4271/2011-24-0026
Payri F, Desantes J, Corberaan J (1988) A study of the performance of an SI engine incorporating a hydraulically controlled variable valve timing system. SAE Trans J Engines 97 (SAE Paper No. 880604):6.1133–6.1145. https://www.jstor.org/stable/44547444
Quoilin S, Lemort V (2009) Technological and economical survey of organic Rankine cycle systems. In: European conference on economics and management of energy in industry. Vilamoura, Portugal
Rakopoulos C, Giakoumis E (2006) Second-law analyses applied to internal combustion engines operation. Prog Energy Combust Sci 32:2–47. https://doi.org/10.1016/j.pecs.2005.10.001
Reitz RD (2013) Directions in internal combustion engine research. Combust Flame 160:1–8. https://doi.org/10.1016/j.combustflame.2012.11.002
Reitz RD, Ogawa H, Payri R, Fansler T, Kokjohn S, Moriyoshi Y, Agarwal AK, Arcoumanis D, Assanis D, Bae C, Boulouchos K (2020) IJER editorial: the future of the internal combustion engine. Int J Engine Res 21:3–10. https://doi.org/10.1177/1468087419877990
Reitz R, Rutland C (1995) Development and testing of diesel engine CFD models. Prog Energy Combust Sci 21:173–196. https://doi.org/10.1016/0360-1285(95)00003-Z
Rissman J, Kennan H (2013) Advanced diesel internal combustion engines. Case Studies on the Government's Role in Energy Technology Innovation, American Energy Innovation Council
Rizzoni G, Guzzella L, Baumann BM (1999) Unified modeling of hybrid electric vehicle drivetrains. IEEE/ASME Trans Mechatron 4(3):246–257. https://doi.org/10.1109/3516.789683
Robert Bosch GmbH (2021) High-pressure injectors-solenoid valve. https://www.bosch-mobility-solutions.com/en/products-and-services/passenger-cars-and-light-commercial-vehicles/powertrain-systems/gasoline-direct-injection/high-pressure-injector/
Shyani R, Caton J (2009) A thermodynamic analysis of the use of exhaust gas recirculation in spark ignition engines including the second law of thermodynamics. Proc Inst Mech Eng Part D J Automobile Eng 223:131–149. https://doi.org/10.1243/09544070JAUTO935
Singh AP, Agarwal AK (2019a) Performance and emission characteristics of conventional diesel combustion/partially premixed charge compression ignition combustion mode switching of biodiesel-fueled engine. Int J Engine Res. https://doi.org/10.1177/1468087419860311
Singh AP, Agarwal AK (2019b) Characteristics of particulates emitted by IC engines using advanced combustion strategies. In: Advanced engine diagnostics. Springer, Singapore, pp 57–71

Singh AP, Sharma N, Satsangi DP, Kumar V, Agarwal AK (2019) Reactivity-controlled compression ignition combustion using alcohols. In: Advanced engine diagnostics. Springer, Singapore, pp 9–28

Singh AP, Sharma N, Kumar V, Agarwal AK (2020a) Experimental investigations of mineral diesel/methanol-fueled reactivity controlled compression ignition engine operated at variable engine loads and premixed ratios. Int J Engine Res. https://doi.org/10.1177/1468087420923451

Singh AP, Sharma N, Satsangi DP, Agarwal AK (2020b) Effect of fuel injection pressure and premixed ratio on mineral diesel-methanol fueled reactivity controlled compression ignition mode combustion engine. J Energy Res Technol 142(12):122301. https://doi.org/10.1115/1.4047320

Song R, Hu T, Liu S, Liang X (2008) Combustion characteristics of SI engine fueled with methanol gasoline blends during cold start. Front Energy Power Eng China 2(4):395–400. https://doi.org/10.1007/s11708-008-0081-7

Splitter D, Wissink M, DelVescovo D, Reitz RD (2013) RCCI engine operation towards 60% thermal efficiency. SAE Technical Paper 2013–01–0279. https://doi.org/10.4271/2013-01-0279

Srinivasan K, Mago P, Zdaniuk G, Chamra L, Midkiff K (2008) Improving the efficiency of the advanced injection low pilot ignited natural gas engine using organic Rankine cycles. ASME J Energy Resour Technol 130:022201-1–022201-7. https://doi.org/10.1115/1.2906123

Subramani DA, Dhinagaran R, Prasanth VR (2020) Introduction to turbocharging—a perspective on air management system. In Design and Development of Heavy Duty Diesel Engines (pp. 85–193). Springer, Singapore.

Sugiyama T, Hiyoshi R, Takemura S, Aoyama S (2007a) Technology for improving engine performance using variable mechanisms. SAE Trans J Engines 116 (SAE Paper No. 2007-01-1290):803–812. https://doi.org/10.4271/2007-01-1290.

Sugiyama T, Hiyoshi R, Takemura S, Aoyama S (2007b) Technology for improving engine performance using variable mechanisms. SAE Trans J Engines 116 (SAE Paper No. 2007-01-1290):803–812. https://www.jstor.org/stable/44699316

Sun C, Kang D, Bohac SV, Boehman AL (2016) Impact of fuel and injection timing on partially premixed charge compression ignition combustion. Energy Fuels 30(5):4331–4345. https://doi.org/10.1021/acs.energyfuels.6b00257

Takeda Y, Keiichi N, Keiichi N (1996) Emission characteristics of premixed lean diesel combustion with extremely early staged fuel injection. SAE Trans J Fuels Lubricants 105 (SAE Paper No. 961163):938–947. https://www.jstor.org/stable/44729107

Taymaz I (2006) An experimental study of energy balance in low heat rejection diesel engine. Energy 31:364–371. https://doi.org/10.1016/j.energy.2005.02.004

Taşkiran ÖO, Ergeneman M (2011) Experimental study on diesel spray characteristics and autoignition process. J Combust 2011:1–20. https://doi.org/10.1155/2011/528126

Teng H, Regner G, Cowland C (2006) Achieving high engine efficiency for heavy-duty diesel engines by waste heat recovery using supercritical organic-fluid Rankine cycle. In: SAE 2006 commercial vehicle engineering congress & exhibition. SAE Technical Paper 2006–01–3522. https://doi.org/10.4271/2006-01-3522

Tritt TM, Subramanian MA (2006) Thermoelectric materials, phenomena, and applications: a bird's eye view. MRS Bull 31:188–198. https://doi.org/10.1557/mrs2006.44

Tuttle J (1980) Controlling engine load by means of late intake-valve closing. SAE Trans 89 (SAE Paper No. 800794):2429–2441. https://doi.org/10.4271/800794

Viswanathan B (2017) Energy sources: fundamentals of chemical conversion processes and applications. Petroleum. Elsevier, Amsterdam, pp 29–57

Wall J (2012) Presentation about possible future Cummins engine technologies, Cummins.

Wirbeleit F, Binder K, Gwinner D (1990) Development of pistons with variable compression height for increasing efficiency and specific power output of combustion engines. SAE Trans J Engines 99 (SAE Paper No. 900229):543–557. https://doi.org/10.4271/900229

Chapter 7
Evolution of Catalytic Converters for Spark Ignition Engines to Control Emissions

Sulav Kafle, Hardikk Valera, and Avinash Kumar Agarwal

Abstract Engine exhaust species such as nitrogen oxides (NOx), carbon monoxide (CO) and unburnt hydrocarbons (HC) are hazardous and pose a significant threat to the environment and the human health. Catalytic converter is installed in the engine exhaust manifold of modern vehicles for emission reduction. This device plays a significant role by simultaneous catalytic oxidation and catalytic reduction reactions. Catalytic converters themselves do not take part in chemical reactions but catalyse these reactions. There is a drastic reduction in CO, HC and NOx emissions in the exhaust, and these species are converted to harmless products exiting through the tailpipe. Catalytic converters are among the most developed and matured technologies to control exhaust emissions. Emissions concentrations of different emission species can be analysed upstream and downstream of the catalytic converter to evaluate their conversion efficiencies. Also, real-time catalyst temperatures can be determined by 1-D thermal modelling of the three-way catalytic converters. This chapter gives insights into the flow behaviour along with the transient temperature field across the catalytic converter, which helps understand overall working of the catalytic converter. Also, catalyst temperature distributions are discussed based on engine operating conditions in order to understand the effect of exhaust gas temperature on the catalytic converter efficiency.

Keywords Catalytic converter · Exhaust emissions · Catalyst temperature · Conversion efficiency

---

S. Kafle · H. Valera · A. K. Agarwal (✉)
Engine Research Laboratory, Department of Mechanical Engineering, Indian Institute of Technology Kanpur, Kanpur 208016, India
e-mail: akag@iitk.ac.in

A. P. Singh and A. K. Agarwal (eds.), *Novel Internal Combustion Engine Technologies for Performance Improvement and Emission Reduction*, Energy, Environment, and Sustainability, https://doi.org/10.1007/978-981-16-1582-5_7

Abbreviations

1-D	One **d**imensional
ATDC	**A**fter **T**op **D**ead **C**entre
BMEP	**B**rake **M**ean **E**ffective **P**ressure
BTDC	**B**efore **T**op **D**ead **C**entre
CC	**C**atalytic **C**onverter
CCC	**C**lose **C**oupled **C**atalyst
EFI	**E**lectronic **F**uel **I**njection
HCNG	**H**ydrogen Enriched **C**ompressed **N**atural **G**as
NMHC	**N**on-**M**ethane **H**ydro**c**arbons
PGM	**P**latinum **G**roup **M**etals
PI	**P**roportional **I**ntegral
SI	**S**park **I**gnition
TC	**T**hermo**c**ouples
THC	**T**otal **H**ydro**c**arbons
TWC	**T**hree **W**ay **C**atalyst
UCC	**U**nderfloor **C**atalytic **C**onverter
ZDP	**Z**inc-**d**ialkyl-dithio**p**hosphate

Nomenclatures

Al	**Al**uminium
CO	**C**arbon **M**onoxide
CO_2	**C**arbon **D**ioxide
Cr	**Cr**omium
Fe	**I**ron
HC	**H**ydrocarbon
Ni	**Ni**ckel
NO_x	**N**itrogen **O**xides
Pt	**P**latinum
Pd	**P**alladium
Rh	**R**hodium
RPM	**R**evolution **P**er **M**inute
Si	**Si**licon

7.1 Introduction to Catalytic Converters

Air pollution and global warming are the two major threats to the global environment. Contribution of the transport sector to the overall global pollution cannot be ignored because automotive vehicles emit deleterious exhaust species such as carbon monoxide (CO), hydrocarbons (HC), nitrogen oxides (NOx), etc. These are significant contributors to global environmental threats. Since last 40 years, catalytic converters are being used in the engine/vehicle exhaust system to minimize the harmful emissions. Catalytic converters also reduce the particulates mass by destroying their organic fraction (Searles 1998). They typically consist of an intake section, comprising a pipe of uniform cross-sectional area followed by a diverging region, which contains a coated catalytic substrate in the middle and then the exhaust after catalysis passes through a converging section, eventually leading to the atmosphere. In catalytic converters, significant improvements and modifications have been made with time, in order to comply with stricter emission norms.

Further, innovations and modifications can be helpful in meeting the future environmental norms. Three-way catalysts (TWCs) are considered as one of the significant advancements in the automotive catalytic converters. Currently, the TWC converter, along with the electronic fuel injection (EFI) system, is used to reduce hazardous emissions by up to ~95%, if the engine operating conditions are conducive enough. The driving mode of the vehicle affects the exhaust species concentration, temperature and mass flow rate at the intake of the catalyst. These parameters affect the conversion efficiency of catalysts and therefore the emissions at the catalytic converter outlet. Other factors affecting the conversion efficiency include type of catalyst, size of catalyst and its location. Therefore, thermo-mechanical stresses, vibrations and efficiency degradation through ageing as well as flow distribution should be taken into consideration for optimum design of catalytic converters (Tartakovsky et al. 2012; Mitianiec 2001).

Globally, energy policies are focussing on minimizing excessive dependence on petroleum products and introduce an economical, eco-friendly and easily available energy resource for transport sector. Therefore, alternative fuels like alcohols, ethers, biodiesel and alternative propulsion such as fully electric/hybrid vehicles are gaining importance. The main objective of these fuels and propulsion systems is to minimize emissions and achieve a performance similar to conventional fuels driven vehicles. The problem associated with renewable fuels is that although they reduce regulated emissions, they tend to increase unregulated emissions compared to fossil fuels. Chao et al. (2000) reported higher formaldehyde, acetaldehyde and acetone emissions with ethanol-gasoline blends in comparison to gasoline. He et al. (2003) reported that acetaldehyde emissions surged with increasing ethanol content in the test fuel blends. Therefore, extensive research is required to be undertaken for minimizing emissions by modifications in catalytic converters, development of newer technologies for overall emission control or by selection of suitable alternative fuels.

Catalytic converters are heterogeneous catalytic reactors consisting of catalyst in solid phase and reactants and products in vapour phase. Physical and chemical phenomena undergoing in the catalytic converters control their performance. For instance, the efficiency of the converter is related to the transport of mass, energy and momentum in radial and axial directions and complicated chemical kinetics. In hybrid vehicles, thermal management of catalytic converters is very important. Hybrid vehicles suffer from the cold-start emission problems because electric engines are usually switched off during low speed-torque requirement. The IC engine catalytic converters cannot attain light-off temperature quickly, hence it decreases the conversion efficiency of pollutant species (Simon et al. 2015). Walls of the parallel channels of catalytic converters consist of a wash-coat, where catalyst pellets are present. This results in the formation of a boundary layer through which heat and mass transfer takes place between fluid and fluid/solid interface. The reaction occurs within the porous medium in the wash-coat due to the catalyst. On the active site of catalysts, reactants get adsorbed, and products get desorbed and diffused back to the fluid in the channel (Bertrand et al. 2012; Hayes and Kolaczkowski 1997).

7.2 Internal Structure of Catalytic Converters

The catalytic converter comprises different essential components, namely, monolith, wash-coat, catalyst, intumescent mat, covering case and heat shields. Monolith is the ceramic or metallic block with a large number of parallel channels with a large heat transfer area. Wash-coat increases the active surface area as it coats the substrate and facilitates application of catalyst. Precious metals coated over the wash-coat of monolith are referred to as 'catalysts'. The intumescent mat acts like thermal insulation and prevents mechanical shock. Covering case encloses all these essential components and heat shield is provided over the periphery to protect it from thermal shocks. The internal structure of the catalytic converter is quite complicated, as shown in Fig. 7.1. Detailed description of the composition and properties of monolith, wash-coat and catalyst is given in the following sub-sections.

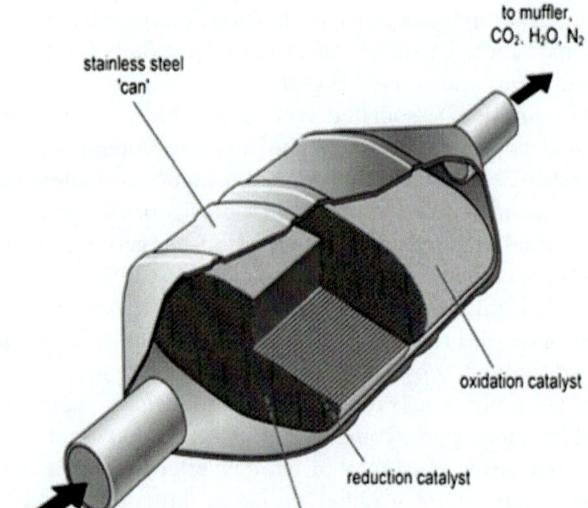

Fig. 7.1 Internal structure of a typical three-way catalytic converter (Cleveland and Morris 2013)

7.2.1 Monolith

Monolith is a block consisting of a large number of parallel channels, which enables a high flow rate, has large geometric surface area and gives low pressure drop to a fluid flow across it. Monoliths are manufactured by extrusion process. Ceramic and metallic are the two types of monoliths used for these applications. Ceramic monoliths are more commonly used in the catalytic converters. Metallic substrates are relatively less suitable because they have low porosity; hence only a thin layer wash-coat can be applied over its surface. Monolith structures are found in various shapes. The diameter of the monolith structure can be ~15 cm and length can reach up to 1 m. The opening dimension of parallel channel varies from 0.5 to 10 mm (Shelef and McCabe 2000; Heck and Farrauto 2001; Catalytic Converter (http://wiki.ask.com/Catalytic_converter).

Monolithic substrates possess two important mechanical properties: (i) low coefficient of thermal expansion, and (ii) high resistance to thermal shock. Ceramic monoliths are formed from synthetic Cordierite ($2MgO \cdot 2Al_2O_3 \cdot 5SiO_2$-14% MgO, 35% Al_2O_3, 51% SiO_2). Cordierite possesses a high melting point of ~1450 °C. It also has a low coefficient of thermal expansion and provides resistance to thermal fracture and oxidation (Williams 2001). Metallic monoliths have a higher coefficient of thermal expansion compared to ceramic monoliths; thus, they need special bonding treatment for protecting the wash-coat layer. For high temperature applications such as in the close coupled catalysts (CCC), FeCr alloys (73% Fe, 20% Cr, 5% Al with traces of Ni and Si) are used, which melt ~1500 °C (Williams 2001). Due to its thermal and mechanical properties, a metallic monolith is used in a CCC to increase the conversion efficiency during cold start. Both monoliths can be made with protrusions in their channels to increase the turbulence. Therefore, the mass transfer from these channels is enhanced. In this regard, metallic substrates have an advantage as corrugated foils can be used for the next generation catalysts called 'turbulent catalysts' (Lotti et al. 2005). Metallic substrates have the advantage of lower back pressure because of their larger frontal area (Pfalzgraf et al. 1996). Ceramic substrates increase the conversion due to its larger geometric surface area through cell density enhancement and thermal mass reduction within the walls (Schmidt et al. 1999; Umehara et al. 1997). It is also observed that the maximum conversion of pollutants is obtained by increasing the geometrical surface area, where the rate of conversion is controlled by the bulk mass transfer in the substrate (Umehara et al. 2000).

In an exhaust manifold, flow is turbulent (Chan and Hoang 1999), while it is laminar in parallel channels of a monolith (Koltsakis and Stamatelos 1997). The transition takes place from turbulent to laminar flow in these channels after a few millimetres from the entrance because flow becomes fully developed in all parallel channels after the entrance region. The surface transport of chemical species is governed primarily by diffusion, while axial transport follows a convection phenomenon (Santos and Costa 2008). Channel geometry is one of the significant design variables that affect the performance of the catalytic converter. Residence time and reaction time both decrease in extreme operating conditions; however, the reaction time reduces more in comparison to residence time since it is exponentially related to the temperature for the catalytic reactions. It implies that the conversion is mainly due to external mass transfer, hence channels with a higher mass transfer coefficient are desirable. Space velocity is a parameter used for analysing the conversion efficiency of the catalytic converters. It is the function of the mass flow rate of the exhaust gas. With an increase in engine load, space velocity increases linearly. Higher engine speeds give higher space velocities. For high space velocities, conversion efficiencies are high for longer channels, which have longer residence time. The same trend is observed with higher cell density (channels with smaller hydraulic diameters). However, pressure loss increases with increasing length of the channel and the cell density, which ultimately affects vehicle's performance as well (Santos and Costa 2008).

7.2.2 Wash-Coat

Wash-coat is the coating of inorganic oxides like aluminium oxide, silicon dioxide, titanium dioxide, or a mixture of silica and alumina applied over the surface of channels of the monolith. It increases the active geometric area for the dispersion of catalysts. This coating forms a coarse and uneven rough surface, which leads to the formation of most active catalytic surface layer and increases the conversion. Wash-coat also prevents sintering of the catalysts into catalytic pellets even at high temperatures up to 1000 °C (Catalytic Converter (www.catalyticconverters.com)).

7.2.3 Catalyst

Catalytic converters use precious metals as the catalysts and these metals are combination of platinum group metals (PGMs), namely, platinum, palladium and rhodium. Platinum is very active; however, it is not suitable for all applications due to being very expensive and its use leading to some unnecessary reactions. Noble metals have excellent thermal resistance for degradation, superior cold-start performance and low deactivation efficacy toward sulphur. The amount of these noble metals in the catalytic converters is ~1–2 g only. The particle size of noble metals in the fresh catalytic converter is less than or equal to 50 nm. At high temperatures, these particles get sintered and the particle size grows up to 100 nm. Palladium is commonly used as an oxidation catalyst and rhodium as a reduction catalyst. Platinum can be used as both, an oxidation and a reduction catalyst. Other than these metals nickel, iron, copper, manganese and cerium are also used as a catalyst; however, some of these are banned for use in different parts of the world. For instance, nickel is restricted for use in the European Union due to its reaction with CO to form nickel tetra-carbonyl. Also, copper is unlawful to use in North America due to formation of dioxin (Shelef and McCabe 2000; Catalytic Converter (www.catalyticconverters.com); Marsh and Acke 2001).

7.3 Evolution of Catalytic Converters

In the past few decades, several technologies related to catalytic converters have been developed for meeting emission standards such as a CCC for cold-start HC emission reduction, an electrically heated catalyst for shortening the light-off time, hydrocarbon trap for storing unburned hydrocarbons, exhaust gas igniter for reducing the light-off time, chemically heated catalyst, secondary air injection for emission reduction via thermal oxidation and turbulence generating devices for maximum conversion efficiency (Farrauto and Heck 1999; Lee and Heywood 2010; Tien et al. 2020; Agrawal et al. 2019). Use of additional combustion devices, higher idle speed, retarded ignition timing and variable valve timing are the other ways to mitigate harmful emissions via shortening light-off duration during the warm up phase. Thermal management of these converters is also essential to maintain high exhaust gas temperatures. High catalyst temperature increases the thermal sintering of substrates, which shortens the durability of the catalyst (Gao et al. 2019). The catalyst light-off performance also depends on the fuel used in the engine. The use of alternative fuels and blends with gasoline are common as they reduce the emissions and increase the efficiency of the engine. Among various alternative fuels, alcohol blends have higher oxygen content. This inherent oxygen increases the exhaust gas temperature and reduces the light-off duration. Hydrogen has high flame velocity, which brings relatively fast light-off temperature in case of hydrogen enriched compressed natural gas (HCNG) (Gao et al. 2019).

7.4 Limitations of Catalytic Converters

At high exhaust gas temperature (~1000 °C), metals present in the catalytic converter are subjected to deactivation because of sintering, which leads to reduction in catalytic activity due to a reduction in the surface area covered by the catalyst in the catalytic converter. Further, the effectiveness of the catalytic converter decreases via two ways: (i) thermal deterioration and (ii) poisoning. Thermal deterioration takes place once the catalyst is exposed to high temperatures. This results in the sintering of platinum group metals (PGM), loss in functional area of the catalytic converter and the phase transformation of catalysts. Catalytic poisoning refers to reduction in catalytic activity due to blocking of the pores, which form active sites or even blocking the active sites themselves (Shelef et al. 1979; Beck et al. 1997). Lead, sulphur and zinc poisoning are some permanent chemical effects, through which catalyst efficiency gets diminished (Tartakovsky et al. 2012). When the lubricating oil is consumed, zinc-phosphorus compounds present in the oil form non-combusted zinc-di-alkyl-di-thio-phosphate (ZDDP). In the catalytic converter, ZDDP forms a glaze of zinc-phosphate over the catalyst, which deteriorates its ability. For a long time, ZDDP was used as an anti-wear and anti-oxidant in the automotive lubricants, however, nowadays, its content is lowered due to its adverse impacts (Tartakovsky et al. 2012; Ball et al. 1997).

During the warm up period in the first five minutes, majority of pollutants are emitted by the vehicle. However, the catalytic converter is not activated because it is unable to reach sufficiently high light-off temperature in such a short span. In the past, heating coils were used in the catalytic converter for its accelerated heating. They are electrified just after starting the engine, which helps in achieving the light-off temperature of the catalyst quickly (Glander and Zidat 1998). Catalytic converters are most effective in reducing the hydrocarbons and CO emissions; however, they cannot reduce CO_2 emissions. CO_2 is a major contributor to the global warming. Also, modern three-way catalytic converters run close to the stoichiometric region; hence the engine consumes more fuel compared to lean-burn engines. Thus, CO_2 emissions are higher from these vehicles though catalytic converter fitted engines are cleaner compared to lean burn engines. Catalytic converters use precious metals such as palladium or platinum, which are available scarcely in only few parts of the world and the extraction process of these metals involves significantly polluting processes (Catalytic Converter (www.catalyticconverters.com)). One of the main problems for the future three-way catalytic converter use will be NOx control with lean fuel/air mixtures. Since the future trend of the engines is towards developing lean burn engines, a three-way catalytic converter would be ineffective in lean burn engines compared to stratified charge engines and diesel engines (De Nevers 2000).

7.5 Three-Way Catalytic Converters

The most commonly used catalytic converter configuration in a modern engine emission control system is three-way catalytic converter. Precious metals, namely platinum, palladium and rhodium, are used for HC and CO oxidation and NOx reduction simultaneously. With the adaptation of stringent emission norms nowadays, emission control technology is required to be developed to address these environmental challenges. It requires development of a three-way catalytic converter (TWC) along with refinements in the electronic fuel injection (EFI) system. An efficient TWC design reduces significant harmful emissions by almost ~90% (Ribbens 2017). Its effectiveness is a complex phenomenon and requires a deep understanding of chemical reactions, species conversion efficiency and feedback loop control. These are discussed further in detail in the following sub-sections.

7.5.1 Chemical Reactions in TWC

The first set of chemical reactions includes the oxidation and reduction reactions of CO, HC and NOx, respectively. However, there are other side reactions such as water gas shift reaction and steam reforming reaction that also takes place. N_2O and NO_2

are formed as intermediate products. There is also the concept of storing NOx during lean conditions for some time.

Oxidation reactions

$$2CO + O_2 \rightarrow 2CO_2$$
$$2HC + 5/2O_2 \rightarrow 2CO_2 + H_2O$$

Reduction reactions

$$2CO + 2NO \rightarrow 2CO_2 + N_2$$
$$4HC + 9NO \rightarrow 4CO_2 + H_2O + H_2 + 9/2N_2$$
$$2NO + 2H_2 \rightarrow 2H_2O + N_2$$

Water-gas shift reaction

$$CO + H_2O \rightarrow CO_2 + H_2$$

Steam reforming reaction

$$HC + 2H_2O \rightarrow CO_2 + 5/2H_2$$

7.5.2 Conversion Efficiency of TWC

The effectiveness of a catalytic converter is defined by its conversion efficiency of the exhaust species. For determining the conversion efficiency of the TWC, measurements of exhaust gas samples at upstream and downstream of the catalytic converter are required. This measurement is generally performed, when an engine is running at different loads and speeds. Conversion efficiency is defined by the following equation:

$$\eta = 1 - \frac{\text{emission concentration after TWC}}{\text{emission concentration before TWC}}$$

Catalytic conversion efficiency depends on inlet conditions of the catalyst, its position, size and type. Threshold temperature at which catalyst becomes activated to convert particular exhaust species is called 'light-off temperature'. If the catalytic converter does not reach this temperature, then the harmful exhaust species do not get sufficiently oxidized. Thus, the purpose of the catalytic converter doesn't fulfil and its presence in the exhaust circuit does not make any difference to the emissions.

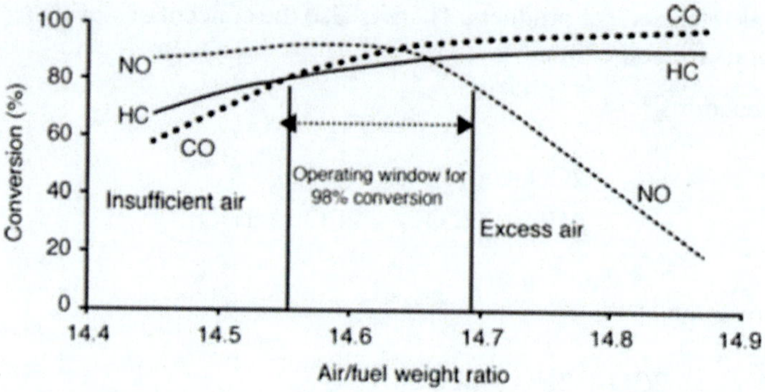

Fig. 7.2 Typical TWC conversion efficiency versus air/fuel ratio trends (Wade and Farrauto 2012)

This typically happens during the cold-start period. Conversion efficiency depends on air/fuel ratio primarily and its variation for the three exhaust gases is shown in Fig. 7.2. At low air/fuel ratio, NOx conversion efficiency is quite high but, CO and HC conversion efficiencies are relatively smaller but at high air/fuel ratio, NOx conversion efficiency decreases while CO and HC conversion efficiencies increase. Hence, there is a region of trade-off between NOx and HC/CO conversion efficiencies and that is an acceptable window that can help meet the emission standards.

7.5.3 *TWC with Feedback Loop Control*

TWCs were introduced in 1980 and it required the engine to run at stoichiometric air/fuel ratio ($\lambda = 1$). This could be achieved by placing an oxygen sensor at the intake of the catalytic converter. It provides a signal to the engine control module; this further directs a signal to the fuel delivery system to maintain the air/fuel ratio. Oxygen storage component consisting of ceria/zirconia oxides is also a part of TWC. When the exhaust stream is in lean conditions ($\lambda > 1$), it adsorbs oxygen and utilizes it during fuel rich conditions ($\lambda < 1$). This is useful in conversion efficiency fluctuations generally encountered between rich and lean conditions. It maintains nearly stoichiometric conditions. The modern TWC consists of mostly noble metals Pt, Pd and Rh and CeO_2-ZrO_2 deposited on γ-Al_2O_3 over the channels of the metallic or ceramic monoliths. In commercial catalysts these days, palladium has replaced platinum. To improve the HC conversion efficiency during initial cold-start period, a close coupled catalyst is used as an additional catalyst near the exhaust manifold along with TWC. This is very important for cold-start emission reduction by the conversion of un-combusted hydrocarbons emissions. This is very effective for increasing the overall conversion efficiency above 98%, up to 150,000 miles of vehicle travel.

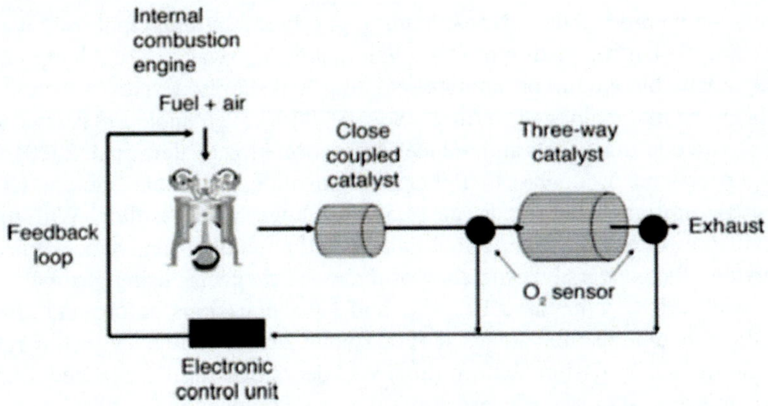

Fig. 7.3 Feedback loop control for TWC with close coupled catalyst (Wade and Farrauto 2012)

The electronic engine control system optimizes the quantity of fuel delivered subjected to complex constraints that are to be maintained for compliance with emission regulations and for improving the fuel economy. The lambda/oxygen sensors placed in both inlet and outlet of the catalytic converter assist in maintaining the stoichiometric air/fuel ratio (Ribbens 2017; Wade and Farrauto 2012). Measurement errors, errors in sensors and actuators are corrected by feedback controller using the oxygen sensors. To match the dynamic offset of the upstream lambda sensor, the feedback controller uses a switch type downstream lambda sensor. The most common feedback control strategy these days includes the use of PI controller to compensate for deviations in the signal from the lambda sensor placed downstream of the TWC. This maintains stoichiometric conditions without taking into account the dynamics of oxygen storage of the used catalyst. This, in turn, leads to unnecessary control actions during fuel cut-off stages. Thus, the feedback controller is kept off during fuel cut-off and it remains off until the signal from the downstream lambda sensor is nearly stoichiometric. Therefore, the feedback controller minimizes the recovery time of signals from a downstream lambda sensor compensating the oxygen storage dynamics of the catalytic converter (Kiwitz et al. 2012). Typical schematic of feedback loop control with CCC is shown in Fig. 7.3.

7.6 Catalytic Converter for Gasohols

Increasing demand for petroleum products and restrictive environmental norms warrant use of renewable fuels. These renewable fuels should be economical, eco-friendly and easily available. Alcohols, mainly ethanol and methanol, seem to be strong contenders because their chemical and physical characteristics are similar or superior to conventional fuels. Also, they can be produced from renewable as well as non-renewable resources. Ethanol can be produced from sugar cane and biomass.

Methanol can be produced from coal, natural gas, biomass/municipal solid waste as well as catalytic hydrogenation of CO_2 (Valera and Agarwal 2019). Alcohol blends are easily adaptable via minor tuning of existing fuel supply devices. Alcohols have higher latent heat of vaporization (methanol: 1098 kJ/kg; ethanol: 838 kJ/kg), which provides extra cooling effects and reduces NOx formation (Valera et al. 2020). Since the oxygen content in alcohol is higher than gasoline, emissions such as HC and CO from alcohol-gasoline blends (gasohols) are lower than gasoline. With the use of catalytic converters, these emissions can be further reduced, and hence, it is better to understand the conversion efficiency of the exhaust species using gasohol.

Kim et al. (2019) compared HC, CO and NOx emissions before and after the TWC using ethanol/gasoline blends at 1500 rpm at 2 bar BMEP by adjusting relative air/fuel ratios. HC emissions before the TWC decreased until λ reached 1.05 and then HC emissions increased. These emissions were reduced in the TWC, hence HC emissions decreased continuously and observed to be very low under stoichiometric conditions. HC conversion efficiency was almost ~100% after attaining stoichiometry. It was due to the presence of O_2 concentration in leaner and richer conditions. Similarly, CO emission reduced with an increasing relative air/fuel ratio before the TWC. It is because of an increase in O_2 concentration leading to improved CO oxidation and consequent reduction in CO emission. With a rise in air/fuel ratio, the conversion efficiency increased due to active oxidation reactions with higher O_2 concentration. CO was almost totally oxidized after $\lambda \geq 1$. In the region of $\lambda \leq 0.95$, the conversion efficiency of the catalyst was low due to incomplete oxidation with lower O_2 concentration. At lower and richer air/fuel ratios than stoichiometry, NOx emissions were relatively smaller and increased closer to the stoichiometry because their concentration in the exhaust is determined by the peak combustion temperature and O_2 concentration in the mixture (Heywood 1988). In this study, NOx emissions didn't change significantly with increasing air/fuel ratio since achieved peak combustion temperature was not enough to produce NOx at operated air/fuel ratios. For $\lambda < 1$ region, NOx conversion efficiency was very high. However, it plummets down to negligible for lean mixtures. Increase in NOx is related to an increase in oxygen content in the ethanol/gasoline blends and availability of free oxygen in the exhaust, which participated in reactions, instead of oxygen present in NOx.

7.7 Flow in a Catalytic Converter

Internal flow distribution in a monolith significantly affects the light-off temperature, warm up performance and temporal variations in the catalyst. Light-off temperature increases with reduction in uniformity index, which in turn decreases the conversion efficiency of the catalytic converter. This also increases the radial temperature gradient of the monolith (Guojiang and Song 2005; Martin et al. 1998). Upstream flow is highly non-uniform, but this non-uniformity of flow decreases in individual channels due to frictional pressure drop along the flow direction. Also heat losses and chemical reactions in the monolith affect the flow distribution (Agrawal et al.

2012). In many studies, it is implicitly assumed that flow, temperature and species distribution are the same throughout all channels. Still, it is not completely correct due to these two reasons: (i) flow mal-distribution and (ii) heat losses. Flow mal-distribution results due to geometry of inlet diffuser of the catalytic converter, which in turn leads to significant difference in the exhaust flow across various channels. Heat losses from the catalytic monolith lead to a variety of temperature profiles in the central and peripheral channels (Agrawal et al. 2012). Non-uniformity of flow due to recirculation zones affects the monolith performance during cold start (Chakravarthy et al. 2003). Induced flow mal-distribution can be reduced by deploying flow tailoring devices. However, their utilization results in significant additional pressure drop and thermal inertia (Agrawal et al. 2012). Also, it can be reduced by increasing the angle of the diffuser section. This may face space and design constraints in case of a CCC, where it is subjected to non-uniform flows (Salasc et al. 2005).

As flow expands in the diverging section of the diffuser, recirculation zones form. This leads to high and low axial velocities around the centreline and the periphery, respectively. Frictional pressure drop increases with inlet velocity, and a higher flow rate in the channels offers higher frictional resistance. To minimize this resistance, the pressure field redirects flow from the centre towards the periphery and this resistance makes the flow uniform throughout the monolith. From Fig. 7.4, it is clear that recirculation zones in the diverging section bring non-uniformity in the flow. From the velocity vectors shown, it can be seen that the transverse velocity is higher than the centre velocity, showing that flow is directed towards the periphery. Here, recirculation zones and pressure fields oppose each other, where recirculation zones form flow mal-distribution while the latter homogenizes. The flow mal-distribution creates a difference in residence time, which reduces the conversion efficiency. Exothermic

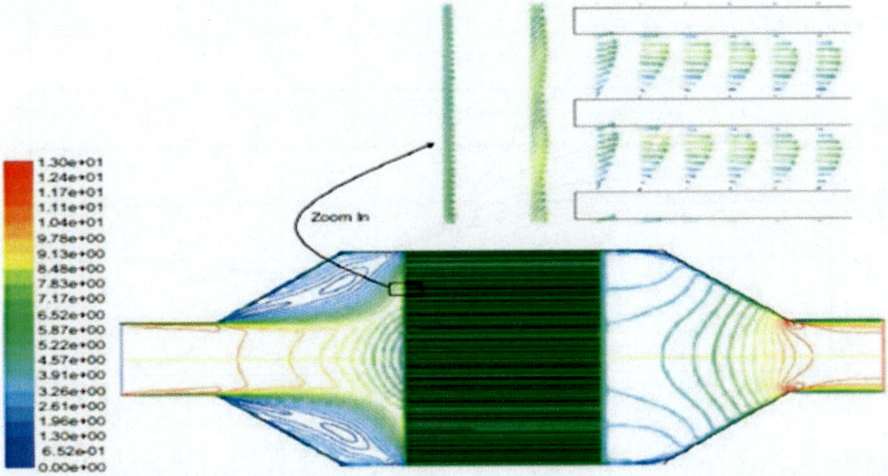

Fig. 7.4 Velocity contours showing recirculation zones and velocity vectors at the inlet of catalyst monolith (Agrawal et al. 2012)

reactions and thermal effects due to heat losses affect the flow distribution. The reason behind this is a significant change in density and viscosity of the fluid w.r.t. temperature and species concentration (Agrawal et al. 2012).

Ideal axial static pressure profile in the catalytic converter consists of two regions. The first region with a higher gradient is the boundary layer region and the second region is a region of fully developed laminar flow (Abu-Khiran et al. 2003). The pressure drop across the catalytic converter using an angled pipe is more significant in comparison to offset pipe as flow through angled pipes is non-uniform. It is also concluded from the studies that pressure drop rises with an increase in mass flow rate because the mass flow rate also enhances non-uniformity in the flow in the substrate. In addition to the above, increased cone angle also increases non-uniformity in the flow, which enhances the pressure drop (Ramadan et al. 2007). Also, non-uniformity of flow increases with increasing Reynolds number of the fluid (Shuai et al. 2001).

Velocity profile varies with increment in monolith temperature. During the cold-start phase, flow mal-distribution occurs at the inlet of the monolith channels. But the velocity profile becomes more uniform after some time, when the converter warms up gradually. After 15–20 s of engine starting, the velocity profiles are quite similar to each other, or no significant changes are observed in the flow distribution, as seen in Fig. 7.5 (Tsinoglou et al. 2004). Besides the chemical efficiency of a catalytic converter, flow resistance in the catalytic converter also plays a vital role because the installation of catalyst in the exhaust system alters the flow behaviour of the system significantly. If the flow resistance is very high, it is quite challenging to bypass the exhaust gas flow from the system, which increases the work consumed during the charge exchange (Postrzednik et al. 2004).

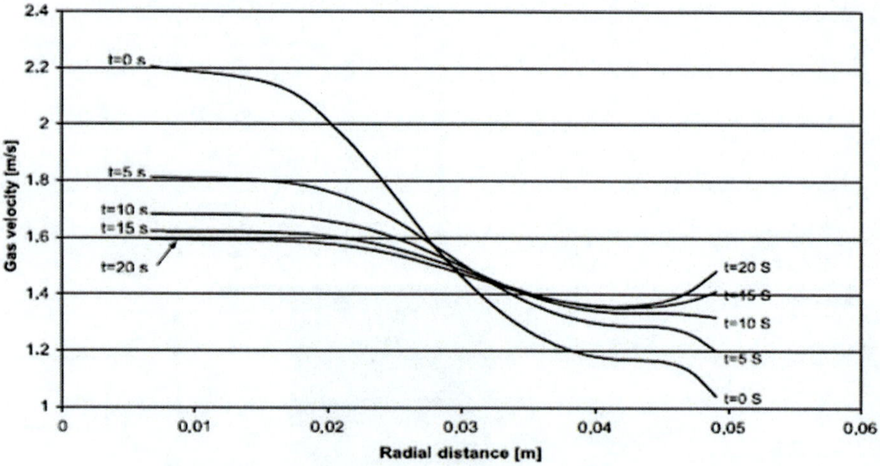

Fig. 7.5 Inlet velocity profiles of monolith during light-off phase (Tsinoglou et al. 2004)

7.8 Thermal Modelling of TWC

It is essential to know the thermal behaviour of the TWC during the warm up phase and in hybrid powertrains. This provides helpful information of effective real-time conversion of harmful exhaust species. 1-D thermal model is considered for a cylindrical catalytic converter to understand the thermal behaviour of the catalytic converter. In this model, the radial temperature gradient is neglected by assuming a well sprayed exhaust gas flow at the inlet. Due to high thermal inertia and low thermal conductivity in the axial direction, an axial thermal gradient is considered. Nomenclature used is shown in Table 7.1.

Using physical and simplified model given by Sabatini et al. (Sabatini et al. 2015) for two-phase models, solid/gas phase, energy balance through the exhaust gas phase is given by:

$$\rho_g \in c_{pg} \frac{\partial Tg}{\partial t} = -\frac{\dot{m}_g}{A_c} c_{pg} \frac{\partial Tg}{\partial z} + \in k_g \frac{\partial^2 Tg}{\partial z^2} + h_{int,cat} A_{geo} \left(T_{cat} - T_g \right) \tag{7.1}$$

Table 7.1 Nomenclature used

Parameter	Description	Units
$\dot{m}_g$	Mass flow rate of exhaust gas	kg/s
ρ_g	Density of the exhaust gas	kg/m^3
ρ_s	Density of the solid phase	kg/m^3
c_{pg}	Specific heat of the exhaust gas	J/kgK
c_s	Specific heat of the solid phase	J/kgK
k_g	Thermal conductivity of exhaust gas	W/mK
k_s	Thermal conductivity of solid phase	W/mK
$h_{ext,cat}$	Heat transfer coefficient TWC/ambient	W/m^2K
$h_{int,cat}$	Heat transfer coefficient TWC/gas	W/m^2K
$h_{int,post}$	Heat transfer coefficient post/gas	W/m^2K
$h_{ext,post}$	Heat transfer coefficient post/ambient	W/m^2K
$\in$	Volume fraction of exhaust gas	–
A_c	Cross-sectional area of catalyst	m^2
A_{geo}	Specific geometric area of the catalyst	m^2/m^3
A_{out}	External surface area of the catalyst	m^2
V_{cat}	Volume of TWC	m^3
η_{conv}	Conversion efficiency	–
s_i	Slope of conversion efficiency function	K^{-1}
T_{lo}	Light-off temperature	K
T_{cat}	Catalyst temperature	K
T_g	Temperature of exhaust gas	K
Q_r	Heat released by conversion	W

The first term in the RHS of the equation represents convective axial heat transport through the gas phase. The second term represents the heat conduction and the last term accounts for the heat exchange between gas and solid phase in the catalytic converter.

Similarly, the energy conservation equation for the solid phase is given by:

$$\rho_{cat}(1-\in)c_{cat}\frac{\partial T_{cat}}{\partial t}=(1-\in)k_s\frac{\partial^2 T_{cat}}{\partial z^2}-\frac{A_{out}}{V_{cat}}h_{ext,cat}(T_{cat}-T_{amb})-h_{int,cat}A_{geo}\left(T_{cat}-T_g\right)+Q_r \tag{7.2}$$

The first term in the right hand side of the equation is heat transfer via conduction through solid. The second term signifies the convective heat exchange between the catalyst and the ambient. The third term refers to internal convection between solid/gas phases and the last term is the heat released from the conversion of pollutants from the exothermic reactions in the catalytic converter.

In Eq. (7.1), gas phase conduction can be neglected as it does not have a significant impact on the catalyst temperature. The storage term in the equation also can be neglected because the dynamics of gas phase is faster than the dynamics of catalyst temperature, which means T_g is at a steady state (Depcik and Assanis 2005). In Eq. (7.2), every term has a significant influence on the catalyst temperature. Therefore, it is clear that heat due to exothermic reactions and heat exchanged with the gas are dominant factors.

Ceramic substrates have low thermal conductivity; thus, conduction across the substrate is negligible (Montenegro and Onorati 2009). From the literature for getting S shaped efficiency behaviour (Kang et al. 2014), as shown in Fig. 7.6, TWC efficiency has been chosen as the hyperbolic function of catalytic temperature.

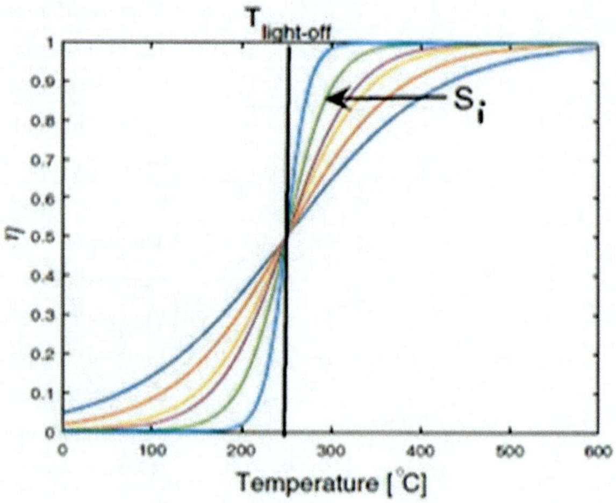

Fig. 7.6 TWC efficiency versus temperature showing S shaped behaviour (Sabatini et al. 2015)

$$\eta_{conv}(T_{cat}) = 0.5\tan\ h(s_i(T_{cat} - T_{lo})) + 0.5 \tag{7.3}$$

Equation (7.1) is further simplified and written as:

$$\frac{\dot{m}_g}{A_c} c_{pg} \frac{\partial Tg}{\partial z} = h_{int,cat} A_{geo} (T_{cat} - T_g) \tag{7.4}$$

Similarly, Eq. (7.2) becomes

$$\rho_{cat}(1- \in)c_{cat} \frac{\partial T_{cat}}{\partial t} = -\frac{4}{D_{cat}} h_{ext,cat}(T_{cat} - T_{amb}) - h_{int,cat} A_{geo}(T_{cat} - T_g) + Q_r \tag{7.5}$$

These partial differential Eqs. (7.4) and (7.5) are discretized using a suitable finite difference scheme (upwind). They can be integrated using computationally less costly numerical methods (explicit Euler forward method), employing appropriate initial and boundary conditions. Thus, this thermal model provides a deep understanding of the thermal behaviour of the catalytic converter.

7.9 Transient Temperature Measurement

One of the main challenges associated with the catalytic converter is to minimize the hydrocarbons released from the vehicles during cold start. Majority of hydrocarbons are emitted during this initial phase of starting of vehicles because of the low exhaust and monolith temperatures (Gao et al. 2019).

Thus, it is crystal clear that the temperature in the catalytic converter plays a vital role in increasing HC conversion efficiency. It is necessary to have a high temperature of catalyst to have excellent conversion efficiency, as shown in Fig. 7.7. There are few studies available on the measurement of the unsteady temperature field inside the catalytic converter. If the temperature of the catalytic converter is

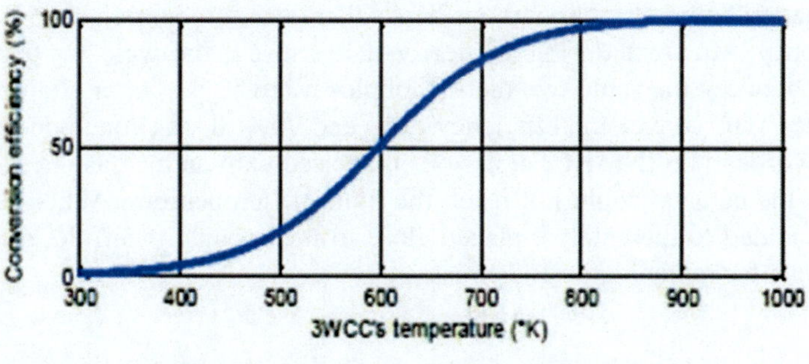

Fig. 7.7 Conversion efficiency versus catalyst temperature (Simon et al. 2019)

known, then it is useful to be aware of the real-time conversion of pollutants. Shuai and Wang (2004) measured unsteady two dimensional temperature fields in coated catalytic monolith and compared it with uncoated monolith using thermocouples. Thermocouples were placed axially and radially at different locations in the catalytic converter. It was concluded that temperature variations and warm up characteristics of the catalytic monolith depends on several thermal/flow factors: (i) convective heat transfer between the monolith and exhaust gas, (ii) heat conduction in the monolith, (iii) heat released from chemical reactions and (iv) flow velocity distribution. Temperature distribution in uncoated channels and catalytic channels are similar when the catalysts in the catalytic monolith are not activated. When the catalysts are enabled, then the maximum temperature is obtained at the rear central part of the monolith. Once the catalyst starts functioning fully, the temperature in the front face of the monolith becomes even higher compared to the exhaust gas temperature. At this instant, exhaust gas acts like coolant and cools the front part. Thus, the front part temperature decreases by convective heat transfer. In contrast, high exhaust temperatures and chemical reactions from upstream accelerate the chemical reactions and increase the temperature in the middle and rear parts of the catalytic monolith. At the centre part of the monolith, flow velocity becomes maximum due to which the highest temperature is observed at the front centre of the monolith (Shuai et al. 2000). Then, this high temperature expands from the centre to the periphery and moves backward with heat transfer in both radial and axial directions. After some time, the temperature tends to become uniform throughout the catalytic converter.

Zammit et al. (2012) studied the effects of catalytic converter location and palladium loading on the tailpipe emissions. Their study was focussed on finding the potential palladium cost penalty due to increased light-off duration required. In this study, the catalyst was placed far away from the engine. At different locations with constant spacing, temperatures were measured for two aged catalytic converters with different palladium loading. In this investigation, EGT and HC conversion efficiency were measured during a standardized cycle with high Pd during the first 500 s of the test. During the first 25–50 s, the temperature difference was around 160 °C between nearest and farthest thermocouples (TC1 and TC24). This was the time required for the catalyst to reach the light-off temperature. When it was second idle in the driving cycle, there was a drop in temperature of the catalyst, which directly affected the oxidation of hydrocarbons hence there was a reduction in THC conversion efficiency. After that, during the steady-state cruise in the cycle, the temperature difference between the same two thermocouples was 150 °C. After attaining Hill 2 in the cycle, THC conversion efficiency exceeded 90%. It was concluded that non-methane hydrocarbon (NMHC) emissions increased with an increase in distance of catalyst as the catalyst could not reach the light-off temperature. When palladium metal was added to the catalyst placed close to the exhaust manifold, it was very cost-effective for reducing the HC emissions.

7.10 Effect of Engine Operating Conditions

It is essential to understand variations in the EGT with engine operating conditions such as engine load, ignition timing, excess air ratio, misfire rates, etc. for understanding the overall effectiveness of the catalytic converter. Lee et al. (2002) concluded that catalyst inlet gas temperatures at the entry of CCC and under floor catalytic converter (UCC) both increased at 2500 rpm with load increment. Also, temperatures of both catalysts increased with load. With an increased load, EGT at the inlet and outlet of both catalytic converters (CCC and UCC) also increased. EGT measured at the entrance of CCC and temperature recorded at catalyst bed, both were maximum at $\lambda = 1$ because the heat produced from exothermic reactions was maximum at the stoichiometry. Temperatures increased up to stoichiometric conditions and then decreased, while the experiment was conducted at constant load at varying engine speeds. Light-off duration decreased with retarded ignition timings because it provided higher exhaust thermal energy. Literature reported a reduction in light-off duration of 34 s with 30 °CA retard ignition timing.

Residence time increased with the length of the catalytic converter, which enhanced the light-off temperature. This enhancement also increased the species conversion, but maximum conversion efficiency did not change significantly. With increasing length, wall temperature decreased because convective heat transfer between the wall and the substrate dominated the heat produced from the local exothermic reactions. Increment in cell densities in monolith through change over in the hydraulic diameter and porosity enhanced the species conversion up to a specific limit. However, it fell due to reduction in particular surface area. Cell density determined the thermal inertia, which was associated with the light-off performance; thus, lower cell density enhanced the light-off due to lower thermal mass (Subramanian et al. 2004).

7.11 Summary

In this chapter, a comprehensive study of a catalytic converter in contemporary spark ignition engines is presented. The main objective of the chapter is to find out the effectiveness of a catalytic converter by developing an overall understanding of multi-physics phenomena undergoing in it. It is vital to understand the innate behaviour, which is key in developing emission mitigation technologies for meeting stringent emission norms for modern day internal combustion engines and hybrid vehicles. With recent developments in the electronic fuel injection system with feedback control along with modified designs of the catalytic converter system, hazardous emissions can be successfully negated. Transient flow behaviour and temperatures in the catalytic converter, variations in conversion efficiency with different catalytic converter geometries, conversion efficiencies with gasohols, changes in exhaust gas temperature and catalyst temperature with engine operating conditions, etc. are analysed to understand the overall complex working of the catalytic converters. This deep understanding would help find out an optimum design for use as catalytic converters in modern engines/vehicles.

References

Abu-Khiran E, Douglas R, McCullough G (2003) Pressure loss characteristics in catalytic converters. SAE International 2003-32-0061

Agrawal G, Kaisare NS, Pushpavanam S, Ramanathan K (2012) Modeling the effect of flow maldistribution on the performance of a catalytic converter. Chem Eng Sci 71:310–320

Agrawal T, Banerjee VK, Sikarwar BS, Bhandwal M (2019) Optimizing the performance of catalytic convertor using turbulence devices in the exhaust system. Springer Science and Business Media LLC. Chapter 31

Ball D, Mohammed A, Schmidt W (1997) Application of accelerated Rapid Aging Test (RAT) schedules with poisons: The effects of oil derived poisons, thermal degradation and catalyst volume on FTP emissions. SAE Technical Paper 972846

Beck DD, Sommers JW, Dimaggio CL (1997) Axial characterization of oxygen storage capacity in closed coupled light of and underfloor catalytic converters and impact of sulphur. Appl Catal B 11:273–290

Bertrand F, Devals C, Vidal D, Préval CS, Hayes RE (2012) Towards the simulation of the catalytic monolith converter using discrete channel-scale models. Catal Today 188:80–86

Catalytic Converter (http://wiki.ask.com/Catalytic_converter)

Catalytic Converter (www.catalyticconverters.com)

Chakravarthy VK, Conklin JC, Daw CS, D'Azevedo EF (2003) Multi-dimensional simulations of cold-start transients in a catalytic converter under steady inflow conditions. Appl Catal A 241:289–306

Chan SH, Hoang DL (1999) Heat transfer and chemical reactions in exhaust system of a cold-start engine. Int J Heat Mass Transf 42:4165–4183

Chao H, Lin T, Chao M, Chang F, Huang C, Chen C (2000) Effect of methanol containing additive on the emission of carbonyl compounds from a heavy-duty diesel engine. J Hazard Mater 13:39–54

Cleveland CJ, Morris CG (2013) Section 26-Environmnet, handbook of energy: diagrams, charts, and tables, vol 1. Newnes, pp 841–876

De Nevers N (2000) Air Pollution control engineering. McGraw Hill, Chapter 13, pp 471–510

Depcik C, Assanis D (2005) One dimensional automotive catalyst modelling. Prog Energy Combust Sci 31(4):308–369

Farrauto RJ, Heck RM (1999) Catalytic converters: state of the art and perspectives. Catal Today 51:351–360

Gao J, Tian G, Sorniotti A, Karci AE, Palo RD (2019) Review of thermal management of catalytic converters to decrease engine emissions during cold start and warm up. Appl Therm Eng 147:177–187

Glander G, Zidat S (1998) Modelling electrically heated converters. Autom Eng Int 2:76–78

Guojiang W, Song T (2005) CFD simulation of the effect of upstream flow distribution on the light-off performance of a catalytic converter. Energy Convers Manag 46:2010–2031

Hayes RE, Kolaczkowski ST (1997) Introduction to Catalytic Combustion. Taylor and Francis

He Q, Wang J, Hao J, Yan X, Xiao J (2003) A study on emission characteristics of an EFI engine with ethanol blended gasoline fuels. Atmos Environ 37:949–957

Heck R, Farrauto R (2001) Automobile exhaust catalysts. Appl Catal A 221:443–457

Heywood JB (1988) Internal combustion engine fundamentals. McGraw-Hill Publications

Kang SB, Han SJ, Nam IS, Cho BK, Kim CH, Oh SH (2014) Detailed reaction kinetics for double-layered Pd/Rh bimetallic TWC monolith catalyst. Chem Eng J 241:273–287

Kim J, Myung CL, Lee KH (2019) Exhaust emissions and conversion efficiency of catalytic converter for an ethanol-fueled spark ignition engine. Biofuels Bioprod Biorefin 13(5):1211–1223

Kiwitz P, Onder C, Guzzella L (2012) Control-oriented modeling of a three-way catalytic converter with observation of the relative oxygen level profile. J Process Control 22:984–994

Koltsakis GC, Stamatelos AM (1997) Catalytic automotive exhaust after treatment. Progr Energy Combustion Sci 23:1–39

Lee D, Heywood JB (2010) Effects of secondary air Injection during cold start of SI engines. SAE International 2010-01-2124

Lee S, Bae C, Lee Y, Han T (2002) Effects of engine operating conditions on catalytic converter temperature in an SI Engine. SAE Technical paper 2002-01-1677

Lotti C, Rossi V, Poggio L, Holzinger M, Pace L, Presti M (2005) Back pressure optimized, metal supported, close coupled PE catalyst-first application on a Maserati powertrain. SAE Paper 2005-01-1105

Marsh P, Acke F (2001) Application guidelines to define catalyst layout for maximum catalytic efficiency. SAE Paper 2001-01-092

Martin AP, Will NS, Bordet A (1998) Effect of flow distribution on emissions performance of catalytic converters. SAE Paper 980936

Mitianiec W (2001) Influence of location of catalytic converter on two-stroke engine performance. SAE Technical paper 2001-01-1820/4241

Montenegro G, Onorati A (2009) 1D thermo fluid dynamic modelling of reacting flows inside three-way catalytic converters. Technical report, SAE Technical Paper

Pfalzgraf B, Rieger M, Ottowitz G (1996) Close coupled catalytic converters for compliance with LEV/ULEV and EG III legislation—influence of support material, cell density and mass on emission results. SAE Paper 960261

Postrzednik S, Zmudka Z, Ciesiolkiewicz A (2004) Basic aspects related to operation of engine catalytic converters. Int J Thermodyn 7(1):9–13

Ramadan BH, Lundberg PC, Richmond RP (2007) Characterization of a catalytic converter internal flow. SAE Technical Paper 2007-01-4024

Ribbens WB (2017) Chapter 4—The basics of electronic engine control. In: Understanding automotive electronics, 8th edn. Butterworth-Heinemann, pp 135–182

Sabatini S, Kil I, Dekar J, Hamilton T, Wuttke J, Smith MA, Hoffman MA, Onori S (2015) A new semi-empirical temperature model for the three-way catalytic converter. IFAC-PapersOnLine 48(15):434–440

Salasc S, Barrieu E, Leroy V (2005) Impact of manifold design on flow distribution of a close-coupled catalytic converter. SAE Technical Paper 2005-01-1626

Santos H, Costa M (2008) Evaluation of the conversion efficiency of ceramic and metallic three way catalytic converters. Energy Convers Manag 49:291–300

Schmidt J, Waltner A, Loose G, Hirschmann A, Wirth A, Mueller W (1999) The impact of high cell density ceramic substrates and washcoat properties on the catalytic activity of three way catalysts. SAE Paper 1999-01-0272

Searles R (1998) Welcome to emission control. Engine Technology International No. 4

Shelef M, McCabe RW (2000) Twenty-five years after introduction of automotive catalysts: What next? Catal Today 62:35–50

Shelef M, Otto K, Otto NC (1979) Poisoning of automotive catalysts. Adv Catal 27:311–365

Shuai SJ, Wang JX (2004) Unsteady temperature fields of monoliths in catalytic converters. Chem Eng J 100:95–107

Shuai SJ, Wang JX, Zhang RJ, Chen JR. Study on flow characteristics of automotive catalytic converters with various configurations. SAE Paper 2000-01-0208

Shuai SJ, Wang JX, Dong QL, Zhang RJ (2001) PIV measurement and numerical simulation of flows in automotive catalytic converters. SAE International 2001-01-3494

Simon A, Michel P, Nelson-Gruel D, Chamaillard Y, Nouillant C (2015) Gasoline-HEV equivalent consumption and pollutant minimization strategy. In:Vehicle power and propulsion conference (VPPC), pp 1–6

Simon A, Nelson-Gruel D, Onori S, Charlet A, Jaine T, Colin G, Nouillant C, Chamaillard Y (2019) An observer looks at the temperature in 3WCC. IFAC PapersOnLine 52–5:531–537

Subramanian M, Gajendra Babu MK, Subrahmanyam JP (2004) Optimization of catalytic converter for cost and effective conversion for spark ignition Engines. SAE International 2004-28-0008

Tartakovsky L, Baibikov V, Veinblat M, Popescu D, Zvirin Y, Gutman M (2012) Mileage influence on conversion efficiency of catalytic converter from in-use vehicles. SAE International 2012-01-1672

Tien ND, Quang KV, Luong NT, Tuyen PH, Khanh ND (2020) Study on improving emission conversion efficiency of three-way catalyst equipped in carburetor motorcycle by air supplement system. Int J Amb Energy, 1–9

Tsinoglou DN, Koltsakis GC, Missirlis DK, Yakinthos KJ (2004) Transient modelling of flow distribution in automotive catalytic converters. Appl Math Model 28:775–794

Umehara K, Yamada T, Hijikata T, Ichikawa Y, Katsube F (1997) Advanced ceramic substrate: catalytic performance improvement by high geometric surface area and low heat capacity. SAE Paper 971029

Umehara K, Makino M, Brayer M, Becker ER, Watson R (2000) Prediction of catalytic performance for ultra thin wall and high cell density substrates. SAE Paper 2000-01-0494

Valera H, Agarwal AK (2019) Methanol as an alternative fuel for diesel engines. In Methanol and the alternate fuel economy. Springer, Singapore, pp 9–33. https://doi.org/10.1007/978-981-13-3287-6_2

Valera H, Singh AP, Agarwal AK (2020) Prospects of methanol-fuelled carburetted two wheelers in developing countries. In: Advanced combustion techniques and engine technologies for the automotive sector. Springer, Singapore, pp 53–73

Wade J, Farrauto RJ (2012) Controlling emissions of pollutants in urban areas. In: Metropolitan sustainability. Woodhead Publishing, pp 260–291

Williams JL (2001) Monolith structures, materials, properties and uses. Catal Today 69:3–9

Zammit M, Wuttke J, Ravindran P, Aaltonen S (2012) The effects of catalytic converter location and palladium loading on tailpipe emissions. SAE International 2012-01-1247

Chapter 8
Engine Emission Control Devices for Particulate Matter and Oxides of Nitrogen: Challenges and Emerging Trends

Utkarsha Sonawane and Avinash Kumar Agarwal

Abstract Internal combustion engines (ICEs) have wide applications in several sectors, which are responsible for boosting the economy. However, engine emissions are of major concern as they significantly contribute to global air pollution. Worsening air quality has negative impact on human beings and nature. Regulated engine emissions include carbon monoxide (CO), unburned hydrocarbon (HC), oxides of nitrogen (NO_x), and particulate matter (PM). Globally, these emissions have been put under strict regulations by various emission regulation bodies. Severe issues related to the engine emissions can be resolved using efficient and advanced after-treatment devices. Use of exhaust gas after-treatment devices is one of the effective ways for engine emission reduction to meet the stringent emissions norms effectively. These devices involve complex chemical reactions. In this chapter, simultaneous reduction of PM and NO_x is focused, as both pollutants are difficult to reduce together because of PM/NO_x trade-off. Several NO_x and PM control devices used in ICEs, such as lean NOx traps (LNT), selective catalytic reduction (SCR) catalysts, and diesel particulate filter (DPF), are discussed in detail. Each after-treatment device has its advantages and limitations. Different integrated exhaust gas systems have been developed in the last couple of decades to address these limitations and enhance emission control efficiency. Various challenges and solutions to meet the emission norms using exhaust gas after-treatment devices have been discussed in this chapter.

Keywords Diesel emission control · Lean NOx Traps (LNT) · Selective Catalytic Reduction (SCR) catalyst · Diesel Particulate Filter (DPF) · Non-Thermal Plasma (NTP)

U. Sonawane · A. K. Agarwal (✉)
Engine Research Laboratory, Indian Institute of Technology Kanpur, Kanpur 208016, India
e-mail: akag@iitk.ac.in

A. P. Singh and A. K. Agarwal (eds.), *Novel Internal Combustion Engine Technologies for Performance Improvement and Emission Reduction*, Energy, Environment, and Sustainability, https://doi.org/10.1007/978-981-16-1582-5_8

8.1 Introduction

Automotive emission regulations are becoming stringent across the globe due to their adverse effects on human health, ambient air quality, and the environment. Mostly, automotive engines are fuelled by conventional fuels, namely gasoline and diesel, for power generation. Combustion of these fuels produces deleterious pollutants such as CO, HC, NO_x, and PM. Diesel engines have higher thermal efficiency vis-à-vis gasoline engines due to relatively higher compression ratio. However, diesel engines emit relatively higher concentrations of NO_x and PM (Dec 2009). Several emission regulatory bodies have taken this issue seriously and adopted stringent emission norms to control PM and NO_x along with several other regulated emissions. However, simultaneous reduction of NO_x and PM is not easy due to PM-NO_x trade-off (Fig. 8.1). The aim of this chapter is to summarize emission control devices used for simultaneous reduction of NO_x and PM.

Various low temperature combustion (LTC) techniques have also been explored by researchers, which have the potential to reduce PM and NO_x simultaneously. These advanced combustion technologies are used widely but they have several limitations associated with their practical implementation. Dec (2009) listed challenges faced by engine manufacturers in implementation of LTC. These challenges could be overcome by improving the part load efficiency, increasing the upper load limit, and accounting for fuel property effects. These are difficult challenges to overcome and are 'work in progress'. However, exhaust gas after-treatment is proven to be effective and reliable for control of NO_x and PM emissions. Engine exhaust emissions could be controlled in three ways: (i) pre-combustion control, (ii) combustion control, and (iii) post-combustion control. Compared to pre-combustion and combustion emission control strategies, post-combustion control strategy has greater

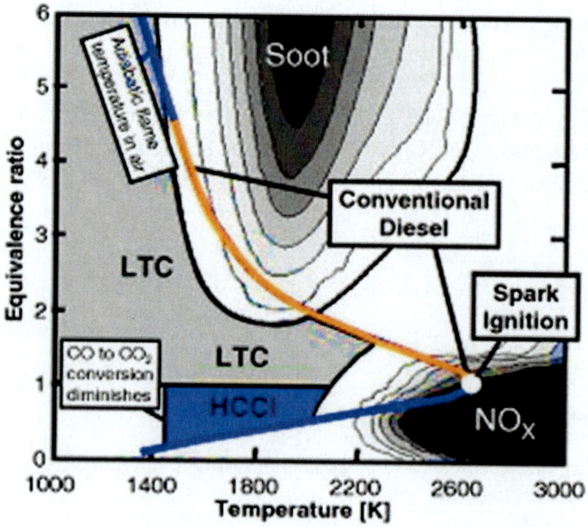

Fig. 8.1 PM-NO_x trade-off curve for IC engines (Dec 2009)

emission control/removal efficiency. Hence, post-combustion control has caught the attention of researchers in last couple of decades. Post-combustion control devices like DOC, SCR catalysts, and DPF are widely used to control regulated emissions such as HC, CO, NO_x, and PM. These after-treatment devices work efficiently when the exhaust gas temperature reaches the light-off temperature of these devices (typically after one to three minutes of engine start). Hence, an extra amount of fuel needs to be injected during cold start, in order to achieve light-off temperatures fast, although this increases the brake specific fuel consumption (BSFC) of the engine. In this chapter, authors have emphasised on NO_x and PM emissions only.

Nitrogen (N_2) is diatomic and inert gas, which constitutes ~79% of the ambient air. N_2 molecule breaks into atomic N at very high temperatures. This atomic N is highly reactive and forms nitrous oxide (N_2O), nitrogen monoxide (NO), di-nitrogen dioxide (N_2O_2), di-nitrogen trioxide (N_2O_3), nitrogen dioxide (NO_2), di-nitrogen pentoxide (N_2O_5), and di-nitrogen tetroxide (N_2O_4). Typically, NO, NO_2, and N_2O form in the engine, which together are referred to as NO_x emissions. Conventional fuels upon engine combustion produce NO_x comprising ~5% NO_2 and ~95% NO with traces of N_2O (laughing gas). N_2O emission leaves higher greenhouse effect compared to carbon dioxide (CO_2). NO_2 has a major contribution to smog formation and is far toxic than NO. NO_x formed in an engine environment can be sub-divided into three types (thermal, prompt, and fuel NO_x), as shown in Fig. 8.2.

Thermal NO_x: It is emitted due to oxidation of N_2 by oxygen molecule (O_2) at high combustion temperatures (typically >1800 K) during combustion. Thermal NO_x is produced normally in lean and stoichiometric premixed flames due to high combustion temperatures. The thermal NO_x formation reactions are as follows (https://www.engineerlive.com/content/using-nox-detectors-burners-and-boilers):

$$N_2 + O = NO + N \quad (8.1)$$

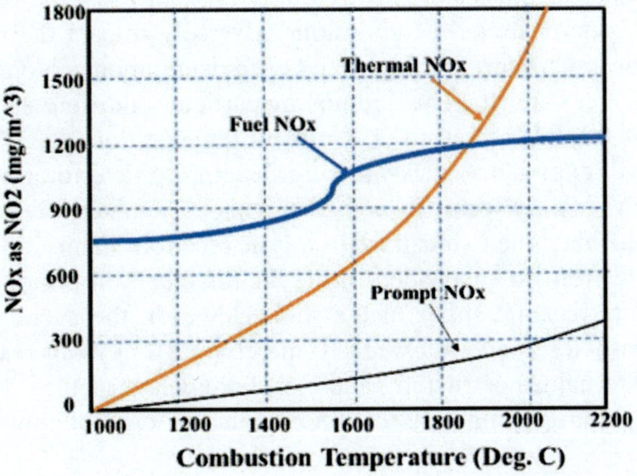

Fig. 8.2 NO_x variations versus peak combustion temperature (Adapted from https://www.engineerlive.com/content/using-nox-detectors-burners-and-boilers)

$$N + O_2 = NO + O \tag{8.2}$$

$$N + OH = NO + H \tag{8.3}$$

The first two reactions of NO_x formation were suggested by Zeldovich and the third reaction was later added by Lavoir, which is referred to as extended Zeldovich mechanism (Dec 2009; https://www.engineerlive.com/content/using-nox-detectors-burners-and-boilers).

Fuel NO_x: It is formed when N containing compounds in the fuel get oxidized partially to NO at low combustion temperatures.

Prompt NO_x: It is formed, when HC free radicals react with N_2 at low combustion temperatures under fuel-rich conditions (Rao et al. 2017). NO_x formation depends on the temperature of gas after combustion, residence time at peak temperature, and localized oxygen availability (Dec 2009). The resident time indicates how long the burnt gases remain at high temperature conditions. Longer resident time leads to formation of higher NO_x. Initial burnt gas packets experience higher pressure and temperature conditions compared to later burnt gas packets. This gives more time for NO formation to the initial burnt gas packets. Therefore, longer combustion duration (CD) provides more time for NO formation at high combustion temperature conditions. Generally, this type of NO_x is formed due to longer CD in the slow-speed engines, compared to high-speed engines.

Increasing concern of PM emissions can be addressed using diesel particulate filters (DPF), which trap and burn-off the soot collected from the engine exhaust. Automotive manufacturers have introduced DPF in diesel vehicles in order to comply with stringent emission regulations. However, now the concern has been shifted to controlling the particulate number emissions from diesel engines, for which, DPFs may not be that effective. Upcoming future emission norms will restrict the particle number emissions to further lower numbers, and more realistic certification will make it difficult to comply with these stringent emission regulations. The engine exhaust consists of particles, which are of the order of few nanometres to hundreds of nanometres. PM emissions adversely impact the human health and cause lung diseases, heart problems, and even death upon prolonged exposure. Advanced DPF devices are efficient in removing particles of different size ranges. The soot deposited in the DPF finally burns out by regeneration. If DPF is not regenerated, engine will experience back-pressure, leading to deterioration of the engine performance. PM mainly forms in non-homogenously mixed fuel-rich zones of the combustion chamber due to localized pyrolytic reactions (Tree and Svensson 2007). These particles remain suspended in the air for a certain period in accordance with size and aerodynamic shape of the soot. However, the average suspension period of these particles is about a week (Cape et al. 2012). Soot reactivity depends on chemical, internal nanostructure, and morphological structure. These parameters change with operating conditions of the engine and type of fuel used (Lapuerta et al. 2020).

PM formation in the engine combustion is a complex phenomenon, mainly dependent upon fuel/air mixture turbulence, chemical kinetics of the process, and thermal processes involved in the engine combustion.

8.2 Conventional NO_x and PM Emission Control Devices

This section covers working mechanisms of different NO_x and PM emission control devices along with their advantages and limitations.

8.2.1 Lean NO_x Traps (LNT)/NO_x Storage and Reduction (NSR) Catalysts

Lean-burn combustion technology has the ability to reduce CO and HC emissions. However, for the engines based on this technology, NO_x reduction is quite challenging using a three-way catalyst (TWC). It is also difficult to reduce NO_x emissions via converting it into N_2 in the presence of excess oxygen. Advanced LNT technology and SCR were developed to handle such challenging situations. LNT catalyst has three main components: (a) A noble metal (platinum (Pt), palladium (Pd), rhodium (Rh)) for performing oxidation and reduction process, (b) storage material (alkali or alkaline earth metal oxide like barium (Ba), strontium (Sr), lithium (Li), calcium (Ca), potassium (K), sodium (Na), etc.), and (c) support with enlarged surface area (cerium oxide (CeO_2), zirconium dioxide (ZrO_2), aluminium oxide (Al_2O_3), etc.) (Liu and Gao 2011). LNT works under lean and rich phases. During the lean phase, NO_x emitted by the engine gets adsorbed by the catalytic substrate. Then the adsorbed NO_x gets converted into nitrogen (N_2) during the rich phase. The catalyst contains active sites for oxidation of NO to NO_2 and reduction of NO_x to N_2 (Pereda-Ayo et al. 2010). Its operational mechanism is schematically shown in Fig. 8.3 (http://web3.bilkent.edu.tr/ozensoy/research-projects). The conversion reactions for NO reduction are as follows (Miyoshi et al. 1995)

$$NO + \frac{1}{2}O_2 = NO_2 \tag{8.4}$$

$$BaCO_2 + 2NO_2 = Ba(NO_3)_2 \tag{8.5}$$

$$Ba(NO_3)_2 = Ba + N_2 + 3O_2 \tag{8.6}$$

In 1990, Toyota proposed LNT technology for the first time (Miyoshi et al. 1995). LNT provided NO_x removal efficiency up to 90% at low engine loads and was mostly used for light-duty applications. It doesn't require reductant fluid for NO_x reduction. However, LNT requires higher noble metal loading, thus increasing the cost of the

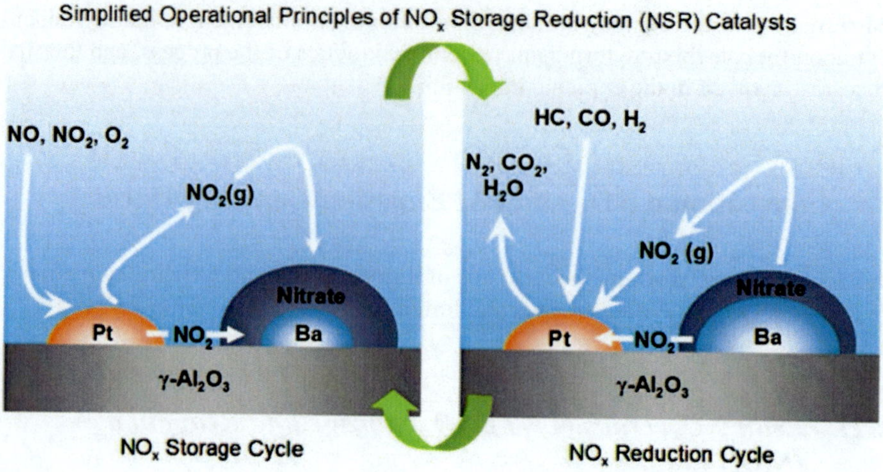

Fig. 8.3 Working principle of LNT/NSR (http://web3.bilkent.edu.tr/ozensoy/research-projects)

system (Kim et al. 2010). LNT has limited capacity to adsorb and lower removal efficiency for heavy-duty applications. LNT forms undesirable NH_3 and N_2O as a by-product during the rich phase. But this problem can be solved by placing SCR downstream of LNT. SCR with up to 90% NO_x reduction efficiency has been widely used in heavy-duty vehicles. Coupling of both technologies has attracted some serious attention in last couple of decades (Onrubia-Calvo and Pereda-Ayo 2020). Hybrid LNT–SCR technology is found to be an emerging solution to overcome the drawbacks and limitations of both stand-alone technologies for future applications (Castoldi and Lietti 2018). It may reduce NO_x emissions effectively with elimination of NH_3-slip (discussed in Sect. 8.3.1).

8.2.2 *Selective Catalyst Reduction (SCR)*

SCR is an advanced NO_x emission control technology. The process used is catalytic reaction, where a catalyst is used to convert NO_x to N_2 and water (H_2O). Also, reducing agents such as ammonia (NH_3), urea, hydrocarbon (HC), carbon monoxide (CO), and H_2 can be used (Fu et al. 2014). The selection of suitable catalyst plays an important role in SCR. Majority of SCR devices that have been commercialized used NH_3 as a reducing agent. In 1957, SCR technology was patented by Englehard Corp., USA. They commercialized vanadium-based catalysts on TiO_2 across the USA in 1960s (Wang et al. 2018). The working schematic of NH_3-SCR is shown in Fig. 8.4. Advantages of NH_3-SCR include flexible working temperature range, good redox nature, high stability, and high NOx removal efficiency. The main reactions associated with NH_3-SCR are as follows (Damma et al. 2019):

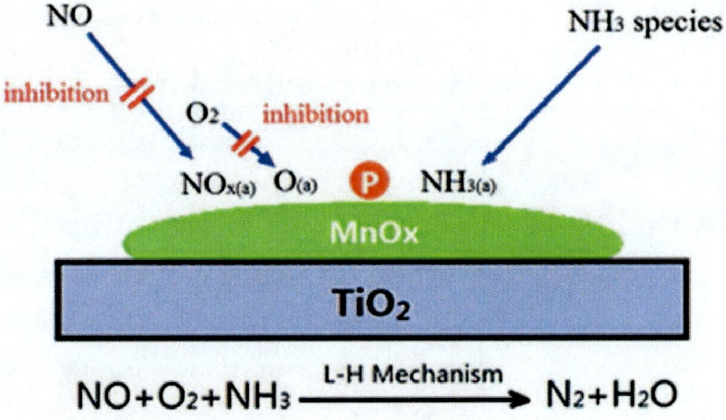

Fig. 8.4 Working principle of NH_3-SCR (Yang et al. 2016)

$$\text{Standard reaction:}\quad 4NH_3 + 4NO + O_2 = 4N_2 + 6H_2O \tag{8.7}$$

$$\text{Fast reaction:}\quad 2NH_3 + NO + NO_2 = 2N_2 + 3H_2O \tag{8.8}$$

$$NO_2\text{reaction:}\quad 4NH_3 + 3NO_2 = \frac{7}{2}N_2 + 6H_2O \tag{8.9}$$

The reactions depend on NO and NO_2 quantity in NO_x emissions (Ko et al. 2019). The problems associated with this process are usage of an expensive reductant, and transportation and storage of NH_3 (Tripathi et al. 2018). Urea can be used as an alternative to tackle the problem of handling NH_3.

The urea solution used for SCR is known by various commercial names (AdBlue, Diesel Exhaust Fluid (DEF), etc.). It decomposes into NH_3 in high-temperature exhaust. The reaction pathways for urea-SCR are shown in Fig. 8.5. Urea is a low cost and comparatively less toxic than NH_3, but it is very challenging to dose and mix urea solution with hot exhaust gases. Urea-SCR is an excellent way to control NO_x emissions, but it has problems such as system complexity, low efficiency at low temperature, and production of solid by-products, and depositions (Peitz et al. 2014).

SCR is still an evolving technology for NO_x reduction, as it has few limitations/drawbacks. Overall, SCR is a promising NO_x control technology for diesel engines (Gholami et al. 2020). In comparison with LNT, SCR offers several advantages such as high NO_x reduction efficiency over a wide range of exhaust gas temperatures and lower cost.

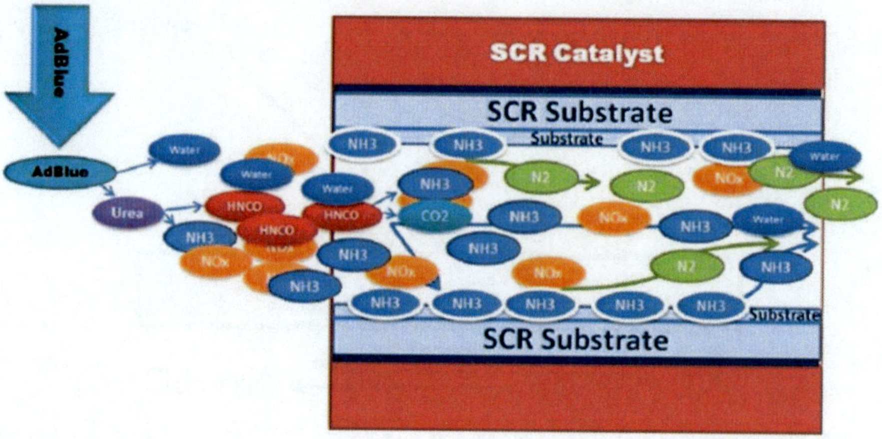

Fig. 8.5 Urea-SCR NO_x conversion mechanism (Wang et al. 2010)

8.2.3 *Ammonia Slip Catalyst (ASC)/Ammonia Oxidation Catalysts (AMOX)*

It is not acceptable to have any ammonia slip (NH_3-slip) through NO_x emission control device into the atmosphere. NH_3 is a dangerous, colourless, nasty gas, which has negative impact on the human health and the environment. Various ways of catalytic oxidation of NH_3 are shown in Fig. 8.6.

ASC is placed downstream of SCR to oxidize unreacted NH_3 into N_2 and H_2O in hot engine exhaust gases. It has SCR catalyst with an oxidation catalyst such as Pt/Al_2O_3 or Pt/SiO_2. The partial oxidation of NH_3 to N_2 and NO_x takes place over Pt catalyst and further conversion to N_2 on a SCR catalyst. The reactions involved in oxidation in ASC are as follows (Jabłońska and Palkovits 2016):

$$2NH_3 + \frac{3}{2}O_2 = N_2 + 3H_2O \tag{8.10}$$

Pt catalyst losses selectivity towards N_2 at high temperature and NO_x gets formed by following the reactions as follows (Jabłońska and Palkovits 2016):

$$2NH_3 + 2O_2 = N_2O + 3H_2O \tag{8.11}$$

Fig. 8.6 Catalytic oxidation pathways for NH_3 reduction

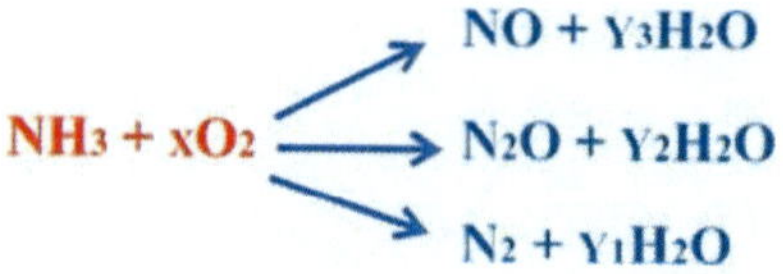

$$NH_3 + \frac{3}{2}O_2 = NO + \frac{3}{2}H_2O \tag{8.12}$$

$$NH_3 + 2O_2 = NO_2 + \frac{3}{2}H_2O \tag{8.13}$$

To tackle this problem, dual-layer ASC is used, which performs the function of Pt/Al_2O_3 oxidation and SCR catalyst for NO_x reduction (Caudle et al. 2010). The NH_3 diffusion occurs through the top layer of SCR wash-coat followed by oxidation as per reactions (8.10)–(8.13) on the oxidation catalyst (bottom layer). Then undesirable products form in reactions (8.11)–(8.13) diffuse back through SCR catalyst and react with NH_3 to produce N_2 via SCR reactions (8.7)–(8.9), as explained in Sect. 8.2.2 (Fig. 8.7).

A vehicle equipped with SCR and without ASC produces high NH_3 emissions, whereas a vehicle equipped with DOC, LNT, SCR, and ASC emits high N_2O and CH_4 emissions (Suarez-Bertoa et al. 2020). Further research in the field of ASC will make it a promising technology for NO_x reduction.

8.2.4 *Diesel Particulate Filter (DPF)*

DPF is a honeycomb structure made of parallel square channels, through whose walls exhaust gases passes, and the soot is collected on the surface and inside the pores of the wall as shown in Fig. 8.8 (Palma et al. 2013). This may lead to blockage of pores. These accumulated particles need to be removed periodically for the proper functioning of the DPF. The soot removal requires a minimum temperature of 600 °C to burn it off. However, normal diesel engine exhaust temperature lies in 200–500 °C range (Neyertz et al. 2014). Therefore, this problem can be resolved through assisted regeneration process (Kuwahara et al. 2013).

$$CO + NO_2 = CO + NO \tag{8.14}$$

$$C + O_2 = CO_2 \tag{8.15}$$

$$C + \frac{1}{2}O_2 = CO \tag{8.16}$$

Ash is another factor, which has a negative effect on the soot oxidation. Ash deposited in the DPF creates back-pressure, reduces soot deposition capacity, and also interferes in the regeneration process because of shifting of temperature balance points (Palma et al. 2013). However, accumulation of a significant ash in the DPF takes longer time and the DPF then needs to be discarded. Research on ash deposition is still limited and not well understood. Hence, it is essential to explore ash formation mechanism and its effects on the engine and DPF performance. The dimensions of

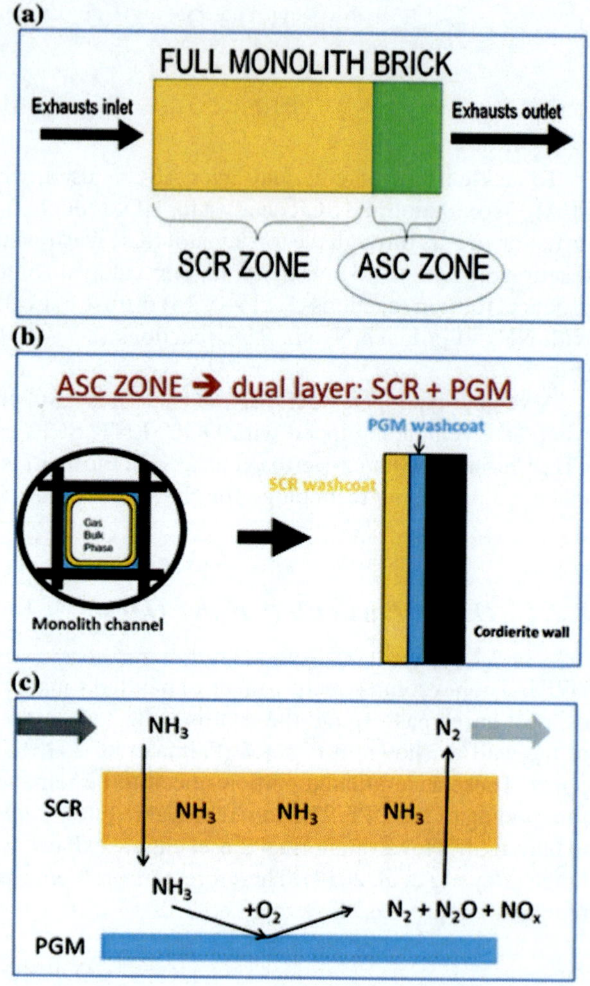

Fig. 8.7 Schematic and working principle of dual-layer ammonia slip catalyst (ASC) (Nova et al. 2014)

DPF channels have a vital role in determining soot particle trapping efficiency. Its size and length to diameter ratio can be changed as per the application requirements. DPF having high PM trapping capability leads to high exhaust back-pressure. Also, particulate number emissions are much higher during regeneration than in normal working conditions. Therefore, the development of economic, fast, and effective regeneration process is a crucial challenge associated with DPF.

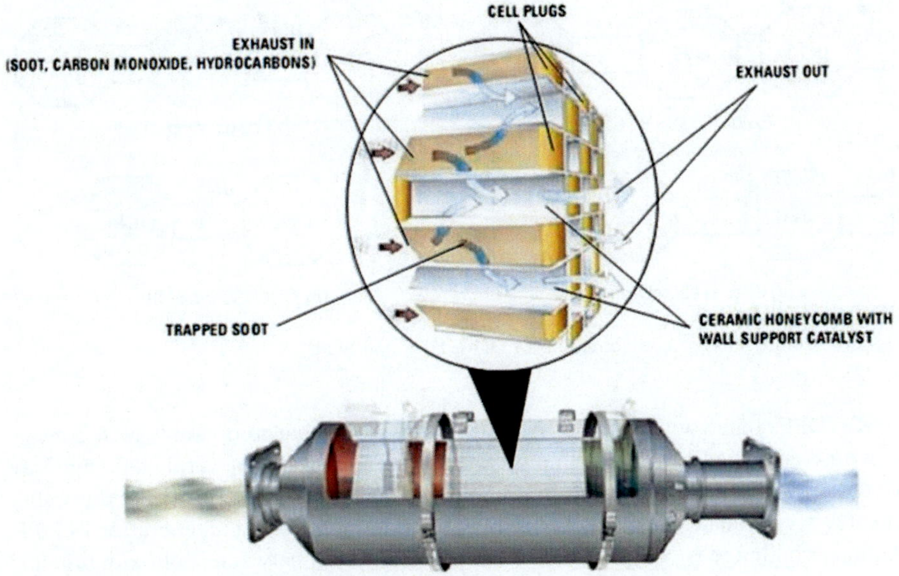

Fig. 8.8 Schematic of DPF structure (Palma et al. 2013)

8.2.5 Combinations of DOC, DPF, and SCR

Different combinations of DOC, DPF, and SCR after-treatment devices are reliable and effective in reducing PM and NO_x emissions simultaneously. Chen et al. (2019) performed experiments to study PM and NO_x removal with various combination of DOC, DPF, and SCR in a heavy-duty diesel engine. They reported that upstream DOC and DPF can remove PM up to 92.7% and downstream SCR can decrease NO_x by 82.51% in case of DOC + DPF + SCR combination. For SCR + DOC + DPF, the same efficiency of PM and NOx conversion was observed. By placing DOC + DPF upstream of SCR, passive regeneration of the DPF is possible due to NO_2 availability. Figure 8.9 shows commonly used combinations of DOC, DPF, and SCR. In Fig. 8.9a, DOC and DPF are placed such that it changes upstream NO/NO_2 ratio to 1:1 and also enhances fast SCR reaction. DPF placed upstream of SCR utilizes NO_2 for regeneration. Figure 8.9b shows DPF followed by SCR. It was found that NO_2 produced from DOC may replace fast SCR reaction with slow SCR reaction. The SCR upstream of DPF reduces warm-up time of DPF. It also reduces passive regeneration due to insufficient availability of NO_2 after the SCR. Figure 8.9c shows a configuration with DOC placed in-between SCR and DPF. In this case, SCR catalyst does not face high PM regeneration temperature and hence, any SCR catalyst can be selected. DOC and DPF take care of unwanted products such as N_2O, NO_x, and NH_3 slip. Drawback of this configuration is poor DPF regeneration due to reduced NO_2 concentration downstream of SCR. Figure 8.9d shows DOC placed upstream to

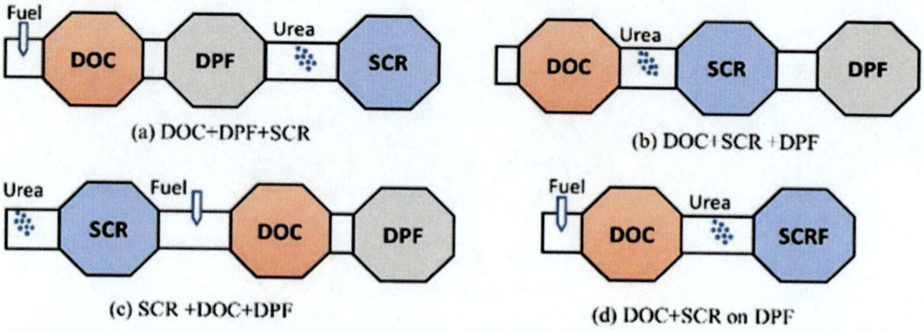

Fig. 8.9 Commonly used combinations of DOC, DPF, and SCR

SCRF (DPF coated with SCR catalyst). It was observed that comparable NO_x conversion efficiency was possible with this configuration. However, several challenges are associated with it. One of the challenges associated with SCRF is thermal durability of catalysts because it gets exposed to high temperatures during regeneration of DPF. Another challenge is competition for NO_2 consumption between soot oxidation and NO_x removal activity. This configuration is explained comprehensively in Sect. 8.3.2.

Figure 8.10 shows various combinations of conventional after-treatment with respect to DEF injection. SCR is equipped with an ammonia oxidation catalyst (AOC) to avoid NH_3-slip. The lower NOx performance can be found in the after-treatment system having DOC with SCRF or DPF downstream of DEF injection. The possible

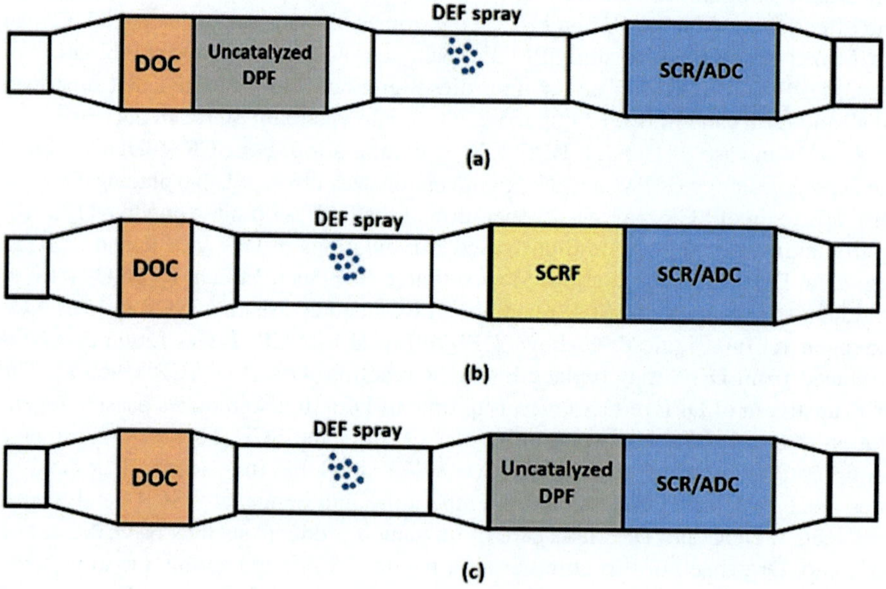

Fig. 8.10 Various combinations of DOC, DPF, and SCR/AOC

reason for degradation of NOx removal could be oxidation of NH_3 by Pt/alumina particles, migrating from the DOC. Therefore, configurations shown in Fig. 8.10b and c exhibit inferior NO_x reduction compared to Fig. 8.10a.

As discussed above, different combinations of after-treatment devices are an effective way, but it has few limitations in implementation due to interaction amongst them (Yuan et al. 2015). Apart from these combinations, advanced after-treatment systems have been explored in last one or two decade.

8.3 Emerging NO_x and PM After-Treatment Technologies

This section provides an overview of various advanced after-treatment technologies such as integrated NSR and SCR catalysts, SCRF, and non-thermal plasma. The working principles, advantages, and limitations of each technology have been discussed briefly in the following sub-sections.

8.3.1 Integrated NSR and SCR Catalyst/Lean NO_x Converters/Lean NO_x Catalyst (LNC)

Integrated NSR–SCR is a combination of two NOx reduction technologies to overcome the drawbacks of both individual technologies. Most common combination is NSR catalyst upstream and NH_3-SCR downstream. This integration of NSR–SCR not only enhances overall NO_x reduction efficiency but also reduces NH_3 emission produced during the rich phase operation of NSR (Shakya et al. 2014). Reaction mechanisms of integrated NSR–SCR are shown in Fig. 8.11, which can be understood by three steps mentioned below (Olsson et al. 2010).

Step 1: In the lean phase, the basic storage compound adsorbs NO. The adsorption becomes easier, when NO converts into NO_2 via oxidation in the presence of precious metal catalysts.

Step 2: In the rich phase, adsorbed NO_x is desorbed and reacts with H_2 present in the exhaust to form NH_3 on surface of precious metals. Then generated NH_3 gets stored by acidic sites on the SCR catalyst.

Step 3: In the next lean phase, NH_3 stored by SCR catalyst reacts with NO_x and O_2 in NH_3-SCR to produce N_2. At this time, reactions referring to Step 1 take place simultaneously

This integrated technology has commercial applications and research interests. Daimler fitted LNT–SCR catalyst technology in Mercedes-Benz E-Class in 2007 (Liu et al. 2012). Recently, Ford Motor patented a combined LNT–SCR system, where LNT was placed upstream of SCR and produced ammonia with partial NO_x removal (Gandhi et al. 2009). John Cavataio et al. (2004) invented an integrated NSR–SCR catalyst and achieved higher NO_x reduction, lower NH_3 slip, and lower

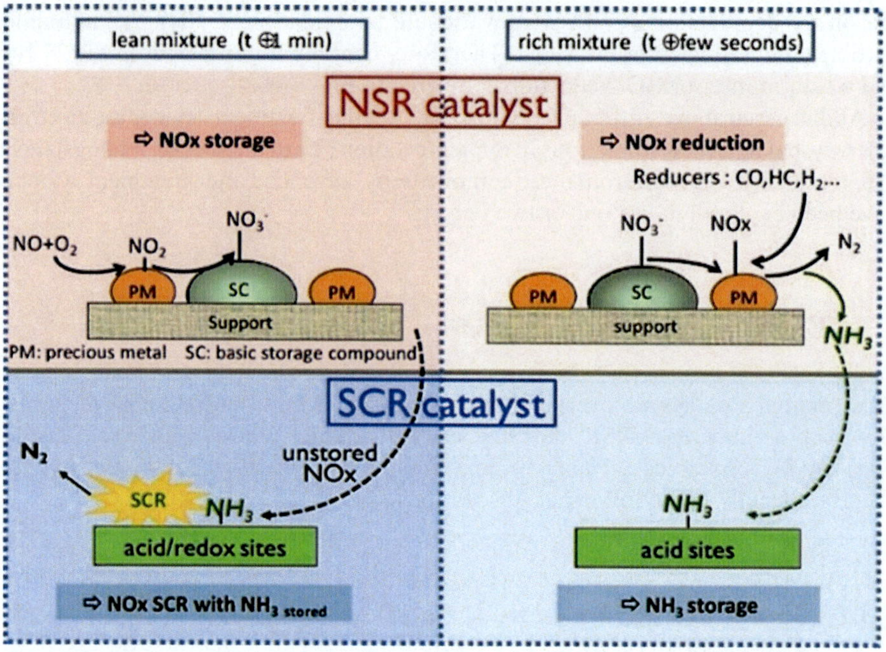

Fig. 8.11 Working mechanism of NSR–SCR (Can et al. 2012)

PGM loading in NSR. It also eliminated the use of urea dosing unit, thus making the overall system lightweight and less complex. De-La-Torre et al. (2018) studied NO_x reduction efficiency of SCR, NSR separately, and compared them with combined LNT–SCR system. They compared thermal stability, NH_3 slip, conversion of NO, and N_2O and NO_2 emissions out. Ageing of NSR catalyst degrades the system performance by reducing adsorption capacity and catalyst textural characteristics. Results showed that addition of ceria to NSR catalyst (Pt–BaO/Al_2O_3) can improve the oxidation of NO to NO_2, NO_x storage, NO conversion, and reduce NH_3 slip.

You et al. (2019) experimented with dual-bed reactor of the HC–SCR and LNT catalyst under lean-burn conditions. HC–SCR catalyst (Cu/Al_2O_3) was placed upstream of LNT catalyst ($La_{0.7}Sr_{0.3}CoO_3$) having same temperature range as that for NO_x reduction. HC–SCR uses C_3H_6 (hydrocarbon) as reductant to partially reduce NO_x emissions before it enters LNT catalyst. N_2O was not formed by LNT catalyst at rich phase using C_3H_6 as reductant. Figure 8.12 shows reaction mechanism of combined SCR (upstream) and LNT (downstream) with C_3H_6 as reductant. It was observed that dual-bed reactor HC–SCR–LNT showed lower NOx leak than individual HC–SCR and LNT under both lean and rich-burn phases. SCR with Cu/Al_2O_3 having 5 wt% Cu was optimum for LNT operation in the temperature range of 300–350 °C. This combined HC–SCR–LNT technique has the ability to overcome limitations of both LNT and SCR. Also, this hybrid technology increases the temperature

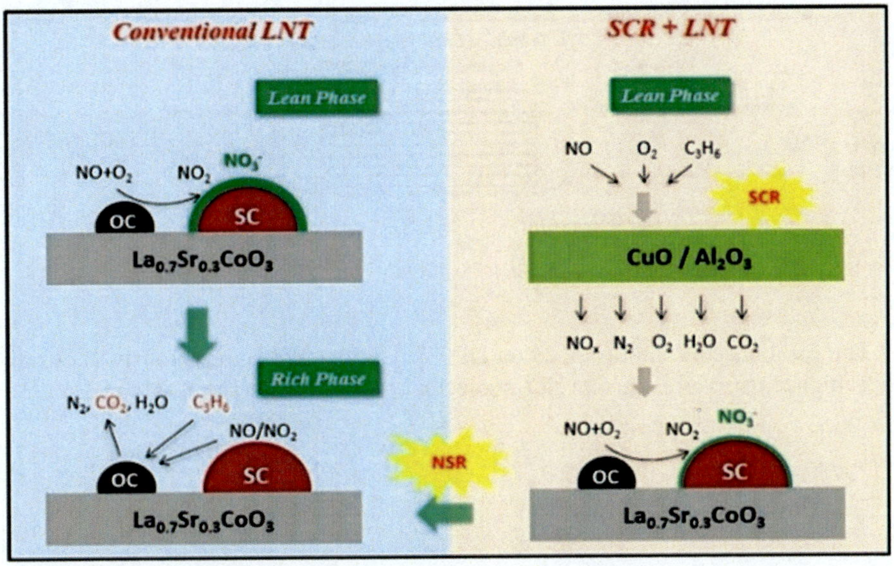

Fig. 8.12 Comparison of conventional LNT with SCR-LNT (You et al. 2019)

operating window and NO conversion, which eliminates the urea dosing module (Onrubia-Calvo and Pereda-Ayo 2020).

It is possible to achieve zero NO_x emissions with proper formulation of catalysts. Hence, the ratio of LNT and SCR catalyst volume and various working parameters are crucial for improved and effective formulation of catalysts (Cortés-Reyes et al. 2020).

From the above discussion, it can be concluded that integration of LNT with SCR can reduce NO_x emissions as well as NH_3 slip.

8.3.2 SCR Coated DPF (SCR-F)/SDPF

Hybrid SCR-DPF is one such technique, which consists of SCR catalyst material coated on porous wall of DPF. The integration of these both into a single device can reduce packaging volume and cost by reducing the number of components. Both NO_x and PM can be removed simultaneously and the functioning of DPF and SCR would be free from precious metals, leading to reduced cost. Figure 8.13 shows the schematic of two-way SCR-DPF concept. Lee et al. (2009) evaluated NO_x and PM removal performance of Cu/zeolite-based integrated SCR–DPF system for light-duty diesel engines. It was found that NO_x reduction of Cu-based SCR–DPF system was comparable with Cu-based SCR technology and not dependant on soot loading of DPF. During high temperature DPF regeneration, oxidation of NH_3 takes place before SCR, thus reducing NO_x removal efficiency.

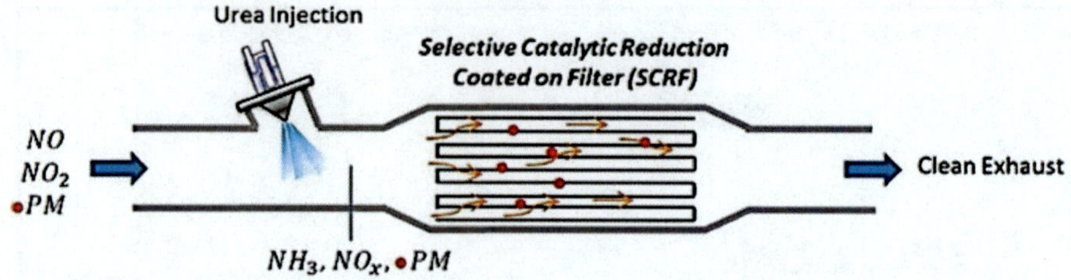

Fig. 8.13 Working mechanism of DPF–SCR (Millo et al. 2017)

The carbonaceous soot trapped on DPF can be oxidized continuously via using O_2 at high temperature or using NO_2 molecule at lower temperatures (Honack 1958).

$$O_2 + C = CO_2 \tag{8.17}$$

$$NO_2 + C = NO + CO \tag{8.18}$$

Generally, regeneration occurs when temperature rises to ~600 °C, which is known as 'active regeneration'. This temperature is achieved by catalytic oxidation of fuel that is injected into the hot engine exhaust. Main drawback of active regeneration is the fuel penalty, which depends on engine running condition. However, passive regeneration does not consume any fuel but it requires NO_2, which can be produced from NO present in the engine exhaust gases. Passive regeneration occurs above ~250 °C temperature.

$$2NO + O_2 = 2NO_2 \tag{8.19}$$

NO_2 not only promotes soot oxidation at low temperature but also de-NO_x reactions and reduces the light-off temperature of SCR (known as fast NO_x reduction reaction).

$$NO + NO_2 + 2NH_3 = 2N_2 + 3H_2O \tag{8.20}$$

In case of an integrated SCR–DPF system, if NO_2 gets consumed in fast NO_x reduction (reaction 8.20) then it will not be available for passive regeneration. So, active regeneration can be done with CO_2. To tackle this problem, oxidation of soot and NO_x reduction process is performed separately. For this purpose, only outgoing side of DPF is coated with SCR catalyst coating, as shown in Fig. 8.14. First, a comparatively faster NO_x reduction reaction takes place at the clean gas side and sufficient quantity of NO_2 is generated at the inlet side for passive soot regeneration. It is necessary to make sure that entire urea injected gets converted into NH_3 and CO_2 before reaching the DPF section. Otherwise, DPF may get blocked due to the deposition of by-products (Doering et al. 2012).

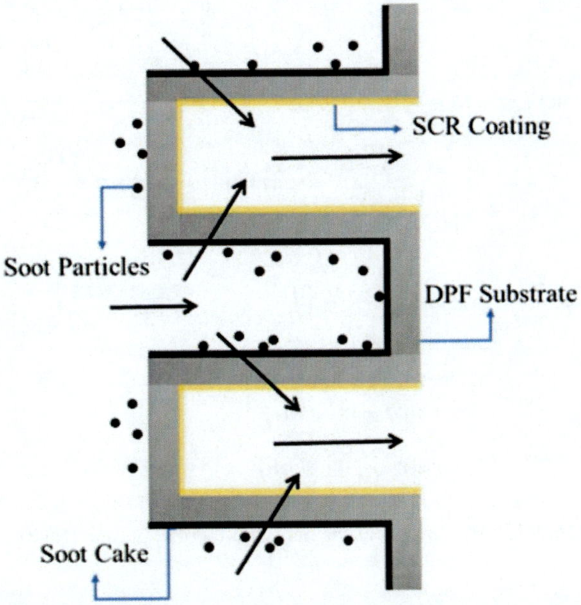

Fig. 8.14 SCR catalyst coated DPF (Adapted from Millo et al. 2017)

Light-duty vehicles have already incorporated SCRF/SDPF but additional aspects need to be evaluated for heavy-duty and off-road applications. Integrated SCR–DPF have shown potential for higher NO_x reduction in future heavy-duty vehicles with optimization of urea dosing strategy. This allowed heavy-duty engines to produce higher NO_x during combustion while meeting the emission norms. However, more NO_x reduction can be achieved by deploying larger SCR volume in SCR–DPF technology (Naseri et al. 2011). It is necessary to understand the cross interactions between SCR and DPF to optimize the system performance and design. Watling et al. (Watling et al. 2012) investigated the interactions between SCR and DPF working. It was observed that the presence of soot had no significant effect on NO_x conversion, but SCR activity retarded soot oxidation rate. Integrated SCRF technology offers better performance compared to alternate arrangements of SCR and DPF in light-duty vehicles. However, it is essential to explore its application in heavy-duty vehicles for effective removal of PM and NO_x. Intense research in the field of catalyst formulation and catalyst design along with adequate loading will make SCRF more reliable.

8.3.3 Non-thermal Plasma (NTP)

Many control measures have been developed to reduce PM emissions (Lee and Huh 2014) such as DOC, and DPF. DPF has a severe limitation of requirement of regeneration process. EGT is less than the soot oxidation temperature under most

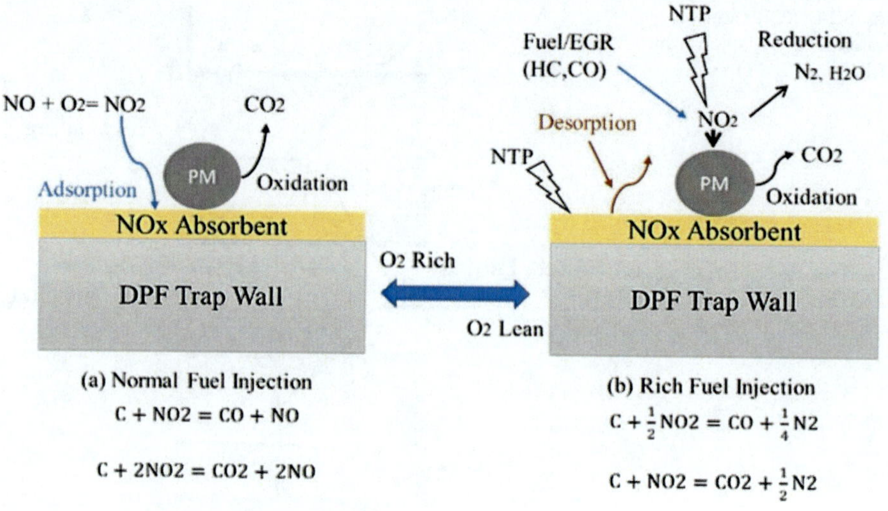

Fig. 8.15 Working mechanism of non-thermal plasma (NTP)

operating conditions in a diesel engine. DPF regeneration by heating may damage the monolith. Catalytic regeneration also puts limitations on sulphur content and NO_x/C ratio in exhaust gas (Yamamoto and Sakai 2015). NTP is an alternative control technique, which has the capacity to reduce HC, NO_x, and PM simultaneously. It also offers high removal efficiency with low back-pressure (Vinh et al. 2012). Plasma is a combination of atoms, ions, free radicals, and electrons, which is known as the fourth state of matter. NTP has a higher electron temperature (more than 10,000 K) compared to gas temperature (<573 K). NTP can promote higher chemical activities/reactions using high-speed electrons, which are not possible under normal conditions. Therefore, NTP requires a huge amount of electrical energy for operation. Figure 8.15 shows the principle of NTP working for simultaneous reduction of NO_x and PM emissions. NO_x absorbents are placed over the DPF trap wall and NTP can be supplied to surface. During normal engine operation, adequate fuel quantity is injected and exhaust gas contains more oxygen. NO gets converted to NO_2 and adsorbed by NO_x adsorbent (Okubo et al. 2017). Carbonaceous soot trapped by filter partially burns under normal fuel injection condition by NO_2 as shown below (reactions 8.21 and 8.22) (Okubo et al. 2017).

$$C + NO_2 = CO + NO \tag{8.21}$$

$$C + 2NO_2 = CO_2 + 2NO \tag{8.22}$$

In oxygen-rich condition, oxidized NO_2 get converted to NO due to the presence of carbonaceous soot. However, during rich fuel injection, i.e., oxygen lean condition, simultaneous reduction of NO_x, and PM takes place based on the following reactions

(Okubo et al. 2017):

$$C + \frac{1}{2}NO_2 = CO + \frac{1}{4}N_2 \quad (8.23)$$

$$C + NO_2 = CO_2 + \frac{1}{2}N_2 \quad (8.24)$$

Generally, diesel engine exhaust has an oxygen-rich condition, which emits harmful NO emission (reactions 8.21 and 8.22). However, it is necessary to have an oxygen lean condition regularly to reduce NO_x into N_2 as per reactions (8.23) and (8.24). This oxygen lean condition is achieved by increasing fuel/air ratio or by providing exhaust gas recirculation (EGR). By using NTP, stored NO_2 is desorbed and converted into N_2. Also, PM and HC are oxidized into CO_2 and H_2O, respectively. Therefore, NTP technology uses combination of CO, HC, O, and NO_2 to reduce engine exhaust emissions. This technology is inspired by LNT. For efficient simultaneous reduction of PM and NO_x, optimized values of plasma condition and PM loading exists (Okubo et al. 2017). Okubo et al. (2017) observed that NO_x and NO removal efficiency decreased and NO_2 increased with an increase in oxygen content. While PM removal efficiency increased with an increase in oxygen content. Kuwahara et al. (2019) studied combined after-treatment system for NOx and PM removal in marine diesel engine. NO_x removal with simultaneous DPF regeneration using NTP was performed experimentally. High amount of absorbent was used in the experiment, which provides higher NO_x reduction efficiency. Also, another study reported that plasma-assisted MnO_2 (manganese dioxide) filter can absorb not only SO_2 but also NO_2. Absorption can be improved via adding ozone in the engine exhaust. Availability of NO_2 can prevent SO_2 absorption and vice versa, but removal of both can be improved by introducing ozone (Osaka et al. 2019). Also, NTP can be used in combination with LNT for activation of reductant in the rich phase and to obtain higher NO_x removal at low temperatures. LNT catalyst shows good catalytic removal efficiency at temperature above 300 °C. However, the exhaust gas temperature of light-duty diesel engines remains in the range of 180–350 °C. Bai et al. (2017) employed H_2 plasma in the rich phase for providing NO_x release and removal at low temperatures (<250 °C). The ability of LNT catalyst to store NO_x is also important at low temperatures. Results showed that $Pt/Co/Ba/Al_2O_3$ catalyst provides higher NO_x removal efficiency and good SO_2 resistance. Zhang et al. (2017) compared $Pt/Ba/Al_2O_3$ (PBA), perovskite-type LMF ($LaMn_{0.9}Fe_{0.1}O_3$) with physically mixed PBA + LMF catalysts. They observed that PBA + LMF with NTP showed high NO_x conversion over a wide range of temperatures.

PM removal increased with an increasing input energy, temperature, and residence time along with the use of catalyst (Ye et al. 2005). Direct non-thermal plasma (DNTP) is a technology, which allows engine exhaust to pass directly through NTP reactor for PM removal. However, indirect non-thermal plasma (INTP) is another technology to remove PM via injection of active ions produced from the NTP reactor. INTP doesn't need any changes in engine exhaust system design. Shi et al. (2014)

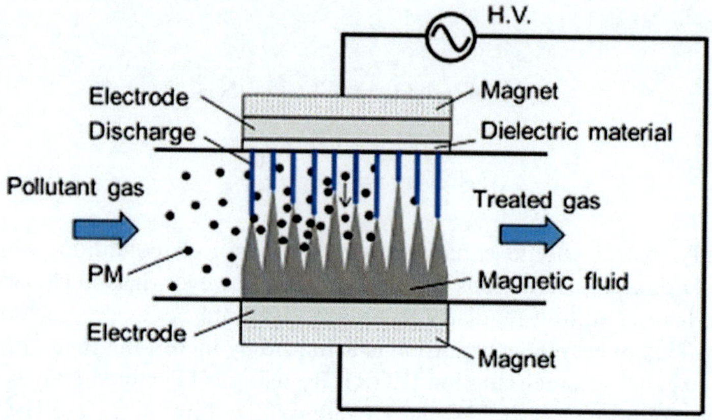

Fig. 8.16 NTP-assisted magnetic fluid filter (Kuwahara 2020)

performed DPF regeneration in a constant temperature chamber using INTP technology. It was found that regeneration of DPF loaded with soot is possible without the use of catalysts at low temperatures. Exhaust gas heat can also be used for INTP technology-based DPF regeneration. CO and CO_2 are the by-products in DPF regeneration. The amount of CO_2 is more than CO, due to vigorous soot oxidation at high temperature caused by DPF regeneration. Overall, NTP injection technology showed advantages for DPF regeneration (Pu et al. 2018). Kuwahara (2020) proposed NTP-based magnetic fluid filter for PM removal without any back-pressure. Further experiments were performed to evaluate the efficiency, and the results confirmed higher PM removal efficiency with magnetic fluid filter. The magnetic fluid consists of tiny ferromagnetic particles (magnetite) of sizes around 10 nm. These particles can be magnetized using a magnetic field. The magnetic field produces spikes, which can be controlled by magnetic flux density applied by the magnet. The exhaust gas passes through this magnetic fluid filter, as shown in Fig. 8.16. PM particles get trapped in the magnetic fluid without any back-pressure. The contact surface area for PM removal increases due to spikes. PM collected in the magnetic fluid filter can be removed by magnetic fluid replacement, injection of ozone, OH radical oxidation, and using ozone produced from the NTP. It is possible to have NTP discharge from spikes and O_3 production using dielectric barrier discharge for PM removal.

As discussed, NTP has been getting attention of researchers as an innovative technology for exhaust gas after-treatment. It is a promising technology for NO_x and VOC control amongst all exhaust control devices. It can be produced from many methods such as using an electron beam, dielectric barrier discharge, and pulsed corona discharge. These methods have been implemented practically for after-treatment presently. However, the performance and efficiency of emission control by using this technology are still questionable. Air pollution control using NTP has found its way for practical implementation. There is a great opportunity in the development of NTP methods for less costly and environment-friendly after-treatment devices. NTP can meet emission requirements in the near future. This technology does not need noble and precious metal catalysts, or dangerous NH_3/urea solution. This method of emission control has significant advantages over conventional exhaust after-treatment methods.

8.4 Summary

Increasingly stringent emission regulations are being adopted for automotive engines by emission regulatory bodies. Diesel engine emission control strategy is expected to comprise of after-treatment devices. Diesel engines are major sources of NO_x and PM emissions. Conventional after-treatment systems are quite expensive and complex. Despite many outstanding research studies, understanding of engine exhaust after-treatment remains limited. This chapter summarizes the research studies and trends in the field. NO_x and PM emission control devices are covered mainly in this chapter. Limitations and advantages associated with practical implementation have also been highlighted for each method/technology. To tackle limitations and drawbacks of conventional after-treatment methods, integrated technology has been used widely. The objective of integrated control technology is to achieve maximum PM and NO_x removal efficiency. The emphasis has been on adapting integrated emission control technology with an aim to achieve compact and less expensive engine exhaust after-treatment systems. However, integrated emission control technologies have few challenges as well. Intense research in the field of catalyst formulation, catalyst structure, catalyst preparation method, regeneration methods, and reactions would help tackle these challenges. Overall, it is recommended to use integrated after-treatment technology for engine exhaust gas after-treatment and control of pollutants. However, integrated emission control technologies must be further developed/evolved in order to achieve simultaneous removal of harmful emission species without compromising engine performance and without any fuel penalty.

References

Bai Z, Zhang Z, Chen B, Zhao Q, Crocker M, Shi C (2017) Non-thermal plasma enhanced NSR performance over Pt/M/Ba/Al_2O_3 (M = Mn Co, Cu) catalysts. Chem Eng J 314:688–699. https://doi.org/10.1016/j.cej.2016.12.034

Can F, Courtois X, Royer S, Blanchard G, Rousseau S, Duprez D (2012) An overview of the production and use of ammonia in NSR + SCR coupled system for NOx reduction from lean exhaust gas. Catal Today 197(1):144–154. https://doi.org/10.1016/j.cattod.2012.07.032

Cape JN, Coyle M, Dumitrean P (2012) The atmospheric lifetime of black carbon. Atmos Environ 59:256–263. https://doi.org/10.1016/j.atmosenv.2012.05.030

Castoldi L, Lietti L (2018) NOx trap catalysts and technologies: fundamentals and industrial applications. http://hdl.handle.net/11311/1072072

Caudle MT, Deiterle M, Roth SA, Xue W-M (2010) U.S. Patent 7722845. BASF Corporation, investors

Chen X, Bai S, Li G, Wang G (2019) Experimental study on the emission characteristics of heavy-duty diesel engine with combination system of doc/dpf/scr. FEB-Fresenius Environ Bull 4591

Cortés-Reyes M, Herrera C, Larrubia MÁ, Alemany LJ (2020) Hybrid technology for DeNOxing by LNT-SCR system for efficient diesel emission control: influence of operation parameters in $H_2O + CO_2$ atmosphere. Catalysts 10(2):228. https://doi.org/10.3390/catal10020228

Damma D, Ettireddy PR, Reddy BM, Smirniotis PG (2019) A review of low temperature NH_3-SCR for removal of NOx. Catalysts 9(4):349. https://doi.org/10.3390/catal9040349

Dec JE (2009) Advanced compression-ignition engines—understanding the in-cylinder processes. Proc Combust Inst 32(2):2727–2742

De-La-Torre U, Pereda-Ayo B, Onrubia JA, González-Velasco JR (2018) Effect of the presence of Ceria in the NSR catalyst on the hydrothermal resistance and global DeNOx performance of coupled LNT–SCR systems. Top Catal 61(18–19):1993–2006. https://doi.org/10.1007/s11244-018-1016-0

Doering A, Emmerling G, Rothe D (2012) Downsizing of the EuroVI exhaust aftertreatment components to fit into the EuroIII silencer; Downsizing der EuroVI-Abgasnachbehandlung auf das Niveau des EuroIII-Schalldaempfers

Fu M, Li C, Lu P, Qu L, Zhang M, Zhou Y, Yu M, Fang Y (2014) A review on selective catalytic reduction of NOx by supported catalysts at 100–300 °C—catalysts, mechanism, kinetics. Catal Sci Technol 4(1):14–25. https://doi.org/10.1039/C3CY00414G

Gandhi HS, Cavataio JV, Hammerle RH, Cheng Y (2009) Method for the reduction of NOx and NH_3 emissions. U.S. Patent 7,485,273. Ford Global Technologies LLC

Gholami F, Tomas M, Gholami Z, Vakili M (2020) Technologies for the nitrogen oxides reduction from flue gas: a review. Sci Total Environ 136712. https://doi.org/10.1016/j.scitotenv.2020.136712

Honack E (1958) Process for treating carbon black by oxidation. Britische Patentschrift GB 895:990

https://www.engineerlive.com/content/using-nox-detectors-burners-and-boilers

http://web3.bilkent.edu.tr/ozensoy/research-projects

Jabłońska M, Palkovits R (2016) Copper based catalysts for the selective ammonia oxidation into nitrogen and water vapour—recent trends and open challenges. Appl Catal B 181:332–351. https://doi.org/10.1016/j.apcatb.2015.07.017

John Cavataio RH, Cheng Y, Gandhi H (2004) US20040076565

Kim CH, Qi G, Dahlberg K, Li W (2010) Strontium-doped perovskites rival platinum catalysts for treating NOx in simulated diesel exhaust. Science 327(5973):1624–1627. https://doi.org/10.1126/science.1184087

Ko A, Woo Y, Jang J, Jung Y, Pyo Y, Jo H, Lim O, Lee YJ (2019) Complementary effects between NO oxidation of DPF and NO_2 decomposition of SCR in light-duty diesel engine. J Ind Eng Chem 80:160–170. https://doi.org/10.1016/j.jiec.2019.07.045

Kuwahara T (2020) Fundamental characteristics of low-resistive particulate matter removal using a magnetic fluid and nonthermal plasma. J Magn Magn Mater 498: https://doi.org/10.1016/j.jmmm.2019.166161

Kuwahara T, Nishii S, Kuroki T, Okubo M (2013) Complete regeneration characteristics of diesel particulate filter using ozone injection. Appl Energy 111:652–656. https://doi.org/10.1016/j.apenergy.2013.05.041

Kuwahara T, Yoshida K, Kuroki T, Hanamoto K, Sato K, Okubo M (2019) Pilot-scale combined reduction of accumulated particulate matter and NOx using nonthermal plasma for marine diesel engine. IEEE Trans Ind Appl. https://doi.org/10.1109/TIA.2019.2958276

Lapuerta M, Rodríguez-Fernández J, Sánchez-Valdepeñas J (2020) Soot reactivity analysis and implications on diesel filter regeneration. Prog Energy Combust Sci 78: https://doi.org/10.1016/j.pecs.2020.100833

Lee Y, Huh KY (2014) Analysis of different modes of low temperature combustion by ultra-high EGR and modulated kinetics in a heavy duty diesel engine. Appl Therm Eng 70(1):776–787. https://doi.org/10.1016/j.applthermaleng.2014.05.090

Lee JH, Paratore MJ, Brown DB (2009) Evaluation of Cu-based SCR/DPF technology for diesel exhaust emission control. SAE Int J Fuels Lubricants 1(1):96–101

Liu G, Gao PX (2011) A review of NOx storage/reduction catalysts: mechanism, materials and degradation studies. Catal Sci Technol 1(4):552–568. https://doi.org/10.1039/C1CY00007A

Liu Y, Harold MP, Luss D (2012) Appl Catal B Environ 121–122:239

Millo F, Rafigh M, Fino D, Miceli P (2017) Application of a global kinetic model on an SCR coated on Filter (SCR-F) catalyst for automotive applications. Fuel 198:183–192

Miyoshi N, Matsumoto SI, Katoh K, Tanaka T, Harada J, Takahashi N, Yokota K, Sugiura M, Kasahara K (1995) Development of new concept three-way catalyst for automotive lean-burn engines. SAE Trans 1361–1370

Naseri M, Chatterjee S, Castagnola M, Chen HY, Fedeyko J, Hess H, Li J (2011) Development of SCR on diesel particulate filter system for heavy duty applications. SAE Int J Engines 4(1):1798–1809. https://doi.org/10.4271/2011-01-1312

Neyertz CA, Banús ED, Miró EE, Querini CA (2014) Potassium-promoted Ce0. 65Zr0. 35O2 monolithic catalysts for diesel soot combustion. Chem Eng J 248:394–405. https://doi.org/10.1016/j.cej.2014.03.048

Nova I, Colombo M, Tronconi E, Schmeißer V, Bandl-Konrad B, Zimmermann L (2014) Dual-layer ammonia slip catalysts for automotive SCR exhaust gas aftertreatment: an experimental and modeling study. In: Urea-SCR technology for deNOx after treatment of diesel exhausts. Springer, New York, NY, pp 553–586. https://doi.org/10.1007/978-1-4899-8071-7_18

Okubo M, Yamada H, Yoshida K, Kuroki T (2017) Simultaneous reduction of diesel particulate and NOx using a catalysis-combined nonthermal plasma reactor. IEEE Trans Ind Appl 53(6):5875–5882. https://doi.org/10.1109/TIA.2017.2748925

Olsson L, Fredriksson M, Blint RJ (2010) Kinetic modeling of sulfur poisoning and regeneration of lean NOx traps. Appl Catal B 100(1–2):31–41. https://doi.org/10.1016/j.apcatb.2010.07.004

Onrubia-Calvo JA, Pereda-Ayo B (2020) Perovskite-based catalysts as efficient, durable, and economical NOx storage and reduction systems. Catalysts 10(2):208. https://doi.org/10.3390/catal10020208

Osaka Y, Iwai K, Tsujiguchi T, Kodama A, Li X, Huang H (2019) Basic study on exhaust gas purification by utilizing plasma assisted MnO_2 filter for zero-emission diesel. Separation Purif Technol 215:108–114. https://doi.org/10.1016/j.seppur.2018.12.077

Palma V, Ciambelli P, Meloni E (2013) Catalyst load optimization for microwave susceptible catalysed DPF. Chem Eng 32

Peitz D, Bernhard A, Kröcher O (2014) Ammonia storage and release in SCR systems for mobile applications. In: Urea-SCR technology for deNOx after treatment of diesel exhausts. Springer, New York, NY, pp 485–506. https://doi.org/10.1007/978-1-4899-8071-7_16

Pereda-Ayo B, Duraiswami D, Delgado JJ, López-Fonseca R, Calvino JJ, Bernal S, González-Velasco JR (2010) Tuning operational conditions for efficient NOx storage and reduction over a Pt–Ba/Al_2O_3 monolith catalyst. Appl Catal B 96(3–4):329–337. https://doi.org/10.1016/j.apcatb.2010.02.029

Pu X, Cai Y, Shi Y, Wang J, Gu L, Tian J, Fan R (2018) Carbon deposit incineration during engine flameout using non-thermal plasma injection. Int J Autom Technol 19(3):421–432. https://doi.org/10.1007/s12239-018-0041-0

Rao A, Mehra RK, Duan H, Ma F (2017) Comparative study of the NOx prediction model of HCNG engine. Int J Hydrogen Energy 42(34):22066–22081. https://doi.org/10.1016/j.ijhydene.2017.07.107

Shakya BM, Harold MP, Balakotaiah V (2014) Modeling and analysis of dual-layer NOx storage and reduction and selective catalytic reduction monolithic catalyst. Chem Eng J 237:109–122. https://doi.org/10.1016/j.cej.2013.10.008

Shi YX, Cai YX, Li XH, Chen YY, Ding DW, Tang W (2014) Meachanism and method of DPF regeneration by oxygen radical generated by NTP technology. Int J Autom Technol 15(6):871–876. https://doi.org/10.1007/s12239-014-0091-x

Suarez-Bertoa R, Pechout M, Vojtíšek M, Astorga C (2020) Regulated and non-regulated emissions from Euro 6 diesel, gasoline and CNG vehicles under real-world driving conditions. Atmosphere 11(2):204. https://doi.org/10.3390/atmos11020204

Tree DR, Svensson KI (2007) Soot processes in compression ignition engines. Prog Energy Combust Sci 33(3):272–309. https://doi.org/10.1016/j.pecs.2006.03.002

Tripathi G, Dhar A, Sadiki A (2018) Recent advancements in after-treatment technology for internal combustion engines—an overview. In: Advances in Internal combustion engine research. Springer, Singapore, pp 159–179. https://doi.org/10.1007/978-981-10-7575-9_8

Vinh TQ, Watanabe S, Furuhata T, Arai M (2012) Effects of particulate matter on NOx removal in dielectric barrier discharges. J Energy Inst 85(3):163–169. https://doi.org/10.1179/1743967112Z.00000000015

Wang J, Chia A, Menq H, Rizzoni G, Yurkovich S (2010) Control of diesel engine urea selective catalytic reduction systems. Etd. ohiolink.edu

Wang F, Ma J, He G, Chen M, Zhang C, He H (2018) Nanosize effect of Al_2O_3 in Ag/Al_2O_3 catalyst for the selective catalytic oxidation of ammonia. ACS Catal 8(4):2670–2682. https://doi.org/10.1021/acscatal.7b03799

Watling TC, Ravenscroft MR, Avery G (2012) Development, validation and application of a model for an SCR catalyst coated diesel particulate filter. Catal Today 188(1):32–41. https://doi.org/10.1016/j.cattod.2012.02.007

Yamamoto K, Sakai T (2015) Simulation of continuously regenerating trap with catalyzed DPF. Catal Today 242:357–362. https://doi.org/10.1016/j.cattod.2014.07.022

Yang NZ, Guo RT, Wang QS, Pan WG, Chen QL, Lu CZ, Wang SX (2016) Deactivation of Mn/TiO_2 catalyst for NH_3-SCR reaction: effect of phosphorous. RSC Adv 6(14):11226–11232. https://doi.org/10.1039/C5RA27713B

Ye D, Gao D, Yu G, Shen X, Gu F (2005) An investigation of the treatment of particulate matter from gasoline engine exhaust using non-thermal plasma. J Hazard Mater 127(1–3):149–155. https://doi.org/10.1016/j.jhazmat.2005.06.040

You R, Meng M, Zhang J, Zheng L, Hu T, Li X (2019) A noble-metal-free SCR-LNT coupled catalytic system used for high-concentration NOx reduction under lean-burn condition. Catal Today 327:347–356. https://doi.org/10.1016/j.cattod.2018.03.022

Yuan X, Liu H, Gao Y (2015) Diesel engine SCR control: current development and future challenges. Emission Control Sci Technol 1(2):121–133. https://doi.org/10.1007/s40825-015-0013-z

Zhang ZS, Shi C, Bai ZF, Li MR, Chen BB, Crocker M (2017) Low-temperature H_2-plasma-assisted NOx storage and reduction over a combined Pt/Ba/Al and LaMnFe catalyst. Catal Sci Technol 7(1):145–158. https://doi.org/10.1039/C6CY01900E

Chapter 9
Variation of Soot Structure Along the Exhaust Aftertreatment System—Impact of Oxygenated Diesel Blends on the Soot/Catalyst Interactions

Nahil Serhan

Abstract Particulate matter structural modifications do not only impact the oxidative/mutagenic properties of the particulates but also influence the motion and contact between the aggregates in the exhaust. Those interactions have a direct impact on the porosity, permeability, and packing density of the soot cake deposited in the diesel particulate filter (DPF). This in turn will influence the filtration efficiency and pressure drop in the DPF channels. Morphology of the PM combined with the carbon layer nanostructure also has a direct impact on the DPF regeneration capability. After recognizing the importance of the particulates' structure on the DPF performance, it becomes a subject of interest to understand if the engine-out PM will face any modifications within the aftertreatment units, e.g., diesel oxidation catalyst, selective catalytic reduction catalyst, etc., before being trapped in the DPF. While the dependency of the PM characteristics on its fueling source and engine technology is a well-researched topic, limited work has been carried out regarding the impact of the aftertreatment system on the structure (i.e., morphology and nanostructure) and chemical characteristics of the exhaust PM. This chapter will discuss the different theories and experimental work provided in the literature regarding the impact of aftertreatment systems on the PM characteristics. Special attention will be given on the impact of alcohols and other oxygenated fuels on this mechanism compared to conventional diesel fuel.

Keywords Alcohol · Soot · Particulate matter · DPF · DOC · SCR · Aftertreatment

---

N. Serhan (✉)
Wärtsilä Catalyst Systems R&D, Vaasa, Finland
e-mail: nahil.serhan@wartsila.com

A. P. Singh and A. K. Agarwal (eds.), *Novel Internal Combustion Engine Technologies for Performance Improvement and Emission Reduction*, Energy, Environment, and Sustainability, https://doi.org/10.1007/978-981-16-1582-5_9

9.1 Introduction

Diesel smoke, so-called particulate matter (PM), can be presented as a combination of carbonaceous material (soot) blended with several types of inorganic and organic substances, resulting in mutagenic and carcinogenic elements by nature (Silverman et al. 2012). Soot structural modifications (i.e., morphology) do not only impact the oxidative/mutagenic properties of the soot but also influence the motion and contact (i.e., convective or diffusive flow) between the particulates which in turn impacts the magnitude of the Peclet number (Pe) accordingly. Pe is a dimensionless factor whose magnitude strongly depends on the soot morphological aspects, mainly the primary particulate size (d_{p0}) and to a lesser extent, the radius of gyration (R_g) (Konstandopoulos et al. 2002). The lower the Pe (<1) the more open and porous the soot cake is (i.e., layer of soot deposited in the DPF channels) and the higher the Pe ($1 < Pe < 5$) the more compact it is. In general, an increase in d_{p0} will yield a higher Pe number and vice versa (Konstandopoulos et al. 2002). Factors as the soot-cake permeability, packing density, porosity, pressure drop in the channels and DPF thermal behavior during regeneration can be calculated (i.e., simulated) once obtaining the Pe number (Konstandopoulos et al. 2002; Chiatti et al. 2009; Konstandopoulos and Kostoglou 2004). The reader is referred to (Konstandopoulos et al. 2002; Chiatti et al. 2009; Konstandopoulos and Kostoglou 2004) for a more detailed interpretation of the incorporated formulas that link the Pe number with the soot-cake characteristics. As a matter of fact, recognizing the filtration efficiency of a DPF is important; however, predicting the exact dimensioning of the system to optimize the resulted backpressure is even more critical, especially in the norm of the severe legislations and economic viability of diesel-powered vehicles (Fig. 9.1).

After recognizing the importance of the particulates' structure on the DPF performance, it becomes a subject of interest to understand if the exhaust particulates will face any modifications within the aftertreatment units, e.g., diesel oxidation catalyst (DOC), selective catalytic reduction (SCR), etc., before being trapped in the DPF. While the dependency of the soot structure on its fuelling source and engine technology is a well-researched topic, to our knowledge, limited work (Ma et al. 2014;

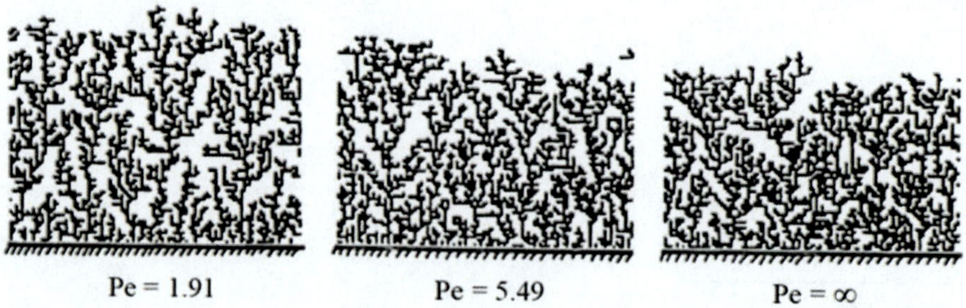

Fig. 9.1 Soot-cake representation with regard to its Peclet number. Reproduced from (Tassopoulos et al. 1989)

Liati et al. 2013; Lapuerta et al. 2007; Fayad et al. 2015; Lee and Zhu 2005; Serhan et al. 2019) has been carried out regarding the impact of the aftertreatment system on the structure of the exhaust PM (i.e., morphology and nanostructure). The chapter will be divided into two sections, research work based on (i) diesel oxidation catalyst and (ii) hydrocarbon-SCR (HC-SCR) catalyst. The impact of the fuel type will be discussed in each section separately.

9.2 Variation in Soot Structure Throughout DOC

Lapuerta et al. (2007) studied the impact of changing the soot sampling point in the exhaust line to check if d_{p0} will be altered by the thermal or fluid-dynamic conditions along the line. However, the study did not involve any aftertreatment unit. It was seen that d_{p0} remains constant throughout all the tested positions since as reported, the exhaust temperature level (110–324 °C) was not enough to launch any chemical reactions with the soot surface. Liati et al. (2013) studied the morphological and nanostructure aspects of soot particulates upstream and downstream the DOC in an effort to understand the impact of this catalyst on the particulate's oxidative reactivity. Concerning the morphological alterations, slight reduction was noted along d_{p0}- when the particulates flow throughout the catalyst. This was due to the devolatilization of the soluble/semi-volatile organic fraction from the particulates' surface; however, these changes were not linked with the existence of any possible carbon oxidation phenomenon. As for the graphitization order, slightly more ordered structure was presented downstream the DOC, and therefore, to some extent, the particulates were considered less reactive toward oxidation. Close edge X-ray absorption fine structure analysis (NEXAFS) was also employed to understand the difference in the particulate's surface functional compounds upstream and downstream the catalyst. It was shown that marginal changes took place downstream the DOC and more specifically, most of the carboxyl functionalities had disappeared and the soot seemed to be more homogeneous in its chemical structure. Ma et al. (2014) studied the impact of the DOC on the particulate's carbon layer arrangement and its corresponding oxidative reactivity. The analysis showed that DOC can alter the primary particulate's inner structure in a way that makes it more ordered toward graphitization, e.g., longer carbon layer length (L_a) and remarkably shorter interlayer spacing (d_{002}). As for the oxidative behavior, the post-catalyst soot seemed to be more reactive, yet a different oxidation mode was recorded compared to that in the pre-catalyst case. To be more specific, the soot mass loss was extremely faster than that of the pre-catalyst test at the early stages of the oxidation (before 50% soot mass loss); however, at later stages (after 50%), the oxidation rate turned to be slower and the residual ash mass significantly higher. Concerning the residuals, they were mainly influenced by the sample collection method and not by any direct soot interactions with the DOC. Longer period was needed to collect the soot downstream the catalyst since the increased concentrations of nitrogen dioxides (NO_2) (resulted from the nitrogen oxide (NO) oxidation in the DOC) continuously oxidized the soot placed on the filter. However,

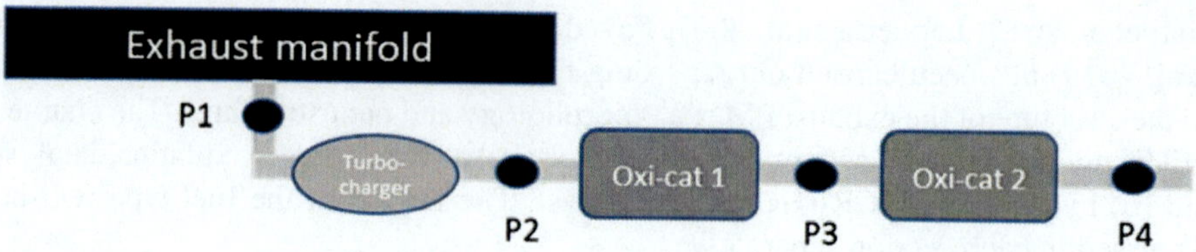

Fig. 9.2 Sampling collection positions for reference (Lee and Zhu 2004, 2005)—self-drawn by author

during the collection time, the ash deposition remained intact and did not oxidize, thus increasing the ash percentage of the sample placed after the DOC. The soot nanostructure variations were also recorded throughout the different stages of the temperature-controlled oxidation to further investigate the oxidation mechanism of the soot samples (pre- and post-catalyst). It was concluded that the soot collected upstream the DOC undergoes a surface burning process; in other words, it keeps a shell-core structure along all the oxidation period. As for the post-catalyst particulates, the same oxidation mode as that of the "biodiesel oxidation" (Song et al. 2006) was seen for this testing case. In summary, despite the more graphitic initial structure, the high-resolution transmission electron microscopy (HRTEM) analysis revealed that the soot undergoes an internal-burning process at the early stages of the oxidation which leads to a capsule-like structure. The significant energy released during this phase helps to thermally restructure the soot silhouette and transform it to a closed outer shell profile. The highly stable configuration of the latter is the main reason for the slower oxidation rates recorded after the first 50% of the soot weight loss. It was speculated that the DOC contributed to the devolatilization of the surface volatile compounds, which in turn generated this unique mode of oxidation. Lee and Zhu (Lee and Zhu 2004, 2005) studied the evolution of the soot morphological parameters along the different positions of the exhaust system (that includes two DOCs) as a function of engine speed and load condition. The different soot sampling positions are presented in Fig. 9.2. The analysis showed that independently from the engine operating settings, the first DOC helped to reduce the particulates R_g and d_{p0} and increased their circularity, e.g., higher fractal dimension (D_f). The reduction in d_{p0} was related to the existence of a soot catalytic oxidation, which was also confirmed through qualitative inspection for the HRTEM micrographs. It was shown that upstream the catalyst, the particulates presented the standard graphitic configuration of diesel soot where the primary particulates are mostly spherical and quite distinct from the other spherules. However, downstream the catalyst, the primary particulates appear to have no definite borders and seem to be fused in each other, indicating the presence of catalytic oxidation. The second DOC (Position P4) also reduces d_{p0}; however, the particulates R_g tend to slightly increase and their geometrical configuration highlights a more chain-like structure (lower D_f). It was expected that the resulted particulates undergo an enhanced aerodynamic mixing after going out of the catalyst (after 200 cm), which in turn favors their aggregation together to yield larger particulates with a more complex geometry. Overall, it was seen that

d_{p0} and R_g decrease progressively throughout the catalysts with R_g presenting a bigger drop compared to d_{p0}. It was speculated that this further reduction in R_g could possibly arise from the breakdown of the particulates throughout the catalyst channels along with reduction in the interstitial distances among the spherules. However, the average number of the primary particulates in the aggregates (n_{p0}) was not evaluated in this study, keeping those findings (i.e., breakdown and reduction in interstitial spacing) as theoretical approaches and not proved by analysis.

Fayad et al. (2015) investigated the influence of the DOC with regard to the oxidation/reduction mechanism of the soot particulates produced from the combustion of diesel, rapeseed methyl ester (RME), and butanol/RME/diesel blend. It was seen that DOC can help in reducing the small-scale particulates (<20 nm) by diffusion mechanism (trapping) and can also result on average with larger soot particulates because of an enhanced collision phenomenon in the catalyst channels. This theory was further verified by morphologically analyzing the particulates, upstream and downstream the catalyst. From the analysis, it was reported that after passing the DOC, the collected soot tended to present a greater n_{p0} and a larger R_g compared to these tested upstream the catalyst, independent of the fuel type. Oxygenated fuel combustion results in particulates which are characterized with smaller primary particles and more skeletal aggregates compared to the diesel case. This, in turn, makes the particles produced from oxygenated fuels more easily trapped in the DOC. However, DOC modified neither the particulates' d_{p0} nor the carbon layers' arrangements, highlighting that a longer soot/catalyst residence time was required to perform those alterations (soot oxidation).

9.3 Variation in Soot Structure Throughout HC-SCR

Serhan et al. (2019) investigated the impact of a silver alumina (Ag/Al_2O_3) HC-SCR catalyst on the morphology and nanostructure of the soot aggregates produced from the combustion of diesel and diesel/tri-propylene glycol ether (TPGME) blends (20% vol. basis, TP20). Numerous studies in the literature proved the advantageous effect of PGMEs in reducing the engine-out PM emissions, particularly TPGME which was recognized for yielding near non-sooting combustion (Burke et al. 2015; Natarajan 2001; González 2001). The catalyst role was evaluated upon three main areas, the exhaust temperature by varying engine load (2 and 4 bars indicated mean effective pressure (IMEP)), fuel type, and hydrogen (H_2) present in the exhaust. H_2 gas has been recognized as a promoter for the Ag/Al_2O_3 de-NOx activity, especially in the low-temperature regions when enough reductants are available (Breen et al. 2007; Theinnoi et al. 2008). This research work present in-depth detail on the catalyst and fuel-type impact on soot particles and will be detailed following two separate sections, the first highlights the impact on soot morphology and the second highlights the nanostructure modifications.

9.3.1 *Catalyst Impact on the Morphology*

9.3.1.1 Low Exhaust Temperature Condition

The different soot morphological parameters including R_g, n_{p0}, and D_f for the different testing conditions at low load operation (2 bar IMEP, 180 °C) are plotted in Fig. 9.3. Also, the normal distribution of the primary particulate's size (d_{p0}) is plotted in Figure.

It is seen that with or without the presence of H_2, aggregates with greater R_g and n_{p0} are seen after the catalyst for both, diesel and TP20 fueling. Therefore, it can be concluded that only a limited portion of the particulates have been trapped inside the catalyst, as it was suggested earlier in reference (Fayad et al. 2015). Besides, it is expected that upon passing the catalyst channels, the soot particulates have faced higher chances of collision with the other neighboring particulates and as a result more mature aggregates (i.e., by size) are spotted downstream the HC-SCR. Furthermore, more spherical particulates were seen downstream the catalyst as suggested from the higher D_f values recorded in Fig. 9.3. This could be also the result of the enhanced collision process suggested earlier. Furthermore, the oxidation/activation of the HCs (exothermic reactions) throughout the catalyst along with the localized thermal effect of the H_2 are likely to increase the local and overall temperature of the catalyst. Higher catalyst temperatures can also help in thermally restructuring the soot shape into a more spherical entity, especially when H_2 is presented. Although it can be assumed that a worsened DPF trapping efficiency has resulted from the presence of more compact and spherical aggregates, the remarkable increase in

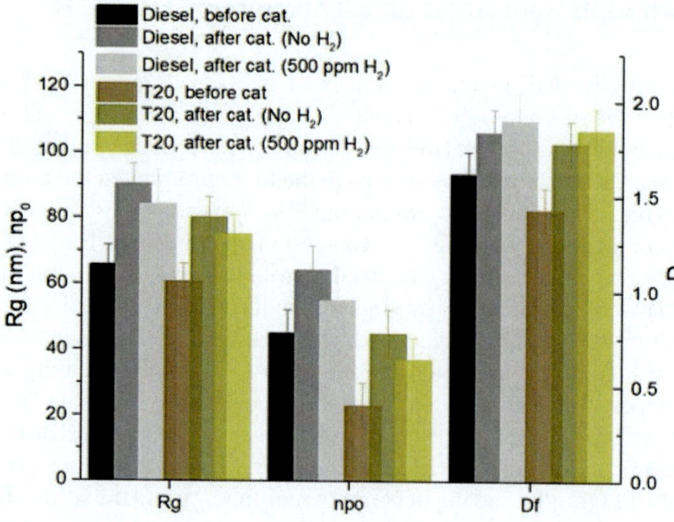

Fig. 9.3 Variation of D_f, n_{p0}, and R_g throughout the HC-SCR in the presence and absence of H_2 at 2 bar IMEP. Reproduced from (Serhan et al. 2019)

the overall particulate's size is expected to compensate their impact (Fayad et al. 2015). This assumption is mainly triggered from the fact the particulate's diffusion/convection pathway in a DPF (i.e., filtration mechanism) is most pronounced from the particulates' size and not their shape.

Concerning d_{p0}, insignificant variations were spotted in the absence of H_2 for both fueling conditions (Fig. 9.4). However, in the presence of H_2, d_{p0} dropped by 2.7% in case of diesel and 3.6% in case of TP20. The only possible theory to clarify these results is that despite the limited residence time of the particulates in the catalyst channels, soot oxidation reactions were launched once H_2 is injected. As for the increased reduction seen in case of TP20 with respect to diesel, it is expected to be the outcome from the enhanced oxidative potential of soot produced from the combustion of TP20 compared to that produced from the combustion of diesel, as reported earlier in Serhan et al. (2018). These possible oxidation reactions are mostly generated from the increased level of NO_2 throughout the catalyst channels (i.e., transition of NO into NO_2 before being converted to N_2 during H_2-SCR-assisted cycle, as it is advised by Breen et al. (2007)) since this oxidizer can launch the soot oxidation reactions in a temperature region as low as 200 °C (Ehrburger 2002). This assumption was partially approved by Houel et al. (Houel et al. 2007) who reported that in low-temperature window (200 °C to 350 °C), H_2 addition will not only enhance the Ag/Al_2O_3 de-NOx activity but also withstand it. The simplified reaction pathways presenting the NO_2-soot-assisted oxidation mechanism are presented below:

$$C(s) + NO_2 \rightarrow CO + NO$$

$$C(s) + 2NO_2 \rightarrow CO_2 + NO$$

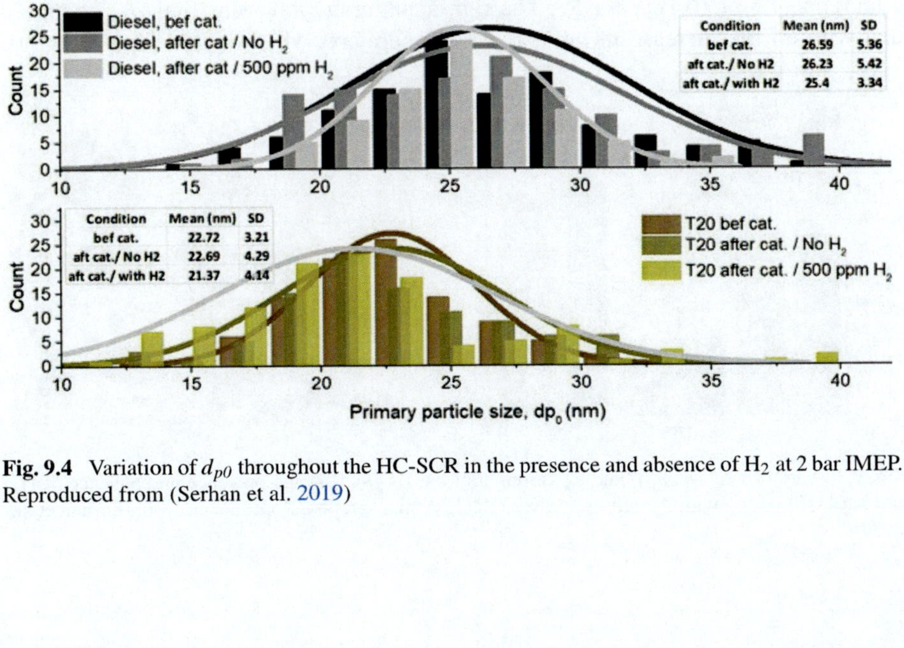

Fig. 9.4 Variation of d_{p0} throughout the HC-SCR in the presence and absence of H_2 at 2 bar IMEP. Reproduced from (Serhan et al. 2019)

Conversely, it is expected that O_2 should not assist the oxidation kinetics in this temperature window since higher temperature level (ranging in 500 to 700 °C) is needed. Furthermore, it is also confirmed in the literature that introducing H_2 gas in conjunction of an oxygen-rich flow (as the diesel exhaust) can change the Ag particulate chemistry to act as an oxidation catalytic agent. This is believed to result from the production of numerous sub-oxide elements (Kim et al. 2013) along with promoting the development of the active O_2- ions that can also assist the soot oxidation process (Vassallo et al. 1995).

In summary, it can be concluded that despite the low catalyst temperature (180 °C), the increase in the local NO_2 and surface O_2 functional species due to H_2 presence can partially oxidize the soot passing throughout the catalyst despite the very limited soot/catalyst residence time (flow-through substrate).

9.3.1.2 Medium Exhaust Temperature Condition

Similar to the earlier section, the same data type has been plotted in 5 and 6 after increasing the engine load to 4 bar IMEP and, respectively, the catalyst temperature to 280 °C (Figs. 9.5 and 9.6).

It is seen that the particulates surviving the catalytic reactions present greater R_g, n_{p0}, and D_f while d_{p0} is kept constant for the diesel case with insignificant drop seen only for the TP20 case. As a result, it is expected that the particulates have survived an enhanced aggregation mechanism throughout the catalyst channels and no possible catalytic oxidation is presented, even though that the catalyst temperature is increased by 100 °C compared to the 2 bar IMEP test. Introducing 500 ppm of H_2 activates the catalyst/soot synergies with significant modifications spotted compared to the 2 bar IMEP testing. As for diesel, a drop of 22% is seen for n_{p0}, 5.29% for d_{p0} and a total reduction of 10.16% for R_g. The significant reduction seen along d_{p0} confirms that with a little increase in the exhaust temperature, which is still representative

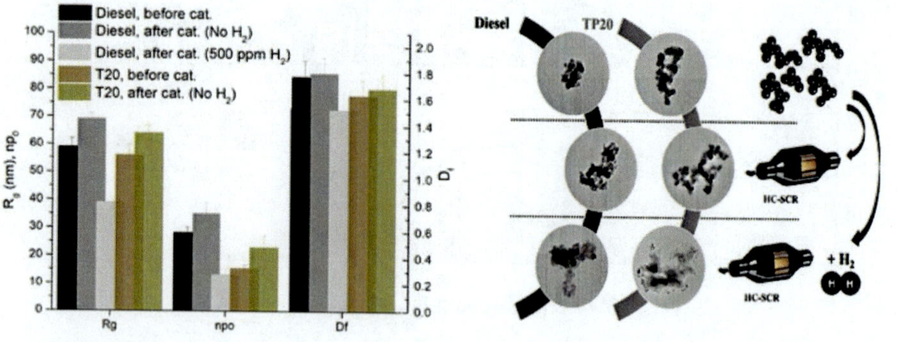

Fig. 9.5 Variation of D_f, n_{p0}, and R_g throughout the HC-SCR in the presence and absence of H_2 at 4 bar IMEP along with the representative HRTEM micrographs. Reproduced from (Serhan et al. 2019)

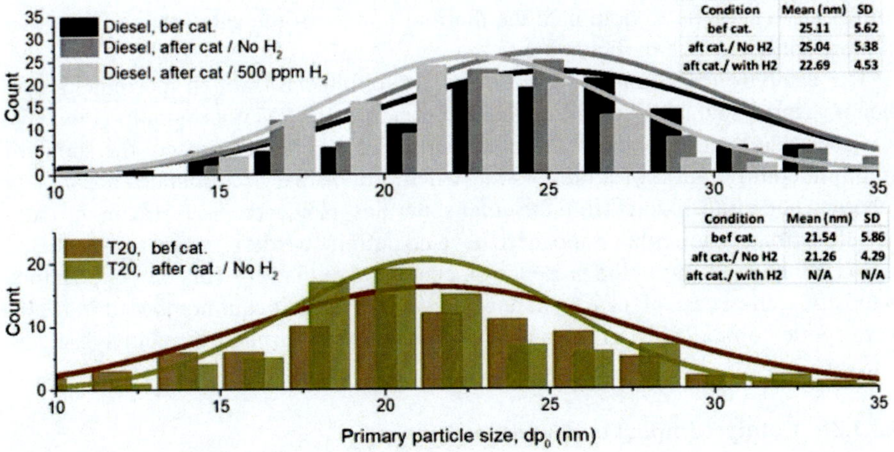

Fig. 9.6 Variation of d_{p0} throughout the HC-SCR in the presence and absence of H_2 at 4 bar IMEP. Reproduced from (Serhan et al. 2019)

of the low-temperature window, minute concentrations of H_2 can yield a powerful oxidation along the soot passing throughout the catalyst channels. It is expected that the NO_2-soot oxidation path is fully activated in this operating margin (~295 °C) and that the localized thermal impact of H_2 can also activate the O_2-soot oxidation pathway (Ehrburger 2002; Theinnoi et al. 2012). As for *Rg*, the catalyst impact was more pronounced with more severe reduction seen along this parameter compared to d_{p0}. A hypothesis could be drawn regarding this trend and can be mainly identified with two different approaches (Lee and Zhu 2004):

(a) Breakdown of the aggregates throughout the catalyst channels.
(b) Reduction in the interstitial distances among the primary particulates constituting the aggregates.

As n_{p0} also faces a noticeable drop downstream the catalyst, it is most probable that hypothesis (a) is the most dominant mechanism leading to this significant reduction in R_g. In addition, it is also speculated that the active oxygen species produced on the Ag particulates can also improve the soot/oxidizers contact and create electronic interaction between the surface of the catalyst and the soot particulates (thus promoting the breakdown phenomena). Concerning the soot fractal like particulates with more chain-like structure (~12% drop in D_f) are spotted downstream the catalyst. This configuration further indicates that hypothesis (b) is not valid for our analysis and highlights that a better DPF functionality can be expected with the application of the H_2-assisted HC-SCR catalyst to the exhaust flow. On the other hand, the increased local and global temperature of the catalyst is expected to thermally restructure the particulates into a more spherical entity (higher D_f). However, this was not the case with this testing condition, which in turn signifies that the catalytic oxidation reactions were remarkably prevailing at this slightly higher exhaust temperature and were sufficiently energetic to dominate the thermal impact of the catalyst (i.e., resulting in more spherical-like particulates).

The catalytic oxidation of the soot was not only seen through the reduction in their d_{p0}, but also through qualitative inspection for the TEM micrographs (Fig. 9.5). It is seen that upstream the catalyst, the particulates collected present the standard graphitic configuration of a diesel soot, where the primary particulates are mostly spherical and quite distinct from the other spherules. However, downstream the catalyst, the primary particulates appear to have no definite borders and seem to be fused in each other, confirming the presence of a catalytic oxidation. This blur appearance is mostly seen in case of TP20 where no statistically significant number of particles were spotted on the TEM grids to produce a confident morphological analysis.

9.3.1.3 Catalyst Impact on Nanostructure

Since the soot morphological analysis highlights the presence of possible oxidation mechanism, it is of interest to check if any variation will be seen in the particulates' nanoscale arrangement. Such kind of information will be valuable to predict the reactivity of the aggregates surviving the catalytic reactions and can also give better insights to validate the oxidation reactions speculated from the reduction seen in d_{p0}. As no noticeable changes were seen with d_{p0} when H_2 is not introduced to the flow, the below analysis will account for the impact of the catalyst only in presence of the H_2. To simplify the results, Figure summarizes the resulted nanostructure parameters as a ratio of the values recorded upstream the catalyst compared to that seen downstream the catalyst (i.e., referred as post/pre in Fig. 9.7).

Concerning the low exhaust temperature condition (2 bar IMEP), similar trend was seen between diesel and TP20: the catalyst helps in reducing the length of the carbon layers (-8.5% L_a for diesel and -13% L_a for TP20) and further twist

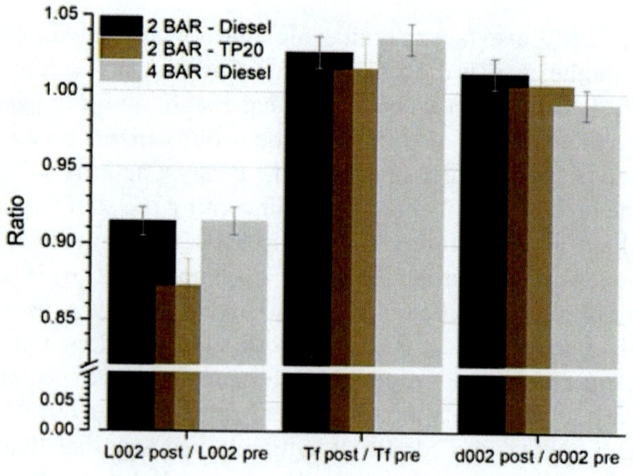

Fig. 9.7 Nanostructure parameter ratios including L_a, T_f, and d_{002}. Reproduced from (Serhan et al. 2019)

their skeleton (+2.8% T_f for diesel and +1.5% T_f for TP20). However, inconspicuous variation ranging within the data-error bars is seen along the interlayer spacing ($d_{002\ \text{after catalyst}}/d_{002\ \text{before catalyst}}$ ~ 1) for both fueling. All the resulted parameters highlight a better oxidative reactivity for the particulates collected downstream the catalyst.

As for the medium exhaust temperature condition (4 bar IMEP), only diesel particulate will be evaluated since as discussed in the earlier section, no statistically significant number of soot aggregates survived the catalytic reactions in case of TP20. Starting with L_a, similar results are seen compared to the 2 bar IMEP condition while T_f show a further increase reaching 3.5%. However, d_{002} present a small drop of 1% compared to the pre-catalyst case, highlighting that more graphitic particulates are presented. It is expected that at this relatively higher exhaust temperature (~300 °C) the higher HC oxidation phenomena and increased NO_2-soot oxidation reactions can help in thermally restructuring the neighboring carbon layers into more ordered series, resulting in slightly more graphitic configuration (Ma et al. 2014). However, accounting all the nanostructure parameters, it can be speculated that despite this shorter interlayer spacing, the particulates are expected to be more reactive since the layers' curvature has been increased by 3.5% and L_a reduced by 8.5%. Carbon layer curvatures and length have been earlier proved to be more influential in dictating the particulate's reactivity compared to d_{002}, as reported in (Serhan et al. 2018).

Furthermore, to relate the nanostructure alterations to the possible soot/catalyst oxidation mechanism, the rate of soot burn-off in the catalyst will be calculated according to the difference in d_{p0} before and after the catalyst. Song et al. (2006) reported that during the first 40% stage of the soot burn-off cycle, the oxidation reactions should follow the surface burning mode mechanism. As a result, the following formula from reference (Song et al. 2006) can be used to assess the soot burn-off rate:

$$dp0\text{post SCR}/dp0\text{pre SCR} = (1 + \text{z})^{1/3}$$

where z is the burn-off rate.

As a result, the soot burn-off is estimated to 8% in case of diesel and 10% in case of TP20 at 2 bar IMEP. Increasing the load to 4 bar IMEP increases the soot burn-off rate to 15% in diesel case while for TP20 there was no clear structure for the soot to measure the morphology/nanostructure parameters; however, it is expected to be higher than diesel case as it is in the 2 bar IMEP condition.

According to Song et al. (2006), surface burning mode mechanism is generally defined with slow combustion reactions translated with a notable reduction in L_a and no more than 3% increase in the particulate graphitic ordering (using Raman analysis) during the first 20% burn-off phase. However, Ma et al. (2014) presented a more detailed approach concerning the above. The study covers the progress of the soot nanostructure alterations during the oxidation process of separate particulates collected upstream and downstream a DOC catalyst. Different trends were seen along L_a and d_{002} between both samples, thus no clear conclusion can be beneficial for our analysis; however, it was confirmed that T_f should increase by ~2% during the first 20% burn-off phase, independently from the soot collection point (downstream or upstream the DOC). Summarizing the findings of reference (Song et al. 2006) and (Ma et al. 2014), during the 20% burn-off phase, L_a is expected to decrease, while T_f is expected to slightly increase with moderate variation along d_{002}. Those findings positively correlate with the nanostructure variation shown in the analysis, thus confirming that these alterations are resulted from an enhanced oxidation mechanism in the catalyst channels.

9.4 Summary

This chapter focuses on the impact of the aftertreatment system on the soot aggregates regarding their morphology and nanostructure characteristics. The particulate trapping efficiency in DOC is linked to the parent fuel type. Oxygenated fuel combustion, e.g., butanol/RME/diesel blends, results in particulates which are characterized with smaller primary particles and more skeletal aggregates compared to the diesel case. This, in turn, makes the particles produced more easily trapped in the DOC. Also, it was seen that independent of the fuel type or exhaust temperature condition, soot particulates tend to aggregate with the other neighboring particulates leading to larger and more circular particulates downstream the DOC. The same hypothesis can be also drawn in case of HC-SCR when H_2 is not introduced upstream the catalyst. However, injecting minute concentration of H_2 helps in significantly modifying the morphology of the particulates which seem to be more dependent on the exhaust temperature when H_2 is presented. At low exhaust temperature (180 °C), particulates' aggregation remains the most dominant mechanism; however, it was seen that the soot primary particle diameter faces some reduction, highlighting the presence of oxidation reactions throughout the catalyst. Increasing the catalyst temperature (290°C) enhances the catalytical modification of the aggregates. It was seen that not only d_{p0} was reduced but also the soot radius of gyration and number of primary particles included in soot aggregates. Those alterations highlight that at this relatively higher exhaust temperature, soot particulates are not only oxidized, but they also tend to breakdown throughout the catalyst channels instead of being aggregated together. This impact was ameliorated when oxygenated diesel blend (TP20) was used. This is mainly due to the higher oxidative reactivity toward oxygen of the particulates produced from TP20 compared to neat diesel. The TP20 soot catalytic oxidation can be qualitatively inspected from the TEM micrographs. It is seen that upstream the catalyst, the particulates collected present the standard graphitic configuration of a soot particle, where the primary particulates are mostly spherical and quite distinct from the other spherules. However, after passing through catalyst, the primary particulates appear to have no definite borders and seem to be fused in each other, confirming the presence of strong catalytic oxidation in case of TP20 despite the limited residence provided throughout the catalyst channels.

References

Breen J et al (2007) A fast transient kinetic study of the effect of H2 on the selective catalytic reduction of NOx with octane using isotopically labelled 15NO. J Catal 246(1):1–9. https://doi.org/10.1016/j.jcat.2006.11.017

Burke U, Pitz WJ, Curran HJ (2015) Experimental and kinetic modeling study of the shock tube ignition of a large oxygenated fuel: tri-propylene glycol mono-methyl ether. Combus Flame 162(7):2916–2927. https://doi.org/10.1016/j.combustflame.2015.03.012

Chiatti G, Chiavola O, Falcucci G (2009) Soot morphology effects on DPF performance. SAE technical paper, 2009-01-1279

Ehrburger P et al (2002) Reactivity of soot with nitrogen oxides in exhaust stream. SAE Technical Paper, 2002-01-1683. https://doi.org/10.4271/2002-01-1683

Fayad MA et al (2015) Role of alternative fuels on particulate matter (PM) characteristics and influence of the diesel oxidation catalyst. Environ Sci Technol 49(19):11967–11973. https://doi.org/10.1021/acs.est.5b02447

González MA et al (2001) Oxygenates screening for advanced petroleum-based diesel fuels part 2: The effect of oxygenate blending compounds on exhaust emission. SAE Technical Paper, 2001-01-3632. https://doi.org/10.4271/2001-01-3632

Houel V et al (2007) Promoting functions of H2 in diesel-SCR over silver catalysts. Appl Catalysis B: Environ 77(1–2):29–34. https://doi.org/10.1016/j.apcatb.2007.07.003

Kim PS et al (2013) Effect of H2 on deNOx performance of HC-SCR over Ag/Al2O3: morphological, chemical, and kinetic changes. J Catalysis 301:65–76. https://doi.org/10.1016/j.jcat.2013.01.026

Konstandopoulos AG, Kostoglou M (2004) Microstructural aspects of soot oxidation in diesel particulate filters. SAE Technical Paper, 2004-01-0693. https://doi.org/10.4271/2004-01-0693

Konstandopoulos AG, Skaperdas E, Masoudi M (2002) Microstructural properties of soot deposits in diesel particulate traps. SAE Technical Paper, 2002-01-1015. https://doi.org/10.4271/2002-01-1015

Lapuerta M, Martos FJ, Herreros JM (2007) Effect of engine operating conditions on the size of primary particles composing diesel soot agglomerates. J Aerosol Sci 38(4):455–466. https://doi.org/10.1016/j.jaerosci.2007.02.001

Lee KO, Zhu J (2004) Evolution in size and morphology of diesel particulates along the exhaust system. SAE Technical Paper, 2004-01-1981. https://doi.org/10.4271/2004-01-1981

Lee KO, Zhu J (2005) Effects of exhaust system components on particulate morphology in a light-duty diesel engine. SAE Technical Paper, 2005-01-0184. https://doi.org/10.4271/2005-01-0184

Liati A et al (2013) Variations in diesel soot reactivity along the exhaust after-treatment system, based on the morphology and nanostructure of primary soot particles. Combus Flame 160(3):671–681. https://doi.org/10.1016/j.combustflame.2012.10.024

Ma Z et al (2014) Effects of diesel oxidation catalyst on nanostructure and reactivity of diesel soot. Energy Fuels 28:4376–4382. https://doi.org/10.1021/ef500467a

Natarajan M et al (2001) Oxygenates for advanced petroleum-based diesel fuels: Part 1. Screening and selection methodology for the oxygenates. SAE Technical Paper, 2001-01-3631. https://doi.org/10.4271/2001-01-3631

Serhan N, Tsolakis A, Martos FJB (2018) Effect of propylene glycol ether fuelling on the different physico-chemical properties of the emitted particulate matters: Implications of the soot reactivity. Fuel 219:1–11. https://doi.org/10.1016/j.fuel.2018.01.065

Serhan N et al (2019) Modifying catalytically the soot morphology and nanostructure in dieselexhaust: Influence of silver De-NOx catalyst (Ag/Al2O3). Appl Catalysis B: Environ 241:471–482

Silverman DT et al (2012) The diesel exhaust in miners study: a nested case-control study of lung cancer and diesel exhaust. J Natl Cancer Inst 104(11):855–868. https://doi.org/10.1093/jnci/djs034

Song J et al (2006) Examination of the oxidation behavior of biodiesel soot. Combust Flame 146(4):589–604. https://doi.org/10.1016/j.combustflame.2006.06.010

Tassopoulos M, O'Brien JA, Rosner DE (1989) Simulation of microstructure/mechanism relationships in particle deposition. Amer Inst Chem Eng J 35(6):967–980. https://doi.org/10.1002/aic.690350610

Theinnoi K et al (2008) Hydrogen promotion of low-temperature passive hydrocarbon-selective catalytic reduction (SCR) over a silver catalyst. Energy Fuels 22:4109–4114. https://doi.org/10.1021/ef8004515

Theinnoi K et al (2012) Diesel particulate filter regeneration strategies: Study of hydrogen addition on biodiesel fuelled engines. Energy Fuels 26(2):1192–1201. https://doi.org/10.1021/ef201355b

Vassallo J, Miró E, Petunchi J (1995) On the role of gas-phase reactions in the mechanism of the selective reduction of NOx. Appl Catalysis B: Environ 7:65–78. https://doi.org/10.1016/0926-3373(95)00032-1

Chapter 10
Effect of Hybrid Nanoparticle on DI Diesel Engine Performance, Combustion, and Emission Studies

Elumalai Perumal Venkatesan, Dhinesh Balasubramanian, Olusegun David Samuel, Muhammad Usman Kaisan, and Parthasarathy Murugesan

Abstract The frequent rise in the use of diesel engines in all fields emits harmful gases such as NO_x and CO, which causes significant environmental emissions, global warming, breathing problems, etc. (Sivalingam et al. 2019). In the investigation of the performance, combustion, and emission characteristics, using diesel water emulsion is mixed with hybrid nanoparticles as additives in Direct Injection (DI) diesel engine. Reducing the emission characteristics and increasing engine performance is to introduce emulsion fuels (Parthasarathy et al. 2021). The water content of 5% is added with the diesel fuel as blends of (D94% + W5%). The surfactant used is Span 80 and Tween 80 and mixed with diesel water emulsion using a mechanical stirring

E. Perumal Venkatesan (✉)
Aditya Engineering College, Surampalem 533437, India
e-mail: elumalai@aec.edu.in

D. Balasubramanian
Department of Mechanical Engineering, Mepco Schlenk Engineering College, Sivakasi, India

Mechanical Engineering, Faculty of Engineering, Khon Kaen University, Khon Kaen, Thailand

Center for Alternative Energy Research and Development, Khon Kaen University, Khon Kaen, Thailand

D. Balasubramanian
e-mail: dhineshbala91@mepcoeng.ac.in; dhinesh.b@kkumail.com

O. D. Samuel
Department of Mechanical Engineering, Federal University of Petroleum Resources, P.M.B 1221, Effurun, Delta State, Nigeria

Department of Mechanical Engineering, University of South Africa, Science Campus, Private Bag X6, Florida 1719, South Africa

M. U. Kaisan
Department of Mechanical Engineering, Ahmadu Bello University, Room 11, Shell Chair Office Complex, Zaria, Nigeria

P. Murugesan
Department of Automobile Engineering, Vel Tech Rangarajan Dr. Sagunthala R&D Institute of Science and Technology, Avadi, India

A. P. Singh and A. K. Agarwal (eds.), *Novel Internal Combustion Engine Technologies for Performance Improvement and Emission Reduction*, Energy, Environment, and Sustainability, https://doi.org/10.1007/978-981-16-1582-5_10

process. This reports on the use of cerium oxide (CeO_2), aluminum oxide (Al_2O_3), titanium dioxide (TiO_2) nanoparticles as an additive to diesel fuel. For this study, the fuel tested was prepared by blending three nanoparticles into diesel in a mass fraction of 50 ppm, 100 ppm, 150 ppm with the assistance of an ultrasonic stirrer. The diesel water emulsion is mixed with the nanoparticle and prepared as three different fuel blends such as (D90% + W5% + S5% + HBNP 50 ppm), (D90% + W5% + S5% + HBNP 100 ppm), and (D90% + W5% +S5% + HBNP 150 ppm). Based on experimental results, BTE increased by 8.3% and BSFC is reduced by 14.42% at (D94% + W5% +S1% + HB 150 ppm) blend when compared with the diesel, due to the atomization of the fuel and oxygen content available in the fuel. The emissions of CO is reduced by 10.2%; smoke, oxides of nitrogen emissions, and HC are reduced by 27.5%, 36.58%, and 27.77%, respectively, when compared with clean diesel fuel, because of microexplosion and proper atomization taking place during the combustion process. The cylinder pressure and HRR are increased by 3.2% and 2.8%, respectively, when compared with neat diesel fuel, due to increased combustion temperature and secondary atomization of fuel take places.

Keywords Water · Emulsion · Hybrid nanoparticle · HLB · Aluminum · Titanium · Cerium oxide

10.1 Introduction

The consumption of fossil fuels and petroleum products has rapidly increased due to industrialization. As a non-renewable source of energy, the use of oil as a fuel is exhausted step by step because of the increment in transportation and energy. The usage of diesel to a maximum extent leads to pollution through hydrocarbons (HC), carbon monoxide (CO), and nitrogen oxide (NOx) emissions. Such pollutants are particularly harmful to humans and the environment. To minimize these emissions, the fuel is doped with an additive or catalyst with the base fuel to improve its physiochemical properties. DI diesel engines are generally utilized as a power source in different applications because of their better performance efficiency of characteristics (Mohd et al. 2017). There are several methods used to minimize exhaust emissions, like EGR, fuel additives in nanoparticles, and fuel additives in antioxidants agents, etc. One of the best methods that attempt to increase the performance parameters and dropping of the emission parameters (HC, CO, NOx) is the addition of nanoparticles.

Various experiments have been performed on the water–diesel emulsion in the diesel engine to increase water content in the diesel to improve brake thermal efficiency and decrease carbon monoxide, hydrocarbon, and nitrogen oxides. As the water content in the diesel fuel slightly increases the BTE, BMEP decreases, due to the rapid evaporation of water, and this adds more fuel consumed in the premixed combustion phase. The oxides of nitrogen, carbon monoxide, and hydrocarbon are drastically reduced by 35.12%, 28.54%, and 24.32%, respectively, because of secondary atomization and microexplosion, where the larger droplets are converted into smaller droplets (Nadeem et al. 2006).

> The diesel–water emulsion fuels used in the investigation to conduct brake thermal efficiency of diesel–water emulsion fuel is increased by 4.59% compared to diesel fuels. Due to the oxidation rate of the fuel water droplets are increased There is some considerable reduction in the emission of carbon monoxide, hydrocarbon, smoke emission (by 7.05% 10.51%, 7.54%, respectively), but the reduction in NOx emission is drastic (28.52%) because the addition of water reduced the peak combustion temperature (Pradeepkumar et al. 2013)

The performance and brake thermal efficiency are marginally increased by adding diesel fuel with nanometal oxides as additives, which is compared to conventional diesel fuel. It also found that the emission of hydrocarbons decreases due to the addition of catalytic nanoparticles. Generally, using the nanoparticles with diesel results in a decrease in NOx emission by 30% (Ramlan et al. 2016).

Diesel–water emulsion with 25 ppm, and 50 ppm aluminum oxide nanoparticle were performed in this investigation. The brake thermal efficiency of the blend diesel water mixed with nanoparticles dosage of 50 ppm shown increased by 9.9%, respectively, when compared with neat diesel due to microexplosion and secondary atomization of the fuels, it improved the combustion process. The blend of an emulsion of diesel–water mixtures of 50 ppm hydrocarbon and carbon monoxide is decreased by 10.52% and 7.8% compared to neat diesel fuels because of shortened ignition delay and nanoparticle present in the water molecules to achieve the better combustion process. There is a drastic change in smoke and nitrogen oxides of the 50 ppm diesel–water nano-emulsion blend where 40% and 27.1% were due to secondary atomization and microexplosion of water molecules in the nanoparticles (Sadhik Basha and Anand 2011).

In this investigation, water–diesel was doped with various dosages of 50 ppm, 100 ppm, and 150 ppm with carbon nanotubes, respectively. Because of the calorific value of the fuel, the BTE of nano-emulsion is increased by 8.1% as correlated with clean mineral fuel, and where the fuels evaporated very quickly. Oxides of nitrogen, carbon monoxide, and hydrocarbon are decreased by 42.7%, 41.2%, and 34.25%, respectively, due to water molecules and nanoparticles present in the fuels completely burn the droplet of fuel during the combustion process (Narinder and Bharj 2015).

The diesel–water emulsion added with zirconium nanoparticle of dosage used 50 ppm and 100 ppm. Owing to the microexplosion and ignition delay phenomenon of emulsion fuel, the brake thermal efficiency (BTE) is increased by 5.2% in the diesel–water nano-emulsion blends of 100 ppm of dosage as correlated with clean mineral fuel. The 100 ppm nano-emulsion blend shows nitrogen oxides and smoke emissions decreased by 22.37% and 25.67%, respectively, due to the water molecules present in the fuel to minimize the high peak temperature during combustion. The mixture of 100 ppm nano-emulsion reduced carbon monoxide and hydrocarbon by 15.1% and 40%, respectively, due to the high secondary atomization, enhanced air–fuel mixing, and intensity in microexplosion and attributes of nanoparticles incorporated emulsion fuels (Vellaiyan et al. 2016).

In this investigation of introducing titanium dioxide for different dosages of 50 ppm, 70 ppm, and 85 ppm to diesel fuel emulsion, the following observations were found. The brake thermal efficiency of the diesel–water nano-emulsion is improved by 16.52% due to the availability of oxygen of nanoparticles that enhance the engine performance. The oxides of nitrogen, carbon monoxide, and hydrocarbon were reduced by 6.6%, 14.9%, and 37%, respectively, and correlated to mineral diesel fuels because of the microexplosion of emulsified fuel that decreased the in-cylinder temperature of the combustion process (Rajwinder et al. 2016).

From the above survey, it was found that carbon monoxide and hydrocarbon in the traditional engine were increased using diesel fuel. The critical goals of reducing NO_X and HC are achieved by bringing water into diesel fuel. As a result, the brake thermal efficiency can increase slightly, nitrogen oxides can dramatically decrease, and hydrocarbon and carbon monoxide can slightly decrease. Water is added to diesel fuel to achieve superior performance and emission characteristics in a conventional diesel engine.

Most work has been done to add more dosages of nanoparticles to increase brake thermal efficiency and decrease the emission of hydrocarbon (HC), carbon monoxide (CO), nitrogen oxides (NOx), and smoke opacity. More oxygen available in the nanoparticles and also microexplosion of smaller droplets of fuels resulted in better combustion.

For the current study, diesel was taken as the base fuel and mixed with water with the constant water percentage of 5% with the help of an emulsion setup. The surfactant chosen are Span 80 and Tween 80 of 5% added with the diesel–water emulsion. The hybrid nanoparticle (Al_2O_3, TiO_2, and CeO_2) mixed with the diesel–water emulsion fuel of various proportions of 50 ppm, 100 ppm, and 150 ppm, respectively. Hybrid nanoparticles (HBNP) such as Cerium oxide (CeO_2), Aluminum oxide (Al_2O_3), Titanium dioxide (TiO_2) are doped together equally as a single additive (HBNP). This is the innovation of the proposed work. In the diesel engine, CeO_2 has also been used to minimize ignition delay, variable fuel consumption, exhaust emission smoke, and thermal brake-efficiency (Janakiraman et al. 2019). Using Al_2O_3 and TiO_2 serves as a catalyst for oxygen donation and provides oxygen for CO oxidation, and absorbs oxygen for NOx reduction. The fuel blends prepared in the current investigation is D100, D94% + W5% +S1% + HBNP 50 ppm, D90% + W5% +S5% + HBNP 100 ppm, D90% + W5% +S5% + HBNP 150 ppm. The diesel-water nanoemulsion is a promising alternative fuel used in the diesel engine to reduce the formation of NOx, HC, and CO in the tailpipe emission.

Table 10.1 shows that previous research work was conducted. The experimental investigation on diesel fuel blended with nanoparticle use two different symbols upwards and downwards. Upwards symbols show increase,downwards symbols show decrease and NC—No change value. Mg—Magnesium, Mn—Manganese, Al—Aluminum, Al_2O_3—Aluminum oxide, Fe—Iron, B—Boron, CeO—Cerium dioxide, TiO_2—Titanium Dioxide, D—Diesel.

Table 10.2 shows that previous researchers conducted an experimental investigation on diesel fuel blended with nanoparticles. They have used five different symbols upwards, downwards, NC, = , and —. Upwards symbols show an increase, and downwards symbols show a decrease. NC means no change value compared with diesel value and = that value is equal value to diesel value and—shows that

Table 10.1 Performance and emissions characteristics of diesel engines fuelled by diesel fuel blends containing nano-additives compared with neat diesel

Type of fuel blend	Performance		Emission characteristics				References
	BSFC	BTE	CO	NOx	HC	PM/Smoke	
D + Mg (0.01%)	▲	▲	▼	▲	▼	▼	Chlopek et al. (2015)
D + Mg (0.02%)	▲	▲	▼	▲	▼	▼	
D + Mg (0.03%)	▲	▲	▼	▲	▼	▼	
D + Mg (0.04%)	▲	▲	▼	▲	▼	▼	
D + Mg (0.05%)	▲	–	▼	▲	▼	▼	
D + Mn (8 μmol/L)	▼	–	▼	▲	–	▼	Keskin et al. (2011)
D + Mn (16 μmol/L)	▼	–	▼	▲	–	▼	
D + Mg (8 μmol/L)	▼	–	▼	▲	–	▼	
D + Mg (16 μmol/L)	▼	–	▼	▲	–	▼	
D + Al (0.5 wt%)	▼	▲	▼	▲	▼	▲	Mehta et al. (2014)
D + B (0.5 wt%)	▲	▼	NC	NC	NC		
D + Fe (0.5 wt%)	NC	NC	▼	▲	▼	▲	
D + CeO (50 ppm)	▼	▲	▼	▼	▼	▼	Mei et al. (2016)
D + CeO (100 ppm)	▼	▲	▼	▼	▼	▼	
D + Al_2O_3 (25 ppm)	▼	▲	▼	▲	▼	▼	Aalam et al. (2015)
D + Al_2O_3 (50 ppm)	▼	▲	▼	▲	▼	▼	
D + Al (0.5 wt%)	▼	▲	▼	▲	▼	–	Nishant et al. (2015)
D + TiO_2 (80 mg/L)	▼	▲	▼	▼	▼	–	DSilva et al. (2015)
D + CeO (25 ppm)	▼	▲	▼	▲	▼	▼	Aalam and Alagappan (2015)
D + CeO (50 ppm)	▼	▲	▼	▲	▼	▼	
D + FeO (25 ppm)	▼	▲	▼	▲	▼	▼	Syed et al. (2015)
D + FeO (50 ppm)	▼	▲	▼	▲	▼	▼	
D + Al (75 ppm)	▼	▲	▲	NC	▲	▼	Babu et al. (2015)

Table 10.2 Performance and emissions characteristics of diesel engines fuelled by various diesel in water emulsion fuel blends containing nano-additives compared with neat diesel

Type of fuel blend	Performance		Emission characteristics				References
	BSFC	BTE	CO	NOx	HC	PM/Smoke	
W/D (5% Water)	▲	▲	NC	▼	NC	▼	Basha et al. (2017)
W/D + CNT (25 ppm)	▼	▲	▼	▼	NC	▼	
W/D + CNT (50 ppm)	▼	▲	▼	▼	NC	▼	
W/D (15% Water)	▼	▲	NC	▼	▲	▼	Narinder and Bharj (2015)
W/D + CNT (50 ppm)	▼	▲	▼	▼	▲	▼	
W/D + CNT (100 ppm)	▼	▲	▼	▼	▼	▼	
W/D + CNT (150 ppm)	▼	▲	▼	▼	▼	▼	
W/D (20% Water)	▼	▲	▼	▼	▼	▼	Narinder and Bharj (2015)
W/D + CNT (50 ppm)	▼	▲	▼	▼	NC	▼	
W/D + CNT (100 ppm)	▼	▲	▼	▼	NC	▼	
W/D + CNT (150 ppm)	▼	▲	▼	▼	▼	▼	
W/D (15% water)	▲	▲	NC	▼	NC	▼	Basha et al. (2011)
W/D + Al_2O_3 (25 ppm)	=	▲	NC	▼	NC	▼	
W/D + Al_2O_3 (50 ppm)	=	–	NC	▼	NC	▼	
W/D + Al_2O_3 (100 ppm)	▼	▲	NC	▼	NC	▼	
W/D (10% water)	–	–	▲	▼	▼	▼	Vellaiyan et al. (2016)
W/D + ZnO (50 ppm)	–	–	▲	NC	▼	▼	
W/D + ZnO (100 ppm)	–	–	▲	NC	▼	▼	
W/D (10% water)	▲	▲	▲	▼	NC	–	Rajwinder et al. (2016)
W/D + TiO_2 (50 ppm)	NC	▲	▼	▼	▼	–	
W/D TiO_2 (70 ppm)	▼	▲	▼	▼	▼	–	
W/D (15% water)	▲	▲	▲	▼	▼	–	

(continued)

Table 10.2 (continued)

Type of fuel blend	Performance		Emission characteristics				References
	BSFC	BTE	CO	NOx	HC	PM/Smoke	
W/D + TiO_2 (50 ppm)	▲	▲	NC	▼	NC	–	
W/D TiO_2 (70 ppm)	NC	▲	▼	▲	▲	–	

emission parameters are omitted. CNT—carbon nanotube, TiO_2—Titanium dioxide, Al_2O_3—Aluminum dioxide, ZnO—Zinc Oxide.

10.2 Emulsion

The emulsion is the process of combining two immiscible fluids, water, and oil. There are two kinds of techniques of emulsification: two-stage phase emulsions and three-stage phase emulsion (Adiga and Shah 1990). The three-phase emulsion uses different surfactants and the time consuming process is also high. Another drawback is the cost is also very high compared to two-phase emulsions (Vellaiyan et al. 2016). Many researches have done the experiment in two-phase emulsions the first phase is the dispersed phase, and the other is the continuous phase (Basha and Anand 2011). The role of surfactants in the emulsion cycle is vital, as the surfactants act as a significant part of the oil emulsion in water. The surfactants tend to mix oil and water easily. The choice of surfactants depends on the high HLB value (Lin and Wang 2003). Most of the authors have used the surfactants Span 80 and Tween 80 because they have enough stability and withstand for longer periods compared to other surfactants (Emiro and Mehmet 2017). The stability of emulsion fuel depends upon emulsified methods such as stirrer speed, the concentration of water, the process of duration, and surfactants. The mineral diesel fuel is combined in the bottle containers, adding 5% water and 5% surfactants. For half an hour, the water and oil undergo stirring phase until a final homogeneous stage is reached. Figures 10.1 and 10.2 show the

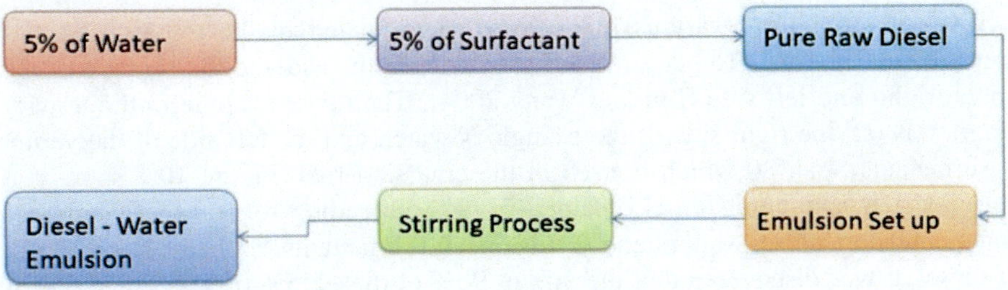

Fig. 10.1 Emulsion process (Ramalingam et al. 2018)

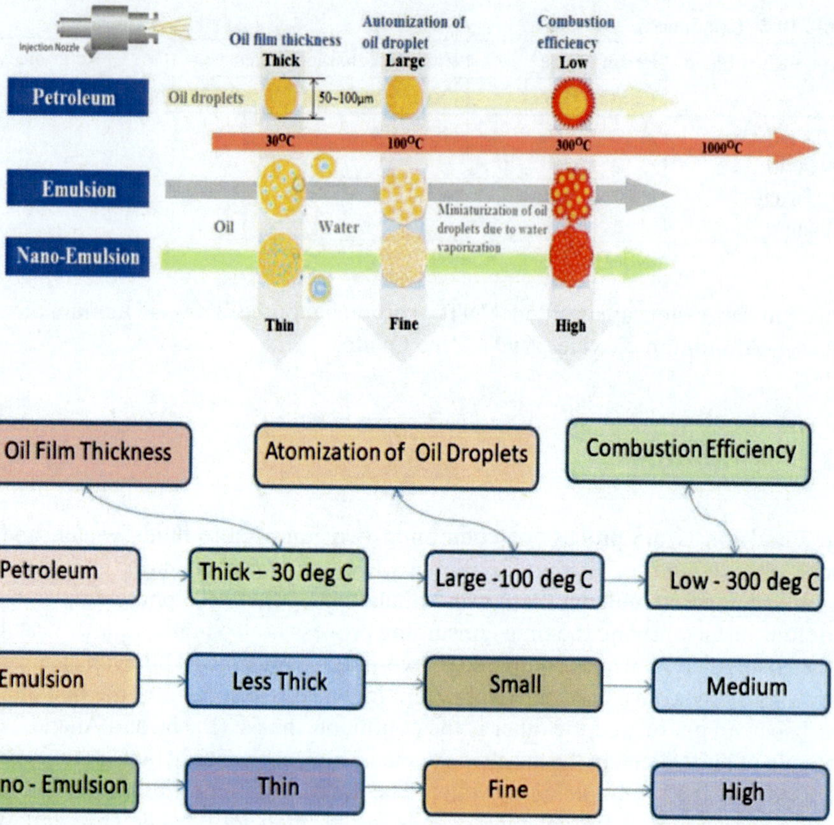

Fig. 10.2 Emulsion process in diesel engine

emulsion setup and emulsion process in a diesel engine.

10.2.1 Ternary Diagram

Results acquired from various plans were encapsulated in the ternary chart as appeared in Fig. 10.3. To get a ternary graph, it was considered the triangle of the upper, right, and left side (Lin and Wang 2004). The upper triangle in the ternary diagram is oil, the right side of the triangle is water, and the left side of the vertex is surfactants Span 80 which is used in the emulsion fuel. Figure 10.3 show that diesel–water nano-emulsion is obtained from the blending of neat diesel, diesel–water emulsion, diesel–water nano-emulsion, and surfactants. From the stability test of diesel, it was discovered that the mix of 90% of diesel, 5% of water, and 5% of surfactant was steady for a more extended period for 64 h. Henceforth the stable of

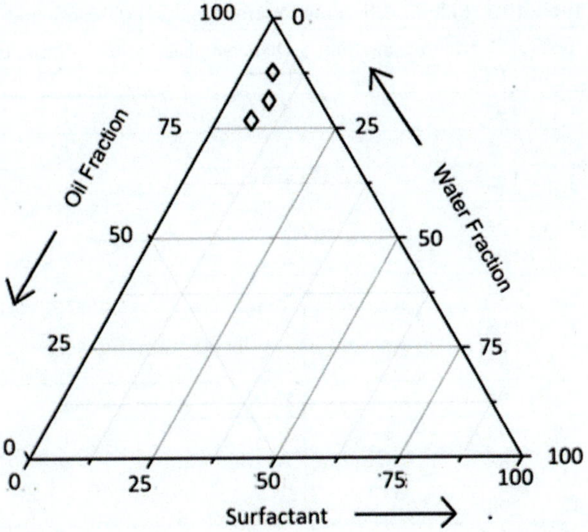

Fig. 10.3 Ternary diagram (Dhinesh and Annamalai 2018)

90% of diesel 5% of water and 5% of Span 80 and Tween 80 by volume is used for the conventional engine. The investigation was further done by utilizing the better emulsion of 90% of diesel 5% of water and 5% of surfactants for experimenting with the engine parameters. Table 10.3 shown that stability checking for emulsion fuel.

10.2.2 Fuel Preparation

The three different nanoparticles are used in the diesel–water fuel blend like Al_2O_3, TiO_2, and CeO_2, the addition of dosage is 50, 100, and 150 ppm, respectively. The blend D100, D90% + W5% + S5% + HBNP 50 ppm, D90% + W5% + S5% + HBNP 100 ppm and D90% + W5% + S5% + HBNP 150 ppm are used as test fuels. All the blends were subjected to the magnetic stirring of 10 min and Ultra-sonication, for 30 min. The combined test fuel nanoparticles were subsequently taken to investigate the stability test, and nanoparticles were not settled by up to 98 h (Elumalai et al. 2021). In addition, the previous analysis of the study into settling, fuel property, and characterization of nanoparticles was carried out. The test fuel properties are shown in Table 10.4 and Fig. 10.4 shown the fuel preparation of the flow chart.

Table 10.3 Stability checking for emulsion fuel (Dhinesh and Annamalai 2018)

Tests	Continuous medium (%)	Disperse phase water (%)	Surfactant (Span 80) (%)	Stability (Hours)
1	97	2	1	10
2	95	4	1	8
3	94	5	1	7.5
4	92	7	1	5
5	96	2	2	18
6	94	4	2	6
7	93	5	2	11
8	91	7	2	13
9	95	2	3	12.5
10	93	4	3	14
11	92	5	3	15.43
12	90	7	3	13
13	94	2	4	6
14	92	4	4	7
15	91	5	4	22
16	89	7	4	25.7
17	93	2	5	32
18	92	3	5	38
19	**90**	**5**	**5**	**64**
20	88	7	5	53

Table 10.4 Fuel property of the emulsion fuel

Sl. No	Property	Diesel	Diesel—water	DWHBNP 50 ppm	DWHBNP100 ppm	DWHBNP150 ppm
1	Density @ 20°C (Kg/m^3)	834	839	859	863	871
2	Viscosity @ 40°C (Cst)	3.16	3.54	3.78	4.12	4.34
3	Fire Point (°C)	58	54	63	68	71
4	Flash Point (°C)	72	69	76	81	84
5	Cetane Number	47	45	48	48.8	49.4
6	Calorific Vaue (MJ/kg)	42	38	39.5	38.3	39

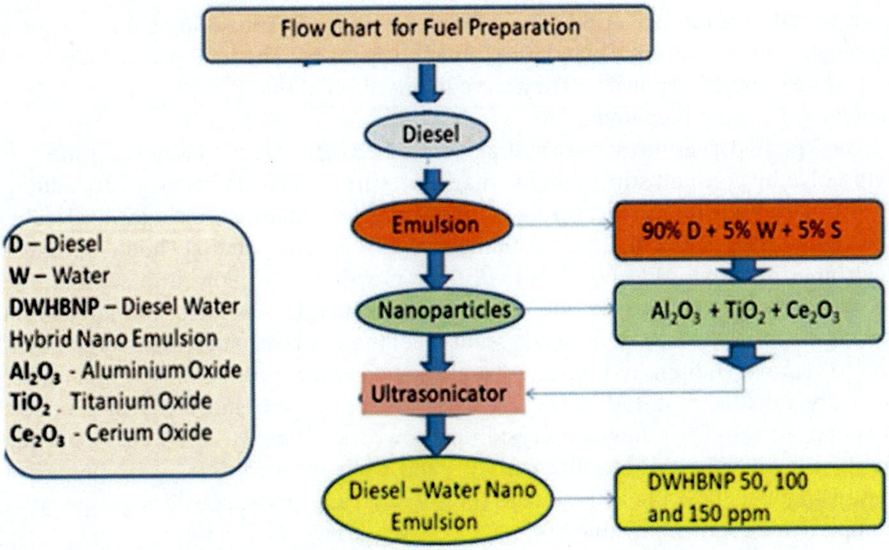

Fig. 10.4 Flow Chart for fuel preparation

10.3 Synthesis of Aluminum, Titanium and Cerium Oxide Nanoparticles

Using aluminum sulfate and sodium hydroxide pellets, the aluminum nanoparticles were prepared by precipitation process. Aluminum sulphate, 34 g was submerged in distilled water and a magnetic stirrer for one hour kept the solution under continuous stirring (Mahesh Kumar et al. 2020). Once the aluminum sulfate was fully dissolved, 90 ml of sodium hydroxide solution was added. The obtained white creamy solution was allowed to settle for an overnight and then allowed to form precipitate carefully. The precipitate was cleaned so many times with refined water and then dried at 80 °C overnight. The full conversion of aluminum hydroxide to alumina takes place during the drying process (Nadeem et al. 2006). At 100 °C, the precipitate obtained is dried in the oven for 5 h. Finally, the obtained precipitate is calculated at 450 °C in a muffle furnace. The dry form of the precipitate is obtained after heating in a furnace. The final products obtained were white nanocrystalline powder. The dry precipitate is eventually ground up into fine particles. Nanoparticles of aluminum were finally prepared.

The nanoparticle TiO_2 was prepared with the solution titanium isopropoxide as the raw material. In a conventional procedure, 90 ml of deionized water and 5 ml of acetic acid at room temperature were extracted to get solution A. In 40 ml of anhydrous ethanol, 14 ml of titanium isopropoxide was extracted with regular stirring to form solution B (Prabu 2018). Then, under intense stirring, solution B was added drop-wise into the solution A within 60 min. Consequently, the collected solution was periodically stirred for 2 h and aged for 48 h at room temperature. TiO_2 gels were stored in a furnace at 80 °C for 10 h as prepared. The solids collected were eventually calculated for 2 h at 450 °C in a muffle oven. The final product obtained was white crystalline powder. They were finally ground into fine particles and TiO_2 nanoparticles were prepared.

The 3 grams of cerium acetate hydrate was added to the distilled water and allowed to dissolve by constant stirring in the magnetic stirrer and after occasionally adding 90 ml of sodium hydroxide solution for 30 min to obtain a well-dissolved solution (Panneerselvam et al. 2015; Patnaik et al. 2017). The solution should be under vigorous stirring at 40 to 60 °C. Finally, the precipitate is allowed to settle at the bottom overnight. After that, the obtained precipitate is separated and kept in an oven at 100 °C for 5 h. The obtained solid precipitate is then made to undergo filtration to remove sodium hydroxide several times and after filtration, the precipitate is finally calculated at 450 °C in a muffle furnace for 2 h and the final products obtained were light yellow nan crystalline powders (Elumalai et al. 2021b). They are then finally ground into fine particles and CeO_2 nanoparticles were prepared. Figure 10.5a–c show the SEM image for Al_2O_3, TiO_2, and CeO_2. The property of nanoparticle is shown in Table 10.5.

10.3.1 Fourier Transform Infrared Spectroscopy (FT-IR)

The hybrid nanoparticle samples were loaded in the FT-IR spectroscopy with relevant peaks generated as seen in Fig. 10.6a–c. They have some feeble peaks from 1700 to 3000 cm $^{-1}$ which was consistent with the –CH band. FT-IR results in the N-H band at 1236 cm $^{-1}$ and this agreement supports the presence of biomolecules in the besides of fuels. The sharp peak 3492 cm^{-1} reveals the existence of the N-H band in the sample, and it was broader than other generated peaks in this region namely, aromatic ring and carbonyl stretching modes. Scissoring in-plane twisting approach of the NH_2 amine group is available at 1746 cm^{-1}. The wide-ranged band 992 cm^{-1} appeared owing to NH_2 mode. The generated peaks 1258 cm^{-1} and 2892 cm^{-1} are related to the absorption band. The peak 1256 cm^{-1} fits the $C = O$ absorption stretch. Some other minor peaks were associated with C–CO–O, C–CH3, –CH3.

10.3.2 X-ray Diffraction (XRD)

Figure 10.7 displays X rays where the electromagnetic radiation frequency is about 1 Å (10^{-10} m), which is nearly a similar size as a particle. X- rays diffraction gives the most authoritative primary data. To provide data about structures, atomic separations need to be tested.– This requires a test frequency of 1×10^{-10} m XRD (X-Ray) diffraction to take a gander at single crystals or polycrystalline materials. XRD analysis of hybrid nanoparticle and the deflection peaks at 2θ of 40, 46 and 66 are related with (1 1 1), (2 0 0) and (2 2 0) planes are associated (Karthikeyan et al.

Fig. 10.5 **a** SEM Image of Al_2O_3. **b** SEM Image of TiO_2. **c** SEM Image of CeO_2

Table 10.5 Property of nanoparticle (Prabu et al. 2018)

Parameters	Specification		
Manufactured company	Alfa Aesar	Sigma Aldrich	Global
Name of Chemical	$Al_2 0_3$	CeO_2	TIO_2
CAS Number	1344-28-1	1306-38-3	13463-67-7.
Weight of Molecular	102.96	173.11	364,64
Diameter Particle Size	52 nm	33 nm	36 nm
Appearance	White	Yellow	White

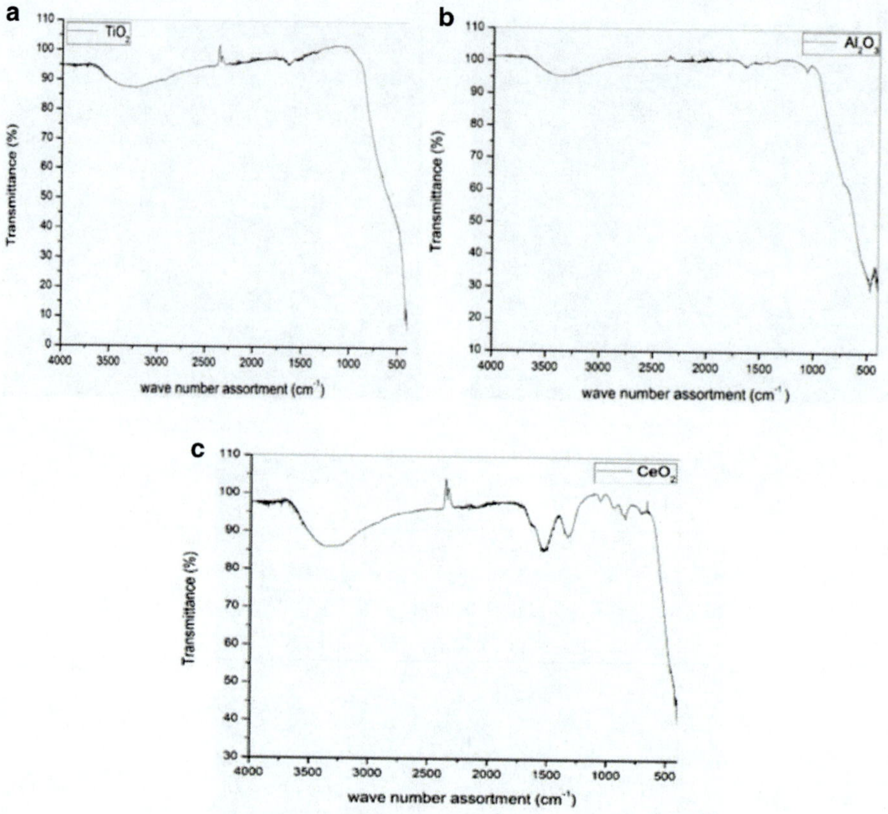

Fig. 10.6 **a** FTIR analysis of TiO_2. **b** FTIR analysis of Al_2O_3. **c** FTIR analysis of CeO_2

2014). The deflection peak (1 1 1) mean particle size was inveterate by Scherrer's method (Venu et al. 2019). Then the chronic mean particle size of the hybrid particle was about 5.2-nanometer range. Finally, the result agreed on the presence of hybrid nanoparticle in the sample with an appropriate nanometer size, and this would aid in upgrading the catalytic activity in the reaction (Karthickeyana et al. 2020).

10.4 Engine Set Up

Figure 10.8 shows the experimental set up of the engine. A 5.2 kW Kirloskar diesel engine with the ability to accelerate speed to about 1500 rpm was used for this study. The engine was associated with an eddy current dynamometer for loading purposes. The specifications of the engine are shown below. A Kirloskar engine consisting of different sensors like pressure sensor, temperature sensor, mass flow sensor, throttle sensor was used along with the numerous instruments as illustrated for

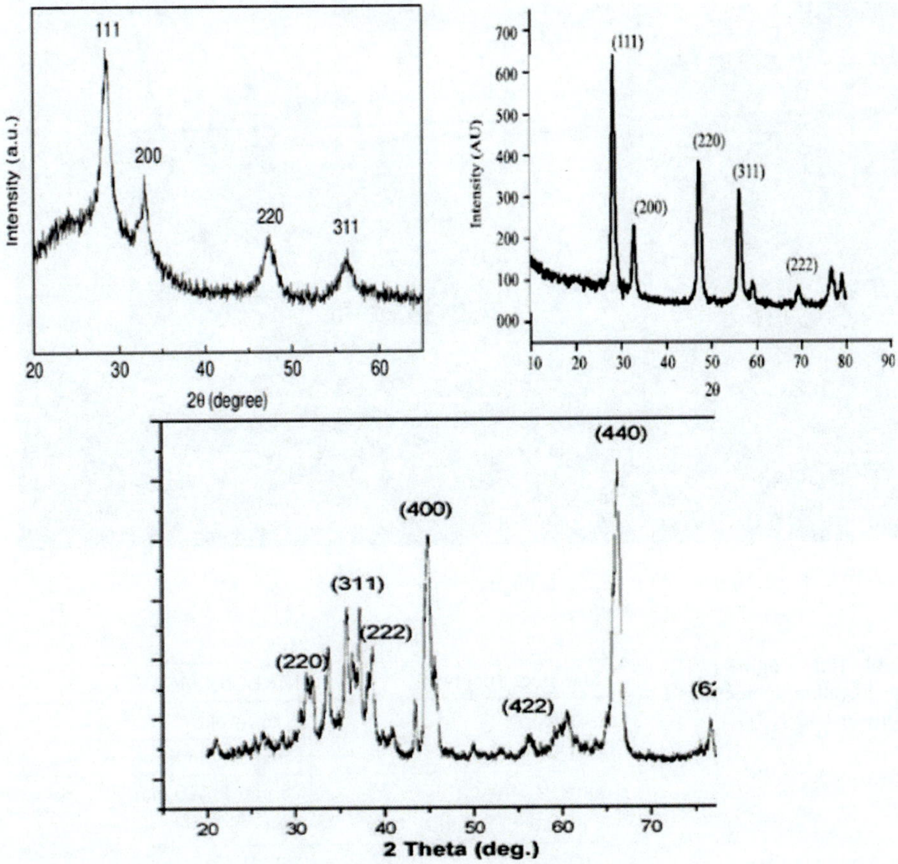

Fig. 10.7 XRD Analysis of Aluminum, Titanium and Cerium Oxide

the analysis of combustion performance and emission characteristics of the diesel–water nano-emulsion blend. The sensors utilized the cylinder pressure information through the DAS (Arun et al. 2014). This DAS took care of information obtained from the sensors that arebeing used for the control systems. Assessment of parameters to be named, cylinder peak pressure and heat release rate were done using an AVL combustion analyzer and a Kistler piezoelectric pressure sensor (Jayakar et al. 2018). The temperatures of water, cylinder, and tailpipe emission were measured using RTD. K chrome type's thermocouple which was present in the cylinder chamber. An AVL five gas analyzer and a smoke meter were used for the analyzer of the CO, HC, Smoke, NO_X, and CO_2 emissions (Saxena and Mauyra et al. 2018). Table 10.6 shows the specifications of the engine.

Fig. 10.8 Experimental setup for diesel engine

Table 10.6 Engine specification (Dhinesh and Annamalai 2018)

Manufacturing types	KIRLOSKAR AV -1
Stroke	Four Stroke
Cylinder	Single Cylinder
Bore & Stroke	88 mm & 110 mm
Compression Ratio	16:1
Power	5.2 kW
Speed	1500 rpm
Injection Pressure	200 bar
Displacement	630 cc
Length of Rod	165.3 mm
Injection Timing	23° bTDC

10.4.1 Uncertainty Analysis

Standard instrument experiment has varying degrees of accuracy, depending on their manufacturers. During this examination, an uncertainty evaluation was performed with the determined measurements and measured values to guarantee the precision level of the values acquired from the devices (Karthickeyana et al. 2020). Many environmental variables that concentrate on the system being tested, calibrated, operating condition, and reliability are due to errors and uncertainties in a product. Analyses were conducted at five cycles, and the moderate values were taken in this study for

Table 10.7 Uncertainty analysis

$$\left(\sqrt{(U_{BTE})2 + (U_{BSEC})2 + (U_{UBHC})2 + (U_{NO_x})2 + (U_{CO})2 + (U_{Smoke})2} = 2.66\%\right)$$

Parameters	Uncertainty analysis (%)
Brake Power	±0.28
BTE	±0.45
BSEC	±0.35
Load	±0.45
CO	±0.21
HC	±0.38
Smoke	±0.68
Oxides of Nitrogen	±0.26

assessing uncertainty (Rodrigues et al. 2013; Sadhik Basha and Anand et al. 2009). The scientists and researchers carried out an uncertainty analysis during the study using the factorial method, which was a highly favored approach. Table 10.7 shows the uncertainties of calculated and evaluated parameters.

10.5 Results and Discussion

10.5.1 Brake Thermal Efficiency

Figure 10.9 shows the fluctuation between brake power and brake thermal efficiency for the fuel types such as D100, DW5, DWHBNP 50 ppm, DWHBNP 100 ppm, and DWHBNP 150 ppm, respectively. The DW5 shows the similar performance of BTE of pure diesel at higher load conditions. The water molecules are present in the pure diesel to reduce the temperature and pressure of the cylinder and it showed inferior combustion (Koc and Abdullah et al. 2013). The operating temperature is high at a higher load point during the combustion process (Suresh et al. 2019). Adding water to diesel is to reduce the temperature range up to 100 degrees where it starts to evaporate and the secondary atomization process is obtained and proper combustion takes place Sadhik Basha and Anand (2012). The brake thermal efficiency of DW5 is slightly increased by 3.3% and then compared with D100. Increment in the dosage of hybrid nanoparticles as additives to the fuel of diesel–water emulsion increases the brake thermal efficiency heat release rate and in-cylinder pressure and reduces the ignition delay (Janakiraman et al. 2019). The brake thermal efficiency of blend DWHBNP 50 ppm, DWHBNP 100 ppm, and DWHBNP 150 ppm of the nanoparticle are increased by 5.87%, 6.12%, and 8.3%, respectively then compared to D100. The highest BTE of a blend of DWHBNP 150 ppm 8.3% is compared with other

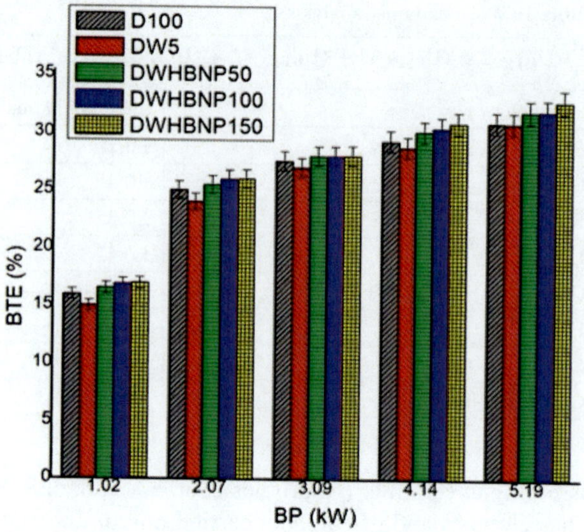

Fig. 10.9 Brake power versus BTE

tested fuels. Because of the oxygen supply of hybrid nanoparticles, it quickly burns the fuel during the combustion cycle and causes economic atomization of the fuel (Ramalingam et al. 2018).

Figure 10.10 shows the diversity between brake specific fuel consumption (BSFC) and load for different fuels such as D100, DW5, DWHBNP 50 ppm, DWHBNP 100 ppm, and DWHBNP 150 ppm, respectively. The brake-specific fuel consumption is to generate the unit power where the energy quantity is required from the diesel engine (Mohit Raj et al. 2018). For DWHBNP 150 ppm, the brake specific fuel

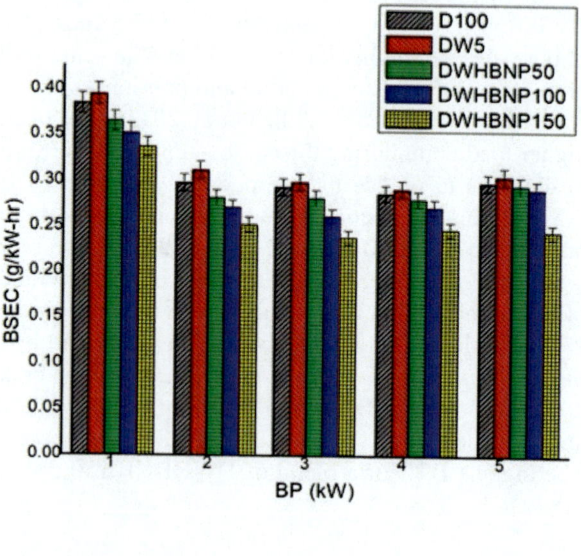

Fig. 10.10 Brake power versus BSFC

consumption is lower than for D100 when compared with other tested fuel. The results obtained are very similar in diesel –water hybrid nano-emulsion, because of reducing the heating value of blend DWHBNP 15O ppm as the diesel fuel and replaced by the water (Alahmer et al. 2010). By blends of DW5, DWHBNP 50 ppm, DWHBNP 100 ppm, and DWHBNP 150 ppm, the BSFC is reduced by 5.68%, 9.03%, 11.7%, and 14.2%, respectively, compared to D100. Sadhik Basha et al. (2013) suggested that the oxygen quality of hybrid nanoparticle has resulted in higher BSFC, a lean oxidation effect that causes an adverse improving effect in the engine performance. When adding the hybrid nanoparticles in the fuels it reduced the ignition delay period and increased the combustion temperature (Elumalai et al. 2020).

10.6 Emission Characteristics

10.6.1 Hydrocarbon

Figure 10.11 shows the variation between hydrocarbon and load for different fuels such as D100, DW5, DWHBNP 50 ppm, DWHBNP 100 ppm, and DWHBNP 150 ppm, respectively. The improper combustion misfired and quenching of the hydrocarbon emission is formed. At full load condition, the emission of hydrocarbon by DW5 is found to be 0.17 g/kw-hr, which is more or less equal to that by D100 (0.18 g/kw-hr). To increase the HC emission of DW5 the reason behind is auto-ignition (Vigneswaran et al. 2018). When adding the hybrid nanoparticle, the hydrocarbon emission is reduced because of higher efficiency and oxygen content available in the nanoparticles (Janakiraman et al. 2019). The lowest hydrocarbon emission of blend DW5, DWHBNP 50 ppm, DWHBNP 100 ppm, and DWHBNP

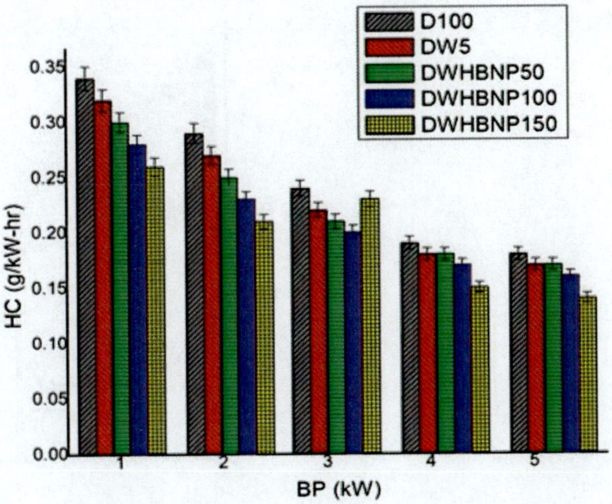

Fig. 10.11 Brake power versus HC

150 ppm is 5.5%, 16.66%, 22.22%, and 27.77%, respectively, when compared with neat diesel. The DWHBNP 150 ppm blend achieved the lowest hydrocarbon emission at full load condition is 27.77% and is then compared to other tested fuel. he hybrid nanoparticles act as a catalyst and this leads to complete combustion and also water molecules are present in the hybrid nanofuel where the microexplosion takes place during the combustion process (Sadhik Basha and Anand 2012) and (C. Z. O et al. 2017).

10.6.2 Carbon Monoxide

Figure 10.12 shows the variation between carbon monoxide (CO) and load for different fuels such as D100, DW5, DWHBNP 50 ppm, DWHBNP 100 ppm, and DWHBNP 150 ppm, respectively. The release of carbon monoxide results from inadequate oxygen content and improper combustion process. It may be decreased by using diesel fuel with hybrid nanowater emulsion. The D100 indicates a small rise in emissions of carbon monoxide as compared to DW5. Due to the incomplete combustion and reduction of oxygenating time in-cylinder temperature during the combustion process (Lin et al. 2006). The carbon monoxide emission is decreased for blends of DW5, DWHBNP 50 ppm, DWHBNP 100 ppm, and DWHBNP 150 ppm by 3.2%, 6.2%, 7.2%, and 10.2%, respectively, as compared with D100. The DWHBNP 150 ppm blends are decreased by 10.2% when compared to other measured fuels at full load level. Due to the hybrid nanoparticle added in the water emulsion, it attributed by secondary atomization in the combustion chamber, and also the availability of oxygen content in the fuel and complete combustion take place (Sadhik Basha and Anand 2012).

Fig. 10.12 Brake power versus CO

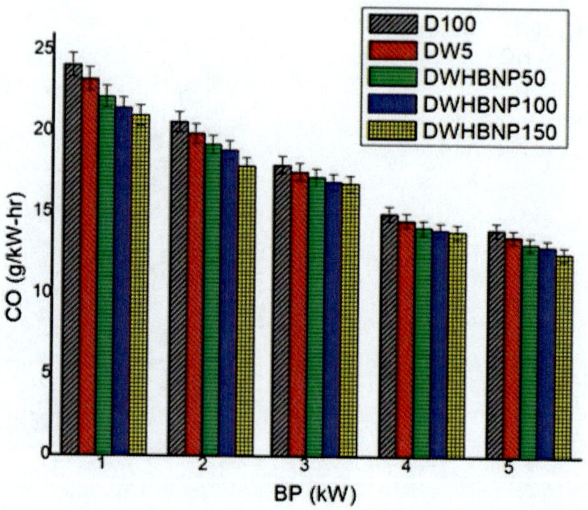

Fig. 10.13 Brake power versus Oxides of nitrogen

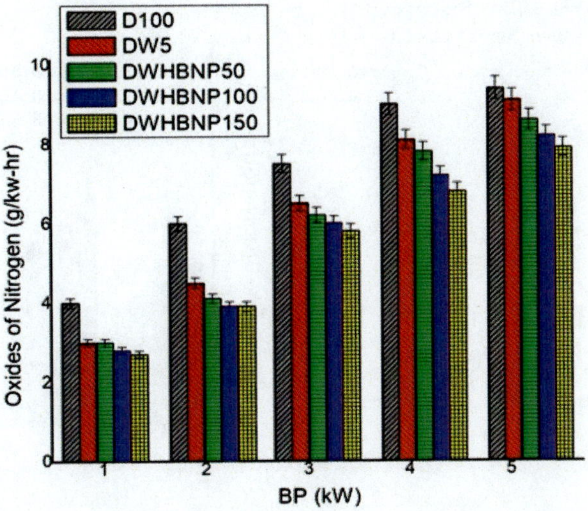

10.6.3 *Oxides of Nitrogen*

Figure 10.13 shows the variation between oxides of nitrogen and brake power for different fuels such as D100, DW5, DWHBNP 50 ppm, DWHBNP 100 ppm, and DWHBNP 150 ppm. During high evaluated temperature and surplus of oxygen content, these two parameters are important to the formation of oxides of nitrogen (Vigneswaran et al. 2018). The D100 (9.4 g/kW-hr) having the highest oxides of nitrogen are then compared with other tested fuel. In order to decrease the oxides of nitrogen to introduce the hybrid nano-water emulsion fuels into diesel fuel (Sharma et al. 2013). The hybrid nano-water molecules have the more significant latent heat of vaporization that reduces its elevated temperature during the combustion process. The oxides of nitrogen emission are decreased for blends of DW5, DWHBNP 50 ppm, DWHBNP 100 ppm, and DWHBNP 150 ppm by 27.37%, 31.36%, 34.55%, and 36.95%, respectively, as compared with D100. DWHBNP 150 is decreased by 36.95% when compared to other measured fuels at a full load level of blends. Due to the less compressible fuels and higher density of emulsion fuels result in shorter ignition delay, less fuel consumption, lower combustion pressure, and microexplosion take place and larger droplets of fuels are converted into smaller droplets, and fuel is quickly burned during the combustion process (Sadhik Basha and Anand 2012).

10.6.4 *Smoke Opacity*

Figure 10.14 shows the variation between smoke and brake power for different fuels such as D100, DW5, DWHBNP 50 ppm, DWHBNP 100 ppm, and DWHBNP

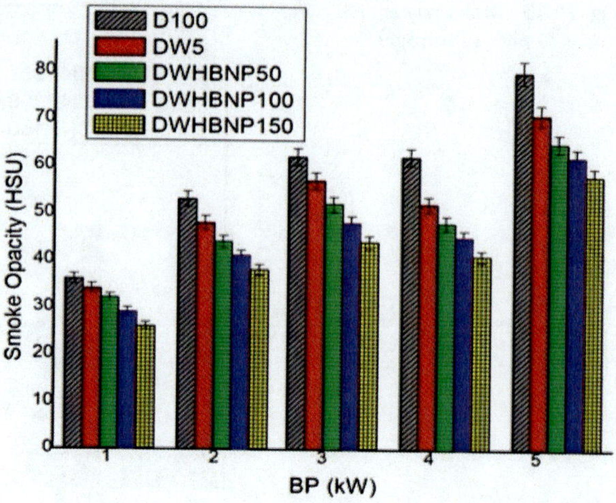

Fig. 10.14 Brake power versus Smoke Opacity

150 ppm, respectively. Lower oxygen levels in the fuel usually have poor combustion, resulting in more significant smoke emissions in the tailpipe (Elumalai et al. 2019). This diesel is observed with the highest smoke opacity emission. Hwang et al. (2016) while addressing this question stated that diesel–water hybrid nanoemulsion and its fuel blends are the ideal solution for reducing smoke opacity because of its intrinsic oxygen molecules. The smoke opacity emission is decreased by combinations of DW5, DWHBNP 50 ppm, DWHBNP 100 ppm, and DWHBNP 150 ppm by11.25%, 18.75%, 22.5%, and 27.5%, respectively, as compared with D100. The visibility of the DWHBNP 150 ppm smoke blends at full load condition are decreased by 27.5% relative to other measured fuels. Owing to the, water molecules and present of nanoparticles in the fuels it makes reduce the temperature in soot particles (Subramanian and Ramesh 2002).

10.7 Combustion Characteristics

10.7.1 Cylinder Pressure, HRR

Figure 10.15 shows the variation between cylinder pressure and crank angle (CA) for different fuels such as D100, DW5, DWHBNP 50 ppm, DWHBNP 100 ppm, and DWHBNP 150 ppm, respectively. The D100 has the higher cylinder pressure was then compared to DW5. In addition, the water molecules were present in the diesel to reduce the temperature in-cylinder pressure during the combustion process (Vigneswaran et al. 2018). The addition of hybrid nanoparticles in the water emulsion fuels leads to improve combustion process the value of which is near to diesel fuel (Sadhik Basha et al. 2012). The cylinder pressure that is decreased by blends of DW5,

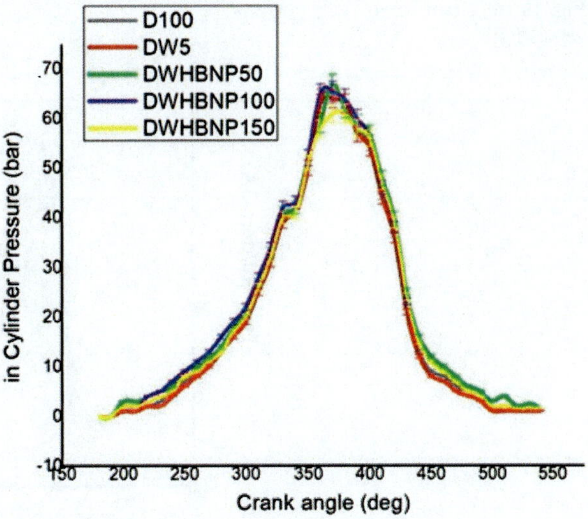

Fig. 10.15 Crank angle versus Cylinder Pressure

DWHBNP 50 ppm, and DWHBNP 100 ppm is 1.56%, 7.8%, and 9.3%, respectively, and is then compared with D100. The cylinder pressure is DWHBNP 150 ppm 3.2% slightly elevated when compared to other measured fuels. Due to the increase in the cetane number the mass of the fuel is lower and it easily burns the fuels during the premixed combustion process because of the shortened ignition period (Perumal Venkatesan et al. 2019). When an accumulation of fuels is the diffusion phase of combustion and quick burning of fuels take places that are the reason to increase the DWHBNP 150 ppm blends. (Subramanian and Ramesh 2002; Lin and Chen et al. 2008).

Figure 10.16 shows the variation between HRR and CA for different fuels such as D100, DW5, DWHBNP 50 ppm, DWHBNP 100 ppm, and DWHBNP 150 ppm, respectively. The net release rate depends on the cylinder pressure during the combustion process (Shukla et al. 2018). The higher HRR rate attributed the better combustion is achieved in the combustion process the lower HRR rate IS obtained due to the improper combustion (Vigneswaran et al. 2018). Because of better combustion in pressure cylinder within the combustion cycle, the DW5 blends showed a higher HRR rate compared to other tested fuels. Also, the DW5 blends in the combustion cycle started early with lower flash point and higher cetane number to boost the combustion characteristics and less delay in ignition. The highest HRR is found in the blend DWHBNP 150 ppm which is then compared to neat diesel fuel due to the shorten ignition delay and increase in the cylinder pressure-temperature (Yamin et al. 2013).

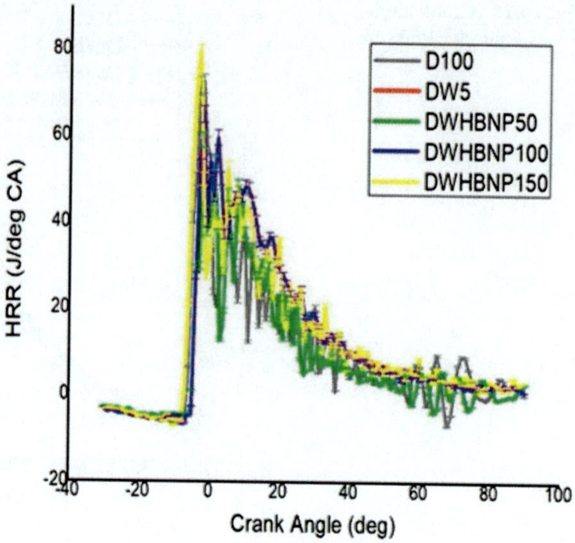

Fig. 10.16 Crank angle versus HRR

10.8 Conclusions

The current empirical research handled the effect of HBNP nanoparticles on the performance emissions and the combustion characterization of the diesel engine. FTIR SEM and XRD were used to analyses the properties of TiO_2, CeO_2 and Al_2O_3 nanoparticles. The following conclusion was drawn By adding hybrid nanoparticles into the diesel–water emulsion to give the superior performance and emission characteristics of the diesel engine. To increase the concentration of nanoparticles to give the optimum results of performance and emission characteristics due to the microexplosion taking place and availability of oxygen content in the fuel. Figure 10.17 shows the comparison results all blended with D100.

1. DWHBNP 150 ppm is obtained with the maximum BTE of 8.3% when correlated with other measured fuels. Due to the oxygen supply of nanoparticles, it quickly burns the fuel during the combustion cycle and causes the economic atomization of the fuel.
2. The DWHBNP 150 ppm blend achieves the lowest hydrocarbon emission at full load condition at 27.77% when compared to other fuel measured. This is because of the available oxygen content in nanoparticles and the activation energy emitted by the nanoparticles.
3. At full load condition, the NOx of blends DWHBNP 150 is decreased by 36.95% and then compared with other tested fuels. Due to the less compressible fuels and higher density of emulsion fuels result in shorter ignition delay, less fuel consumption, lower combustion pressure, and microexplosion taking place the larger droplets of fuels were converted into smaller droplets and the fuel quickly burned during the combustion process.

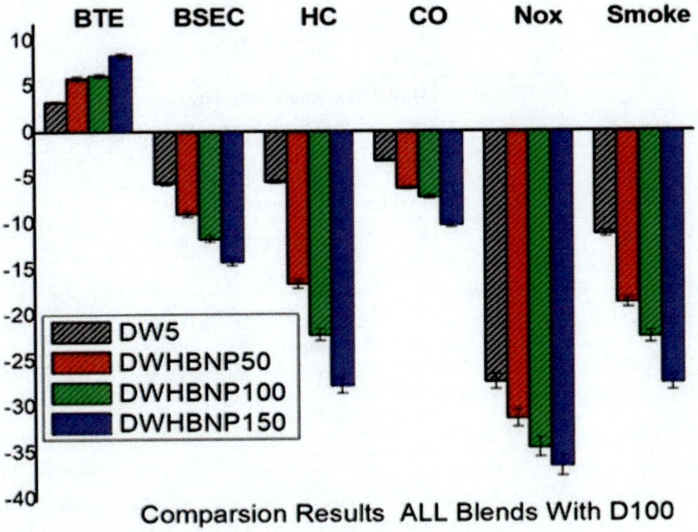

Fig. 10.17 Comparison results all blends with D100

4. At full load condition, the blends DWHBNP 150 ppm smoke opacity is decreased by 27.5% and then compared with other tested fuels. Due to the enhancement in spray volume, an appreciable amount of air entrainment occurs in the emulsion spray, microexplosion effect and causes increase in OH radicals during the combustion in the engine cylinder.
5. The carbon monoxide of DWHBNP 150 ppm blends is then decreased by 10.2% relative to other measured fuels at full load level. Due to the nanoparticles added in the water emulsion, it attributed to secondary atomization in the combustion chamber, and also the availability of air in the atmosphere enters into the cylinder for complete combustion to take place.
6. The cylinder pressure in DWHBNP 150 ppm is 3.2% slightly elevated when compared to other measured fuels. With an increase in cetane number, the mass of the fuel is lower and it easily burns the fuels during the premixed combustion process because of the shortened ignition period.
7. The highest HRR is found in the blend DWHBNP 150 ppm which is then compared to neat diesel fuel due to the shortened ignition delay and increase in the cylinder pressure–temperature.

Abbreviations

DWHBNP50 ppm	D90% + W5% +S5% + HB 50 ppm
DWHBNP100 ppm	D90% + W5% +S5% + HB 100 ppm
DWHBNP150 ppm	D90% + W5% +S5% + HB 150 ppm
ID	Ignition Delay

(continued)

(continued)

DWHBNP50 ppm	D90% + W5% +S5% + HB 50 ppm
BTE	Brake Thermal Efficiency
BSEC	Brake-Specific Fuel Consumption
HC	Hydrocarbon
NOx	Oxides of nitrogen
HLB	Hydrophilic–lipophilic balance
CO	Carbon monoxide
DWS	Diesel–Water Surfactant
HBNP	Hybrid Nanoparticle
TiO_2	Titanium Oxide
Al_2O_3	Aluminum Oxide
CeO_2	Cerium Oxide
CNT	Carbon Nanotube
ZnO	Zirconium Oxide

Acknowledgements The authors thank Dr. V P Ramamurthi, former professor of Anna University and present Chairman of Dhanalakshmi College of Engineering for motivating them in preparing this article. The authors are also indebted to. Dr. K Annamalai, Associate Professor, Department of Automobile Engineering, Anna University for his guidance and support towards the compilation of this book article. The authors proficiently thank their colleagues Mr. J. Jayakar and Mr. M. Nambiraj Assistant Professors, Dhanalakshmi College of Engineering for their esteemed support.

Declaration of Competing Interest

The authors declare that they have no known competing financial interests or personal relationships that could have appeared to influence the work reported in this paper.

References

Aalam CS, Alagappan N (2015) Cerium oxide nano particles as additive with diesel fuel on DI diesel engine. Int J Innov Res Creative Technol 1(2). ISSN: 2454–5988

Aalam SC, Saravanan CG, Prem Anand B (2015) Reduction of emissions from CRDI diesel engine using metal oxide nanoparticles blended diesel fuel. Int J Appl Eng Res 10:3865–3869

Adiga KC, Shah OH (1990) On the vaporization behavior of water-in-oil micro emulsions. Comb Flame 80:412–414

Alahmer A, Yamin J, Sakhrieh A, Hamdan MA (2010) Engine performance using emulsified diesel fuel. Energy Convers Manage 51(8):1708–1713

Arun B (2014) A study on production, performance and emission analysis of Tamanu Oil-Diesel blends in a CI Engine 3(3)

Babu KR, Raja RB (2015) Theoretical and experimental validation of performance and emission characteristics of nano-additive blended diesel engine. Int J Res Aeronaut Mech Eng 3(5):18–31

Basha JS, Anand RB (2011a) An experimental study in a CI engine using nano-additive blended water–diesel emulsion fuel. Int J Green Energy 8:332–348

Basha JS, Anand RB (2011b) An experimental investigation in a diesel engine using carbon nanotubes blended water-diesel emulsion fuel. Proc Inst MechVol 225(3):279–288

Basha JS, Anand RB (2013) The influence of nano additive blended biodiesel fuels on the working characteristics of a diesel engine. J Brazilian Soc Mech Sci Eng 35(3):257–264

C. Z. O, (2017) Assessment on performance, combustion and emission characteristics of diesel engine fuelled with corn stalk pyrolysis bio-oil/diesel emulsions with. Fuel Process Technol 167:474–483

Chlopek Z, Darkowski A, Piaseczny L (2015) The influence of metalloorganic fuel additives on CI engine emission. Polish J Environ Stud 14:559–567

Dhinesh K, Annamalai M (2018) A study on performance, combustion and emission behaviour of diesel engine powered by novel nano nerium oleander biofuel. J Clean Prod 196:74–83. https://doi.org/10.1016/j.jclepro.2018.06.002

DSilva R, Binu KG, Bhat T (2015) Performance and emission characteristics of a C.I. Engine fuelled with diesel and TiO_2 nanoparticles as fuel additive. Mater Today: Proc 2:3728–35

Elumalai PV, Annamalai K, Dhinesh B (2019) Effects of thermal barrier coating on the performance, combustion and emission of DI diesel engine powered by biofuel oil–water emulsion. J Therm Anal Calorim 137:593–60 https://doi.org/10.1007/s10973-018-7948-6

Elumalai PV, Dhinesh B, Parthasarathy M, Pradeepkumar AR, Mohamed Iqbal S, Jayakar J, Nambiraj M (2020) An experimental study on harmful pollution reduction technique in low heat rejection engine fuelled with blends of pre-heated linseed oil and nano additive. J Cleaner Prod. https://doi.org/10.1016/j.jclepro.2020.124617

Elumalai PV, Sivakandhan C, Parthasarathy M et al (2021) Investigation on the mitigation of environmental harmful emissions by incorporating nanoparticles to biofuel water nano emulsion in low heat rejection engine. Heat Mass Transfer (2021). https://doi.org/10.1007/s00231-021-03028-7

Elumalai PV, Dhinesh B, Jayakar J et al (2021b) Effects of antioxidants to reduce the harmful pollutants from diesel engine using preheated palm oil–diesel blend. J Therm Anal Calorim (2021a). https://doi.org/10.1007/s10973-021-10652-2

Emiro AO, Mehmet S (2017) Combustion, performance and emission characteristics of various alcohol blends in a single cylinder diesel engine 212(676):34–40

Hwang J, Bae C, Gupta T (2016) Application of waste cooking oil (WCO) biodiesel in a compression ignition engine. Fuel 176:20–31

Janakiraman S, Lakshmanan T, Chandran V, Subramani L (2019) Comparative behavior of various nano additives in a DIESEL engine powered by novel Garcinia gummi-gutta biodiesel. J Cleaner Prod. https://doi.org/10.1016/j.jclepro.2019.118940

Jayakar J, Elumalai PV, Mailainathan M, Dhinesh B, Harikrishnan G (2018) Macroscopic characteristics of palm oil and palm oil methyl ester using dimensionless analysis. J Oil Palm Res 30:666–73. https://doi.org/10.21894/jopr.2018.0055

Karthickeyana K, Thiyagarajanb S, Edwin Geob V, Ashok B, Nanthagopalc K, Ong Hwai Chyuand, Vignesh K (2020) Simultaneous reduction of NOx and smoke emissions with low viscous biofuel in low heat rejection engine using selective catalytic reduction technique. Fuel 255:115854

Karthickeyana K, Thiyagarajanb S, Ashok B, Edwin Geob V, Azad AK (2020b) Experimental investigation of pomegranate oil methyl ester in ceramic coated engine at different operating condition in direct injection diesel engine with energy and exergy analysis. Energy Convers Manag 205(1):

Karthikeyan S, Elango A, Prathima A (2014) An environmental effect of GSO methyl ester with ZnO additive fuelled marine engine. Indian J Geo-Marine Sci 43:564–570

Keskin A, Guru M, Altıparmak D (2011) Influence of metallic based fuel additives on performance and exhaust emissions of diesel engine. Energy Convers Manage 52:60–65

Koc AB, Abdullah M (2013) Performance and NO x emissions of a diesel engine fueled with biodiesel-diesel-water nano-emulsions. Fuel Process Technol 109(x):70–77

Lin C-Y, Wang K-H (2003) The fuel properties of three phase emulsions as an alternative fuel for diesel engines. Fuels 82:1367–1375

Lin CY, Wang KH (2004) Effects of a combustion improver on diesel engine performance and emission characteristics when using three-phase emulsions as an alternative fuel. Energy Fuels 18:477–484

Lin CY, Chen LW, Wang LT (2006) Correlation of black smoke and nitrogen oxides emissions through field testing of in-use diesel vehicles. Environ Monit Assess 116:291–305 (2006). https://doi.org/10.1007/s10661-006-7402-2

Mahesh Kumar AR, Kannan M, Nataraj G (2020) A study on performance, emission and combustion characteristics of diesel engine powered by nano-emulsion of waste orange peel oil biodiesel. Renew Energy 146:1781–1795

Mehta RN, Chakraborty M, Parikh PA (2014) Nanofuels: combustion, engine performance and emissions. Fuel 120:91–97

Mei D, Li X, Wu Q, Sun P (2016) Role of cerium oxide nanoparticles as diesel additives in combustion efficiency improvements and emission reduction. J Energy Eng 142:142–148

Mohd FO, Abdullah A, Najafi G, Rizalman M (2017) Green fuel as an alternative fuel for diesel engine: a review. Renew Sustain Energy Rev 80:694–709

Nadeem M, Rangkuti C, Anuar K, Haq MRU, Tan IB, Shah SS (2006) Diesel engine performance and emission evaluation using emulsified fuels stabilized by conventional an gemini surfactants. Fuel 85:2111–2119

Narinder S, Bharj RS (2015) Effect of CNT-emulsified fuel on performance emission and combustion characteristics of four stroke diesel engine. Int J Curr Eng Technol 5:477–485

Nishant M, Mayank Sh, Singh RC, Pandey RK (2015). Performance study of diesel engine using nanofuel. Int J Adv Res Innov 3(4):665–8

Panneerselvam N, Murugesan A, Vijayakumar A, Kumaravel D, Subramaniam L, Avinash A (2015) Effects of injection timing on bio-diesel fuelled engine characteristics—an overview. Renew Sustain Energy Rev 50:17–31

Parthasarathy M, Ramkumar S, Elumalai PV, Gupta SK, Krishnamoorthy R, Mohammed Iqbal S, Dash SK, Silambarasan R (2021) Experimental investigation of strategies to enhance the homogeneous charge compression ignition engine characteristics powered by waste plastic oil. Energy Conv Manag 236:114026. https://doi.org/10.1016/j.enconman.2021.114026

Patnaik PP, Jena SP, Acharya SK, Das HC (2017) Effect of $FeCl_3$ and diethyl ether as additives on compression ignition engine emissions. Sustain Environ Res 27(3):154–161

Perumal Venkatesan E, Kandhasamy A, Sivalingam A, Kumar AS, Ramalingam KM, Joshua P james thadhani (2019) Performance and emission reduction characteristics of cerium oxide nanoparticle-water emulsion biofuel in diesel engine with modified coated piston. Environ Sci Pollut PP 26:27362–71. https://doi.org/10.1007/s11356-019-05773-z

Prabu A (2018) Nanoparticles as additive in biodiesel on the working characteristics of aDI diesel engine. Ain Shams Eng J 9:2343–2349

Prabu (2018, Aug) Engine characteristic studies by application of antioxidants and nanoparticles as additives in biodiesel diesel blends. Energy Resour Technol 140(8):082203. https://doi.org/10.1115/1.4039736

Pradeep Kumar AR, Annamalai K, Premkartikkumar SR (2013) An experimental investigation into chosen parameters of diesel-water emulsion on combustion processes 2013, Article ID 764514. https://doi.org/10.1155/2013/764514

Rajwinder S, Sumeet S, Gangacharyulu D (2016) Effect of TiO_2 nanoparticle blended water diesel emulsion fuel on ci engine performance and emission characteristics. Int J Eng Res Technol 5. ISSN: 2278–0181

Rajwinder S, Sumeet S, Gangacharyulu D (2016) Effect of TiO_2 nanoparticle blended water diesel emulsion fuel on ci engine performance and emission characteristics. Int J Eng Res Technol 5(7). ISSN: 2278–0181

Ramalingam KM, Kandasamy A, Subramani L, Balasubramanian D, Paul James Thadhani J (2018) An assessment of combustion, performance characteristics and emission control strategy by adding anti-oxidant additive in emulsified fuel. Atmospheric Pollut Res 9:959–967

Ramlan NA, Yahya WJ, Ithnin AM, Hasannuddin AK, Norazni SA, Mazlan NA (2016) Performance and emissions of light-duty diesel vehicle fuelled with non surfactant low grade diesel emulsion compared with a high grade diesel in Malaysia. Energy Convers Manage 130:192–199

Rodrigues RC, Ortiz C, Berenguer-Murcia A, Torres R, Fernández-Lafuente R (2013) Modifying enzyme activity and selectivity by immobilization. Chem Soc Rev 42(15):6290–6307

Sadhik Basha J (2011) Role of nanoadditive blended biodiesel emulsion fuel on the working characteristics of a diesel engine. J Renew Sustain Energy 3:023106. https://doi.org/10.1063/1.3575169

Sadhik Basha J, Anand RB (2009) Performance and emission characteristics of a DI compression ignition engine using carbon nanotubes blended diesel. In: Proceedings in international conference on advances in mechanical engineering, Surat, India, pp 312–316

Sadhik Basha J, Anand RB (2012) Effects of nanoparticle additive in the water-diesel emulsion fuel on the performance, emission and combustion characteristics of diesel engine. Int J Veh Design 59(2/3):164–181

Saxena MR, Maurya RK (2019) Nanoparticle emissions in reactivity-controlled compression ignition engine. In: Agarwal A., Dhar A., Sharma N, Shukla P (eds) Engine exhaust particulates. Energy, environment, and sustainability. Springer, Singapore. https://doi.org/10.1007/978-981-13-3299-9_12

Saxena MR, Mauyra RK (2018) Performance, combustion, and emissions characteristics of conventional diesel engine using butanol blends, advances in internal combustion engine research, 1st edn, pp 93–110

Sharma A, Kumar N, Vibhanshu V, Deep A (2013) Emission studies on a VCR engine using stable diesel water emulsion. SAE Technical Paper. SAE International. https://doi.org/10.4271/2013-01-2665

Shukla PC, Gupta T, Agarwal AK (2018) Techniques to control emissions from a diesel engine. Air Pollution and Control, 1st edition, pp 57–72

Sivalingam A, Kandhasamy A, Senthil Kumar A et al (2019) Citrullus colocynthis - an experimental investigation with enzymatic lipase based methyl esterified biodiesel. Heat Mass Transf 55:3613–3631. https://doi.org/10.1007/s00231-019-02632-y

Subramanian, Ramesh (2002) Use of diethyl ether along with water-diesel emulsion in a Di diesel engine. SAE Trans 111, Section 4. J Fuels Lubricants, 1361–1367

Subramanian KA, Ramesh A (2002) Use of Diethyl ether along with water-diesel emulsion in a DI diesel engine, SAE Technical Paper No. 2002-01-2720

Suresh K, Aggarwal, Xiao Fu (2019) Using Petroleum and biomass-derived fuels in duel-fuel diesel engines, novel combustion concepts for sustainable energy development. 1st edition, pp 246–276

Syed AC, Saravanan CG, Premanand B (2015) Influence of iron (ii, iii) oxide nanoparticles fuel additive on exhaust emissions and combustion characteristics of CRDI system assisted diesel engine. Int J Adv Eng Res Sci 2(3). ISSN: 2349–6495

Vellaiyan S, Amirthagadeswaran K (2016) Zinc oxide incorporated water-in-diesel emulsion fuel formulation, particle size measurement, and emission characteristics assessment. Pet Sci Technol 34:114–122

Venu H, Subramani L, Raju VD (2019) Emission reduction in a DI diesel engine using exhaust gas recirculation (EGR) of palm biodiesel blended with TiO_2 nano additives. Renew Energy 140:245–263

Vigneswaran R, Annamalai K, Dhinesh B, Krishnamoorthy R (2018) Experimental investigation of unmodified diesel engine performance, combustion and emission with multipurpose additive along with water-in diesel emulsion fuel. Energy Convers Manag 172:370–380. https://doi.org/10.1016/j.enconman.2018.07.039

Yamin JA, Sakhnini N, Sakhrieh A, Hamdan MA (2013) Environmental and performance study of a 4-Stroke CI engine powered with waste oil biodiesel. Sustain Cities Soc 9:32–38